TIN

The Working of a Commodity Agreement

William Fox O.B.E.

Secretary of the International
Tin Council (1956-71)

©1974 MINING JOURNAL BOOKS LIMITED

MINING JOURNAL BOOKS LIMITED
LONDON 1974

First published 1974

©1974 MINING JOURNAL BOOKS LTD.

ISBN 0 900117 05 2

LIBRARY OF CONGRESS 74-78299

This book has been set in 10 on 12pt Plantin printed by Dragon Press Ltd., Luton, Bedfordshire, England for the publishers Mining Journal Books Ltd., 15 Wilson Street, London, EC2M 2TR.

Made and printed in Great Britain

DEDICATED
to my dear wife
Kathleen Frances
in deepest gratitude.

CONTENTS

A Note on Terms and Standards Used

Preface

MAPS AND CHARTS

TABLES

TERMS AND STANDARDS USED

STATISTICS

Statistics of production and consumption of tin are based on the International Tin Research and Development Council: *Statistical Year Book, 1937* for the years prior to 1900; on the International Tin Study Group: *Statistical Year Book, 1949* for the period 1900-48; on the International Tin Council: *Statistical Year Book, 1968: Tin, Tinplate and Canning* for the period after 1948. Other sources used are so stated. Occasional discrepancies between sources are ignored. All figures are quoted in long tons (of 2,240 pounds) to maintain continuity. The International Tin Council for 1970 and thereafter switched in tonnage and in prices to the metric ton (of 1,016 kg); when these are used it is so indicated.

Mining production is given in terms of the tin content of the tin concentrates mined (unless otherwise stated). Consumption is limited to virgin tin metal; secondary tin metal (never a factor of much importance) is ignored.

The U.S.S.R. and the People's Republic of China are excluded (unless otherwise stated) from world total of production or consumption.

PRICES AND CURRENCIES

The basic price used is the price for cash tin (for prompt delivery) on the London Metal Exchange (that is, standard tin with a minimum of 99°75% tin content). The other quotation on the L.M.E. is for forward tin (that is, for contracts maturing three months ahead of the date on which the deal was made and the price fixed).

The quotation used for New York is the price arrived at by *The American Metal Market* for prompt tin, expressed in U.S. dollars and cents per pound.

The quotation for the Malaysian market (Penang) is in Malaysian dollars per picul (=133·3 lb of tin. 16·8 piculs=one long ton).

The £ is always the pound sterling of the United Kingdom. The *dollar*, without any qualification, is always the U.S.A. dollar. The M$ is the Malaysian dollar and includes, before 1950, the S$ (Straits dollar).

COUNTRIES

Nigeria is used throughout. It became independent of the U.K. in 1960.

Siam=Thailand. Siam was changed officially to Thailand in 1939. In dealing with the history of the international tin arrangements prior to 1946 Siam is generally used.

Belgian Congo: Congo: Congo-Zaïre: Zaïre. The Belgian Congo became independent of Belgium in 1960, was the Democratic Republic of the Congo and was renamed Zaïre in 1972. It is referred to as the Belgian Congo up to 1960, as the Congo during the 1960s and as Congo-Zaïre or Zaïre in very recent years.

Netherlands East Indies: Indonesia. The area now known as Indonesia was under the Netherlands government until 1945 as the Netherlands East Indies (sometimes the Netherlands Indies or the N.E.I.). Independence was proclaimed as Indonesia in 1945. For this book Netherlands East Indies (N.E.I.) is used for events prior to 1945 and Indonesia for events thereafter.

Malaya: Malaysia. Malaya is generally used before 1963 to cover the Federated and other Malay States. Malaysia was formed in 1963 as an independent state covering the Federation of Malaya, Singapore, Sabah (North Borneo) and Sarawak. Singapore seceded in 1965. Malaysia is used for events after 1963.

MINING TERMS

Tin concentrate: the material obtained ultimately in the mining and dressing process and sent to the smelter for conversion into tin metal. The tin element in the concentrate is usually cassiterite (stannic oxide, SnO_2). The concentrate is normally around 70 to 75% tin content in Malaysia, Indonesia, Nigeria, Thailand and Congo-Zaïre; it varies from 20 to 65% (but usually averaging about 35%) in Bolivia.

Tin-in-concentrates: the standard for measuring production in terms of the tin content of the tin concentrates.

Alluvial deposits: tin-bearing rocks and veins, disintegrated and weathered by natural agencies, are transported and concentrated by running water. The tin-bearing material then appears as residual deposits on hillsides, as alluvial deposits on valley slopes, old or submerged river valleys, beach and marine deposits or as buried deposists.

Dredging: the use of a dredge, mechanically propelled, on an artificial pond made by itself. The dredge uses mechanical buckets to dig into tin-containing ground; screens and washes on board this material to retain the tin (cassiterite) content; and rejects the waste material from the rear as it moves. The dredge may be sea-going (in Thailand and Indonesia).

Gravel pumping: the tin-containing open face at the mines is broken up by water from the monitors; the disintegrated material is washed into a sump, elevated by gravel pumps and sluiced for the recovery of tin content.

Hydraulic mining: the gravel in the tin-containing face of the mines is disintegrated by water power operating under natural pressure through monitors and is carried through sluice boxes and there washed for the recovery of tin.

Dulang washing: small-scale operations for recovering tin-bearing material from river beds by washing the river gravel in large wooden dishes or pans. It is operated by women holding official licences in Thailand and Malaysia.

Lode mining: lodes or veins of tin-containing ore, normally associated with granitic areas, are mined *in situ*. This usually requires underground mining and a system of working through vertical shafts, inclines, horizontal or sloping adits, drifts, galleries, cross-cuts and stopes to reach the lodes, to extract the tin-containing material (usually with the help of explosives) and to convey it to the surface for crushing and treatment in the mill and processing plant. It is the common form of tin mining in Cornwall, in Bolivia and at Pahang in Malaysia.

Grade of ground: the tin-bearing quality of the ground prospected or mined may be measured in alluvial mines sometimes in terms of the weight of the tin concentrates or the calculated weight of tin content in each cubic yard or cubic metre of ground. In lode mining quality may be expressed as pounds of tin concentrates or tin content per cubic yard of tin-containing ground.

FOOTNOTES

Footnotes relating to each pair of facing pages are set at the foot of the left-hand page. Exceptionally they appear at the foot of the right-hand page—
(i)   where the first page of a new chapter starts on a right-hand page or,
(ii) where the whole of a left-hand page is devoted to an illustration.

PREFACE

The tin mining industry of the world has been subject to some form of international control for many of the past fifty years. The control has usually been intergovernmental in character, first under the International Tin Committee between 1931 and 1942, later, and on a wider basis, under the International Tin Council after 1956. In both periods the general objectives were much the same —over the longer term to redress the imbalance between production and consumption and in the shorter term to even out abnormal price fluctuations. The machinery used for these objectives was the same in both cases—the limitation of exports and the use of a buffer stock.

The principles on which governments should co-operate in seeking to achieve international commodity stabilization were established by the Draft Charter for an International Trade Organisation in Havana in 1947-48. This Havana Charter, although almost universally left unratified, was nevertheless generally accepted. The principles were enlarged upon by the United Nations Conference on Trade and Development (UNCTAD) in the 1960s so as to form a bible which, it was hoped, would take the developing countries out of the old cycle of uncontrolled burst, boom and speculation in commodities, would add to the export earnings of these countries and would help to redress the world trading balance between developed and developing areas.

In the main, the commodity agreements envisaged by the Havana Charter or by UNCTAD have been lamentably slow in coming to life and, once born, sometimes lamentably weak in operation. They have emerged or are emerging, so far, only in tin, coffee, sugar, wheat, olive oil and cocoa. Of these, the tin agreement has been the most effective in its practices; and the initial International Tin Agreement of 1956-61 has been followed by three more five-year agreements.

In writing this book it has been my purpose that it should examine the way in which the tin industry

has reacted to international control since 1921; and
assess the reasons for the success—and sometimes
failure—of the conceptions of buffer stock and
export control. It looks at the abnormal factors
which the agreements have had to take into account.
It also pays considerable attention to the influence
of the post-war policy of the United States through
the creation and disposal of its strategic stockpiles,
a policy which for so many years influenced world
markets very strongly, not only in tin but in other
vital raw materials.

During my involvement for nearly a quarter
of a century with the international tin industry
(first as Secretary-General to the International Tin
Study Group and then as Secretary to the
International Tin Council) I watched civil servants
and their industrial and commercial advisers of
many nationalities and very differing interests study
the complicated and often repetitive problems of an
international commodity agreement. Much of
their work on tin, as in so many other international
discussions, must have been wearisome to the highest
degree; but the world of tin, both producing and
consuming countries, owes these men a deep
debt for the work they did so well. I may be permitted
here to express my particular personal admiration for
Sir Vincent del Tufo (Malaysia), Vija Sethaput
(Thailand), J. van Diest (The Netherlands) and R. van
Achter (Congo and Zaire) and, above all, for Pierre
Legoux (France), who was a masterly influence on all
international tin thinking. Amongst the Americans
working outside the Tin Council I have always
greatly respected Clarence Nichols and Marion
Worthing of the State Department.

It is very pleasant to record my debt to so many former
colleagues on the Tin Council. In particular, I am
happy to thank my personal secretary, Miss Barbara
Stewart (Mrs. Eric Jones), for most invaluable help
both on the International Tin Study Group and on the
Council.

William Fox
London, January 1974

xiii

I

Character of the Tin Industry

THE bulk of the world's tin is mined in a limited number of developing countries in the tropical belts north and south of the Equator. The major mining area, linked to the same basic geological feature, is in the south-east Asian block – Thailand, Malaysia and Indonesia – and produces two out of every three tons of tin entering world trade. In Africa, only two countries (Nigeria and Congo-Zaïre) can be regarded as having tin fields of any world importance. In South America, Bolivia, which is responsible for about one-sixth of world output, is stamped with very sharp geographic, mining and economic problems which make it unique.

Within these producing countries the zones of actual tin mining may be quite localised. In Nigeria almost all mining takes place within the old Plateau province; in Thailand in the southern peninsula; within Malaysia in the two states of Perak and Selangor; in the Congo-Zaïre, in the districts of Kivu and north Katanga; in Indonesia on the tin islands of Bangka and Belitung.

Outside, but sometimes on the edge of, the tropical belt the more important tin producers are in Australia, where a tin mining industry rose into sudden brilliance in the second half of the nineteenth century, almost died in the first half of the twentieth century and has recently been revitalised, in the People's Republic of China and in the U.S.S.R. in Asia. China, although massive in reserves and possibly in output, has kept itself fairly aloof from the international field since the Communist revolution. The U.S.S.R. has not been so successful in following this isolation; on the contrary, it unexpectedly poured a great mass of stockpiled tin on the world market in 1956-60 with devastating effect and so helped in producing the worst price crisis in tin for thirty years.

The six tropical producers, who cover between them normally some 90 per cent of world production outside the U.S.S.R. and China, have been the backbone of all the international arrangements on exports and prices which have been made first by these producers as a group through the International Tin Committee before 1941 and, later, by the producers and consumers in co-operation through the International Tin Council after 1956. The smaller producing areas, especially in Cornwall, have often a fascinating historical background, but in recent times have carried scarcely any weight in the world economic balance of tin.

The art of tin mining

The mining of tin is mainly the mining of alluvial deposits. Less than a quarter of the total mined is obtained by hard rock miners working underground and using drilling and blasting techniques. In either case the key to economic mining is the tin-containing quality of the ground being mined. The key is turned by the world price for tin metal in the world metal markets, although the ease with which that key is turned may depend on a variety of factors such as the existence of a labour force with inherited mining skills, the costs of labour, the costs of transporting mining supplies and equipment to the tin field and of hauling concentrates from the tin field to the smelter, the volume of rainfall (the industry is greedy for water), the degree to which hand labour can be mechanised and the flow of capital. But these factors cannot destroy the over-riding importance of quality in the ground.

In general, an economicalluvial deposit (which itself may show very widely differing values of ground within a relatively limited area) is required to hold an average of around 0·5 pounds of tin per cubic yard of ground (say, 0·017 per cent tin content). This is a very rough and ready figure. Where a large-scale method of working, semi-factory in character, like dredging can be operated this figure may be reduced to 0·4 and even less than 0·3 pounds per cubic yard but with gravel pump mining it may be necessary to raise the economic limit by one-third or one-half more. There is, of course, no upper limit to the tin content. In the early history of any tin field, qualities of ground are often met and initially exploited which are bonanza in character. This is true of the newest tin field in Brazil, and a relatively modern field

(Congo-Zaïre) can still report working ground of about 1·5 pounds tin content per cubic yard. But the bonanzas are a matter mainly of past history and sometimes only romance and do not hide the fact that in alluvial tin mining – as indeed in all non-ferrous mining generally – it is almost an inevitable process that the richer deposits are mined first. The continuation of mining· means a steady decline in the quality of the ground being mined and, in consequence, an increase in the tonnage of material to be handled in order to maintain the same output of metal content and more pressure for a higher metal price to meet the consequent increased costs of production.

In alluvial mining, the over-burden (that is, the immediate surface material which does not contain any tin) is removed and the deposit of tin-containing ground in gravel, clay or even small stones is excavated. The excavated material is washed and cleaned to obtain the tin which emerges not in metal form but as a tin concentrate (usually of cassiterite or tin oxide) for delivery to the smelter for conversion into marketable metal.

Excavation may be in almost any manner, from the most primitive to the most advanced mechanised form. The simplest method is the panning or "dulang washing" followed ·in Malaysia and Thailand. Women pan the tin-containing gravel from river beds and wash it in the stream, a process which separates the heavier cassiterite from the lighter gravel. This method, which has shown no basic change for hundreds of years, still wins in handfuls currently about 3,000 tons of tin a year in Malaysia and 1,000 tons in Thailand (a total equivalent, when translated into tin metal, to meet nearly a year's consumption requirements of India for all purposes). The capital requirements for a dulang washer are the physical ability to stand in the stream, a pan and shovel and an official permit. These permits are usually issued only to women.

A technical step above dulang washing comes in the open paddock of Nigeria where the massive use of hand labour (sometimes seasonal) provides still between one-third and one-quarter of the national output of tin. Higher is the use of gravel pumps and monitors, in which a jet of water is forced on the tin-containing open-cast face of the mine and this material is run or pumped to the cleaning and dressing plant where the cassiterite is isolated and bagged as the tin concentrate. This is the system most common in Malaysia and Thailand and is responsible for half the output of those two countries. If gravel pumps in Indonesia are taken into account this system provides about one-third of total world production. The capital requirement, mainly for pumps, monitors and electrical equipment, is small and well within the reach of local entrepreneurs; the requirement in skill, especially on the recovery side, is high.

The further step which helped to revolutionise the industry in south-east Asia fifty years ago was the dredge – a highly mechanised and floating factory, riding on an often self-created inland pond or at sea, and excavating the tin-containing ground below the water surface. The digging depth of the dredge has been increased until it may now be 120 feet or more. A modern dredge may deal with more than 5·0 million cubic yards of ground a year and its annual output may be as high as 1,000-1,500 tons of tin. It is a complicated and expensive piece of equipment; its capital cost is very high; and it has brought with it into the tin mining industry all the calculations on repairs, maintenance and depreciation as cost elements with which manufacturing industry in the highly industrialised countries is only too painfully aware. Dredging – which is in effect limited to south-east Asia – provides, however, in spite of its technical advantages, less than one-quarter of the world's tin.

Lode mining is the other major form of mining, so different as to seem to belong to another world. In lode mining adits are driven into the hillsides or shafts are sunk down through the granite rocks where the cassiterite lies, normally in thin veins, at depths which may run to 1,500 or more feet below the shaft head.[1]

[1] The Wheal Jane mine, opened in 1971 in Cornwall to treat lodes at a depth of up to 1,500 feet and with an anticipated ultimate output of 1,000 to 1,500 tons of tin a year, cost £5·5 millions (after allowing for capital grants from the government). On the other hand, Selangor Dredging Ltd. expected its new dredge in 1972 to cost $M11·4 millions (around £1·7 millions), to handle about 850,000 cubic yards of ground a month and to produce perhaps 2,000 tons of tin a year.

[2] Pahang Consolidated in Malaya has ground of about 0·9 per cent tin, Wheal Jane in Cornwall about 1·25 per cent, Renison in Tasmania 1·0 per cent and Catavi in Bolivia 0·7 per cent.

The ore is blasted by explosive, hauled to the surface, crushed in the mills and washed in the cleaning plant for its cassiterite content. This was the classic form of mining as practised for two hundred years in Cornwall; it still persists there, but its stronghold now is Bolivia. Almost all the Bolivian production is from lode mining. Normally, although not always, the units of production in lode mining require substantial capital resources to cover the necessary shaft sinking, ventilation, water-pumping, timbering, use of explosives, and milling and crushing plant.[1] There is, however, scope still left in Bolivia for the small mining company or even for the mining co-operative which is prepared to run its lode mine down to about 100 feet on a capital that calls for little more than a winch, pails and shovels, the minimum of explosives and a complete disregard for sudden death.

The rock worked by lode mines at greater costs contains necessarily a much higher tin content than the ground worked in alluvial deposits. There have been many past cases of phenomenal quality in lode mining. In Bolivia the fortunes of Simon Patiño were based on a tin mine - La Salvadora – which contained some tin ores of almost 50 per cent tin content and yielded for many years many ores of well over 20 per cent content. These bonanzas have usually died quickly and the lode veins suffer with the alluvial deposits from a progressive thinning of their quality as the better and more easily reached veins are initially worked out. It would be normal to assume, but not always correctly, that lode ores with less than about 0·8 per cent tin content are not at present economic propositions[2]. The ultimate quantity of reserves of tin in lode mines tends to be greater than the volume of reserves on which an alluvial mine lives, and the most famous of the world's mines – Dolcoath in Cornwall, Llallagua-Catavi and Unificada in Bolivia and Renison Bell and Mount Bischoff in Australia – have been or still are lode underground operations.

For a moment it is worthwhile comparing the metal content of ground in tin mining with the metal content in mining other non-ferrous metals. In tin the metal content in alluvial mining is around or above 0·017 per cent and in lode mining around 1 per cent; but aluminium processes bauxite with a metal content usually of not less than 15 to 25 per cent, for most zinc ores mined the metal content is between 4 and 16 per cent, lead ore ranges between 3 and 12 per cent and copper ores between 0·4 and 6 per cent. Gold-bearing ground in South Africa may carry 0·3 ounces to the ton. Sluicing and dredging are not important in the other metals which are concentrating on underground workings or, still more, on massive open-cast operations; their problem of declining ground is being met by treating still larger tonnages of material. Copper deposits are held to be economic on a copper content of 2·1 to 3·6 per cent in Congo-Zaïre, of 1·2 per cent in Iran, 0·5 per cent in Canada and as low as 0·48 per cent on Bougainville, lead and zinc in Canada on a combined metal content of 8 per cent and nickel on a content of 3 to 4 per cent in Australia[3]. These substantially higher figures for quality of ground in other metals, except gold, are a partial explanation for the relatively high market price of tin.

With such a thin quality of ground or ore to treat, the cleaning and treatment processes in tin mining, and especially in lode mining, become peculiarly significant since any loss of tin at these stages which might be not much more than an embarrassment to other metals may be a financial disaster to tin. The loss of fine tin (cassiterite in such fine grain that it cannot be retrieved by existing treatment methods) is very acute, and it is no consolation to know that this fine tin, beyond collection by present generations, may be building up the tin mines for future and better technicians.

The organisation of the industry

In any general comparison of tin with the other major non-ferrous metals it is necessary to adjust one's views very sharply. Mining in such other metals is normally accepted as a large or very large scale international industry massive in its requirements of capital from multi-national mining houses, massive in its tonnages both in world totals and from individual enterprises, massive in its integration of mines, smelters and fabricators, and mindless of the small producer. In terms of tonnage of metals

[3] See *Mining Journal,* 13 October, 1972 and 26 January, 1973 and *Mining Annual Review,* of *Mining Journal,* 1972.

produced tin is very heavily overshadowed by its non-ferrous brothers. The weight of tin produced annually in the world is only one-fortieth of the weight of aluminium, one-thirtieth of copper, one-twentieth of zinc, one-eighteenth of lead and less than one-half of nickel. Admittedly, in tin mining the quantity of metal in each cubic yard of ground mined is so low that in a comparison of the total volume of ground which is required to be moved tin holds a stronger position. Further, tin is a more valuable commodity selling at a price often three to four times the price of copper, six times the price of aluminium and ten to twelve times the price of zinc or lead, and sometimes even half again above the price of nickel. Measured in terms of the total world value of the product tin emerges in a more favourable light. Currently, the gross value of the tin produced outside the U.S.S.R. and China is of the order of $U.S.750 millions against perhaps $5,500 millions for copper, $4,300 millions for aluminium, $1,600 millions for zinc and under $900 millions for lead (on 1972 prices).

The smaller tin industry has had much of its form of organisation determined by geology, which has given it not only a very small percentage share of all the minerals in the world's surface but also individually smaller deposits to exploit. In the other non-ferrous metals perhaps the most striking feature in organisation, especially in recent years, has been the importance and even domination in mining of a small number of very large owners, many international in character. In nickel, world mining is conceived in terms, at least at the moment, of International Nickel, Falconbridge and Le Nickel; in copper in terms of the enormous properties and quantities of capital – British, South African and U.S. or, perhaps, merely anonymous – in the names of Amax (American Metal Climax), Anglo American Corporation, Rio Tinto-Zinc and American Smelting and Refining Co. (Asarco). Many of these groups do not confine themselves to one metal. Amax is in aluminium, copper, lead, zinc, molybdenum and iron ore; Asarco in copper, lead and zinc; Consolidated Gold Fields in gold, platinum, rutile, iron ore, uranium and zinc; Rio Tinto in copper, lead and zinc, aluminium, uranium and iron ore; Charter Consolidated in copper, potash, iron ore and asbestos.

Even where there has been in copper a full or partial nationalisation within the producing countries, as in Chile, Zambia and the Congo-Zaïre, this has provided a change in ownership and not in the structure of large scale units in production.

The tendency of the international mining houses to use the expertise and profits from the development of one metal in the exploitation of another metal, often in another country, has, however, not spilled over in any very important degree into the tin industry. These groups, so active in recent years in other metals, have been curiously reluctant to be attracted as yet into tin. This is in spite of the existence of an international tin agreement which has shown itself effective enough over many years in ensuring the existence of what all metals are supposed to clamour for – a guaranteed floor price. It is true that the Consolidated Gold Fields group, which was reported as showing an interest in tin internationally forty years ago, has provided the capital for the new Wheal Jane tin mine in Cornwall and for the development of Renison in tin in Tasmania. Cominco of Canada has moved into control of the Aberfoyle tin group in Australia and Charter Consolidated bought its way into the Tronoh tin group in Malaysia and into tin/wolfram at Beralt in Portugal. But the proportion of the assets of these groups devoted to tin is small and the proportion of the world's tin output held by these various groups in combination still remains unimportant. It is probable that, whatever the immediate rate of profit in tin mining, other factors are handicapping the flow into tin of capital from the international houses – the past relatively unstable political position in some producing countries, the fears of the application of nationalisation in the way it has been applied in Bolivia and the Congo-Zaïre and, perhaps more important in the long run, the slowness of the rate of increase in the global consumption of tin. These are all factors which may be counter-balancing whatever favourable climate is fore-shadowed by the stabilising policy of the international tin agreements.

Whatever the reasons for the relative non-entry of the international groups, the tin industry has been allowed generally – and perhaps complacently – to come to terms with the structure which it has known for the last twenty or thirty years.

In the world tin industry there is only one group of great importance operating internationally. This is the London Tin/Anglo-Oriental group. It was formed in the 1920s around British capital, it operated within what was then the British Empire in Nigeria and Malaya, and it was later linked with Patiño money from Bolivia. At its height in the 1930s the group, which seems in practice to have been a much looser organisation than is now customary with international groupings, covered about half the total Nigerian output, about one-sixth of the total Malayan output, and a small proportion of the Thai output as well as, on the Patiño side, the Patiño enterprises in Bolivia (more than half the Bolivian total) and the Consolidated Tin Smelting enterprises in the United Kingdom and in Malaya. This meant that the combined group then covered about one-quarter of the total mine tin production of the world and treated over one-third of the total metal production of the world. It was suspected, at the time, that the London Tin/Anglo-Oriental had many of its properties in Malaya in the high-cost category (at least by Malayan standards) and therefore had every incentive for the very strong support which it then gave to the international export control, buffer stock and price support arrangements before 1941. The power and authority of the group was strong, but was seriously exaggerated by pre-1939 writers (especially by American observers) and was weakened when the Patiño mining interests were nationalised in Bolivia in 1952. The Anglo-Oriental interest has shown little proportionate expansion in the last twenty years; its importance in Nigeria has dwindled as the total output of that country has dwindled; its share in the total Thai or Malaysian output does not seem to have grown; it never developed any further a preliminary interest in the Congo; it was never allowed into Indonesian mining; and it has not been prepared, at least as yet, to risk capital investment in the new or revived tin fields of Brazil or Australia. Currently, it covers perhaps less than one-eighth of total world mine production. The connected Patiño smelting interests have been held together and even extended to cover plants in Australia and Nigeria, thereby giving the Patiño group control of somewhat less than half the world's smelting of tin metal; but the United Kingdom unit in the group is threatened by the potential growth of the new government smelter in Bolivia[4].

There is no other privately-owned group in tin that in size or in international coverage has ever held a candle to the Anglo-Oriental/Patiño connection. In Malaya the old Osborne and Chappel block, perhaps as much technical and administrative as financial, held at one time approaching 10 per cent of Malayan production but has no ownership link with smelters or with tin firms outside Malaya; it was in the past as traditionally associated with opposition to export control as Anglo-Oriental was with favour. Other groups there (Tronoh, Malayan Tin and Austral Amalgamated) are small enough by world standards. In Congo-Zaïre the industry has always had two major groupings —one centring around the Symétain enterprise and one around Géomines (each controlling about one-third of the Congolese output but neither commanding as much as 2 per cent of world production). In Australia the two major mining concerns —Renison and Aberfoyle – are now under foreign control, but their joint output is less than 5 per cent of the world total.

United States capital, so restive elsewhere in the development of the non-ferrous metals, has shown surprisingly little interest in tin, in spite of the almost complete absence of tin mining within its own frontiers. There are isolated U.S. investments in Malaya (through the Pacific Tin Corporation) and the private sector in Bolivia, and the beginnings of interests in Brazil and the Congo-Zaïre. The U.S. capital in mining and smelting (Thaisarco) in Thailand has scarcely given encouraging results. Interest in Indonesia is so far limited to prospecting.

The two largest unified tin mining organisations in the world after London Tin/Anglo-Oriental are now the nationalised bodies which, in Indonesia and Bolivia, cover between them about one-fifth of world production. From the nineteenth century tin was accepted in Indonesia as government-owned with the right to operate being farmed out to private and entirely

[4] In 1973 Williams, Harvey and Co., the United Kingdom subsidiary of the Patiño-owned Consolidated Tin Smelters, went into voluntary liquidation. This reduced the Patiño share of world smelting output to under 30 per cent. In the same year, Haw Par Brothers International acquired 29 per cent of the London Tin Corporation shares.

Dutch capital. The system of farming out ceased finally in 1958 and the industry is now fully nationalised, both in ownership and management. In the case of Bolivia, nationalisation was applied in 1952 to three main groups –Patiño, Hochschild and Aramayo – which had become associated in Bolivian thinking with exploitation and external ownership. About one-third of the Bolivian industry has remained in private hands.

In the world tin industry, even sometimes within the larger private or nationalised groups, there is an astonishingly wide variety of production units. What is certain is that – contrary to the trend in other non-ferrous metal mining – the larger production unit is not dominant; in fact, it appears even to be losing ground. In Malaysia and elsewhere the dredge is generally accepted as the representative of the larger and more efficient mechanised producer. In Malaysia the average dredge produces around 400 tons of tin a year (a figure almost laughably small in comparison with the mechanised open-cast or underground mines of copper, lead, zinc or nickel) and all the dredges provide only one-third of the total Malaysian mine output; in Indonesia the dredges, although they vary very widely in capacity, average under 250 tons of tin a year and provide only half the aggregate national output; and in Thailand dredges of a similar size to those in Malaysia, produce only one-quarter of the Thai output. In Bolivia, where the dredge is almost unknown and underground mining predominates, the nationalised body (Comibol) spreads its substantial total output amongst tin mines that range in size from 250 to nearly 6,000 tons of tin a year.

The smaller unit of production remains very highly resilient. In Bolivia the twenty tin mining companies in the so-called "medium mines" category (some of which are larger than the mines nationalised within Comibol) have an average annual production of 250 tons and have been more successful in maintaining their total production in recent years than has Comibol itself. Malaysia has about 1,000 gravel pump mines. Their average output is no more than 40 tons of tin a year, but their total production is half of the Malaysian mines' and therefore approaching a quarter of total world output (that is, more than the combined output of the nationalised Indonesian and Boli-

vian industries). In Thailand and Indonesia the share held by
gravel pumping is very similar (a little over half), although the
average size of the gravel pump mine is lower in Thailand (40
tons a year against 70 tons in Indonesia).

These are not the limits of decreasing size. In Nigeria (leaving
out the European companies) we find that, in 1969, 101 smaller
companies and private operators averaged only 32 tons of tin a
year each but, in total, nearly one-third of Nigerian production.
In Bolivia there are at the moment 1,000 small mines (that is,
mines selling to the agencies of the Banco Minera de Bolivia).
These miners, working in the cruellest of conditions and on a
standard of living probably lower than in any other mining community in the world, produce from each mine an average of
perhaps less than four tons of tin each year, but still nevertheless
about one-seventh or one-eighth of Bolivia's total output.
Women "dulang washers" in the Malaysian streams and tribut-
ers in Nigeria make contributions, almost in handfuls.

The whole of the world's tin mines employs possibly between
180,000 and 200,000 workers on a basis which is full-time or
which, when it is seasonal, might be regarded as semi-full time.
In addition, there is an indefinite number, perhaps of the order of
50,000 of part-time workers, part-time in the sense that they may
move to mining work if this agrees with the variations in their
agricultural work.

The position in the smelters

Organisation in the tin smelters, which collect the cassiterite,
invest in expensive smelting techniques, turn it into tin metal
and help to finance the miners with advance payments, is necessarily of a different character. There are only nine tin smelters in
the world who can be regarded as important (that is, with an
annual output of over 5,000 tons a year). Of these, four (one each
in the United Kingdom, Malaysia, Nigeria and Australia) are
largely controlled by the Patiño interests, with in total about 40
per cent of the world output. Another in Malaysia (Straits Trad-

⁵ This did not represent a break with foreign capital. The initial capital for
   Thaisarco came from the U.S.A. In 1971 Dutch capital joined in.

ing Co.) makes perhaps another quarter and has steadily been
throwing itself open to local Malaysian and Singapore investment. There are two nationalised smelters – one in Bolivia, one
in Indonesia – but neither smelts the whole of the local production of concentrates and their current joint output is only 10 per
cent of the world total. In Thailand, the Thaisarco firm has a
monopoly in smelting material from the domestic mines and is
responsible for over 10 per cent of the world total.

Prior to 1939, the movement of concentrates to the smelter for
treatment emphasised, but not very heavily, the colonial element
in the tin industry. All Nigerian, all Bolivian, most Indonesian
and most Belgian Congo concentrates were shipped to the
British and western European smelters, but local smelting in
Malaya was established early on (through the Straits Trading
Co. with British money) and a special duty on the export of
cassiterite from Malaya helped to ensure that the smelting of
local material was guaranteed to the Malayan smelters. All the
Thai ore and some smaller amounts from Burma and East Africa
were attracted to the Penang and Singapore smelters. Australia
smelted its own then shrinking domestic mine production.

The pattern of the dependence of Bolivian, African and
Indonesian mines on European smelters, which meant that
Europe mined almost no tin but profited directly from the smelting of about one-third of the world's concentrates, has been
broken in the last twenty years. Local pressure in Nigeria forced
the Consolidated Tin (Patiño) organisation to set up a smelter
there treating all the Nigerian mine production. One
government-owned smelter was rebuilt in Indonesia and another
was built in Bolivia. In Thailand, a new smelter (Thaisarco) was
opened and the smelting of Thai concentrates was shifted there
from Malaysia[5]. The Dutch smelter at Arnhem was closed down
in 1971 as the supply of tin concentrates for smelting from
Indonesia dried up.

There have been few effective instances, although one of them
was very important, of the miners integrating forward into
smelters. The important case was, of course, the use of Bolivian
mining profits by the Patiño organisation to buy up British and
Malayan smelters in the 1920s; the other and very much less

important cases were the creation by mining enterprises of tin smelters in the Congo.

In general, however, the considerable degree of vertical integration seen in other non-ferrous metals has not spread to tin. The smelters may have share interests in some mines (for example, the Straits Trading Co. in Malaysian mines), interests which are likely to influence the continuation of long-term contracts for concentrates. But, in general the smelters appear to have retained what they state to be their traditional roles. The first is as the providers of a process for transforming concentrates into a metal meeting the requirements of the world's metal markets, a process in which they need take no risk on the market and for which they are paid a carefully calculated price; the second role is the provision to the miners, through the system of immediate cash advances on the delivery of concentrates, of the cash flow to cover the necessary working costs in an industry where there is often a long time-lag between the mining of a concentrate and the marketing of a metal. The aloofness of the smelter from market price variations or even from taking a view on the market is not perhaps as complete as the smelter would wish us believe, but it is an attitude that may have helped to keep him from trying to integrate himself in too close a fashion with the ownership of mines.

No serious attempt has been made by the smelters to integrate forward into the ownership of tin-consuming plants; the Billiton interests in white metal subsidiaries and the Consolidated Tin interest in a tin oxide producer guarantee the consumption only of a small proportion of their metal output. Most striking has been the financial independence of the melters from the tinplate users who take so high a proportion of tin consumption.

No tinplate plant in any country is owned or was owned by a tin smelter, nor as yet, does any tinplate plant own any large tin mine.

---

[6] The word "thickness" is used, but it is to be remembered that the steel sheet itself will be less than one-hundredth of an inch thick and that the tin coating will be as little as 15 to 30 millionths of an inch thick. In other words, in hot-dipped tinplate the tin used is 0·7 to 1·25 per cent and in electrolytic coating probably under 0·5 per cent of each ton of tinplate.

The consumption of tin

Concentration of world consumption of tin in the northern temperate belt of developed industrialised countries has been as sharp as the concentration of mine production on the tropical belt of developing countries. The northern belt – the United States, Canada, Japan and Western Europe – takes 80 per cent of all world tin metal used; the developing countries which are not mine producers take only 4 per cent; and the six major mining developing countries which form the backbone of the International Tin Agreement take probably less than 1 per cent.

Within the developed countries there is a heavy dependence on one use – the use of tin in the making of tinplate (that is, of steel sheets coated with a thin, and usually extremely thin, plating of tin). Tinplate in its primary function – as a container in tins or cans of foods and drinks – is dependent on maintaining a share in the whole of the packaging industry, where it is in constant and generally successful enough competition against paper, plastics, glass bottles and aluminium. Tinplate is also in a constant state of internal competition, first, to secure thinner coatings of tin (which is normally regarded by the tinplate manufacturers as a relatively expensive metal) and, secondly, to secure the complete elimination of the tin coating and its replacement by substitute coatings (in the latest instance by tin-free steel, plated with chrome). The problems of technical innovations, of course, are not peculiar to the tin industry. In general, the steel can (tin-plated or not) is so useful in its application for canning that it is likely always to maintain a substantial share of the total world packaging market which is itself growing with modern affluence. The reduction in the thickness of the tin coating[6] over the last thirty years because of the fairly complete transition from hot-dipped, thickly-coated tinplate to electrolytic, thinly-coated tinplate has resulted in pushing down current usage to only one-third of the tin metal per unit of tinplate coated. This drop in the usage on the unit has been counter balanced by the great expansion in the total tinplate produced, so that the total use of tin in tinplate has been more than maintained. But there has been relatively little expansion in that total tin used (as compared for example, with the use of nickel in plating) and the tin industry is well aware of its heavy dependence on this single use (perhaps 40 to 45 per cent of total world tin consumption). It is also well aware that the immediate users of this tin – the tinplate industry – represents only a small proportion of the capital invested in the steel industry and that the primary interest of the steel firms (who own almost all the tinplate firms as a tail end of their steel manufacture) lies not in the manufacture of tinplate sheets but in the manufacture of steel sheets to which there may be applied, in accordance with cost and demand, a tin coating or a coating of a competitive substitute or no coating at all.

Outside tinplate, the consumption of tin falls under four major headings. They are almost all concerned with advanced technological development. Between 20 and 25 per cent of the tin consumed is used in solder[7] where the tin is usually a minor constituent in alloy association with lead. The solder is used for jointing, especially in electrical components. The tendency to use smaller joints or to thin down the proportion of tin in the tin-lead solder is counter-balanced by the growing number of soldered joints used in modern electrical and electronic work. Solder, with tinplate, has been responsible for the bulk of the increase in tin usage in recent years in developed countries.

White metal alloys (including babbitt and anti-friction metal) take 5 to 10 per cent of the total tin used. Here, tin is a component, usually a minor or very minor one, with other non-ferrous metals, and its usage is affected adversely (as in the shift in the United States from journal bearings to roller bearings in the railway industry) or favourably (as in the development of tin-aluminium bearings) by technical developments in industry generally.

Bronze takes about 5 per cent of total tin usage. The tin is a minor component (for example, up to 10 per cent tin in gunmetal, 9 to 11 per cent in phosphor bronze and 4 to 6 per cent in leaded gunmetal) with the copper, but the result of the combina-

[7] 22 per cent in the U.S.A., 8 per cent in the United Kingdom and over 30 per cent in Japan. The variation may be due to different statistical classifications.

[8] This comparison needs some qualification. Early figures would include some tonnage for the U.S.S.R. but nothing in 1971, but the generality of the conclusion is untouched.

tion has been to produce in bronze a material whose mechanical strength, chemical resistance and ease of manufacture have remained constant through scores of civilisations and thousands of years.

The other uses of tin are spread, normally in small quantities, over a host of usages – in tinning electrical wire, as an addition to cast iron, in alloys with metals other than lead and copper, in polyvinyl plastic containers, in tin organic chemicals, etc. There are few uses of tin purely or almost entirely in its undiluted metal form. The only ones of any importance are in the making of organ pipes or in pewter (which is usually over 90 per cent tin content). Unless the world has a religious revival or switches heavily to pewter for its ornaments and tankards, neither is likely to become a substantial user in terms of tonnage.

In a world expanding in every direction in its industrial consumption of metals tin presents no very encouraging picture. Probably the root of the tin problem is not, as is so often held, in over-production but in persistent under-consumption. It is not possible, of course, to enforce on any world economy a rate of tin usage which it does not desire or which it has not bee taught to desire, but it is certain that those interested in the development of tin consumption have seen in their industry a peculiarly slow rate of advance, slow in relation not only to outside general factors but also to other, and sometimes competing, non-ferrous metals.

Since 1900 the world consumption of tin metal has increased only 2·2 times, from 82,000 to 184,000 long tons or a simple increase of 1·7 per cent per annum. Most of this increase was effective in the first third of the twentieth century – at a rate of about 2·2 per cent per annum between 1900 and 1935. Momentum was lost in the second third and between 1935 and 1972 the rate of increase was as little as 0·8 per cent per annum. The peak consumption figure registered in 1937 was not reached again until 1971[8].

Tin consumption, not only in rate of growth but also in trend of growth, has been in direct contrast with the other non-ferrous metals of the world. It is perhaps not reasonable to compare tin with a giant industry like aluminium which has shot, in such a short time, into gigantic uses in new as well as in some old industries. It is also perhaps not reasonable to bring in copper with its massive electrical, communications and transport markets, thirty times the size of the tin market and with a twelvefold growth in usage in the past seventy years. In that same period zinc usage has been multiplied by six and the industry is now in tonnage terms over twenty times the size of tin; lead, which in recent years seems to have been a lagging non-ferrous metal, is now almost four times its 1900 figure and is now sixteen times the size of tin. But, even if these other metals can be dismissed on grounds of comparative size, they cannot be dismissed on grounds of trend. In copper, lead and zinc, but not in tin, the rate of growth in the thirty years before 1938 was lower than the rate of growth after 1938; and this was true of the clearest guide to industrial activity and development – the consumption of steel[9].

The tin industry suffered, especially after 1938, three very severe blows to its actual or potential development in consumption. The first was the technical revolution which shifted the steel industry from the hot-dipped to the electrolytically-coated tinplate. This shift-over took thirty years and is now almost complete. The effects on the non-development of tin usage have been almost startling. Although there was in that period a trebling of the total amount of tinplate produced in the world, the consumption of tin metal for coating that plate has increased by only one-quarter; in the U.S.A. (the real originator of the electrolytic process) tinplate production more than doubled but tin usage fell by one-third. The change-over to the thinner coating was determined in the United States partly by fears of a possible

[9] *Lead and Zinc : Factors affecting consumption* (International Zinc, Lead and Study Group, 1966) gives figures to which are added the tin figures:

| | Compound growth rates per cent per year | | |
	1900-38	1950-60	1960-65
Lead	1·2	2·8	4·3
Zinc	2·8	3·0	6·3
Copper	3·3	3·9	6·1*
Aluminium	11·5	9·2	10·1*
Steel	3·7	6·0	5·8*
Tin	1·58	1·01	0·61

*1964 for copper and aluminium

tin shortage; it is not possible to say how far it was determined by price.

The second severe loss was due not to a smaller use on each unit of product but to the almost complete substitution by another metal. This was in the use of tin metal for foil (in packaging) and in collapsible tubes. In foil the replacement was by aluminium, and in collapsible tubes by aluminium and (in lesser degree) by lead. In 1936 some 17–18,000 tons of tin metal, about one-tenth of total world consumption, was being used in the making of tin foil and tin collapsible tubes; today, the total world usage is negligible, perhaps no more than 2,500 tons. The price of tin, the war-time shortage of tin and the improvement in the extrusion techniques for aluminium combined in the killing.

The third loss was a lost opportunity. In the new major technological revolution of the twentieth century – the coming of the automobile – tin made a most promising start. In 1936 the world automobile industry was estimated as using no less than 19,500 tons of tin a year (much of it in the form of solder). The figure was perhaps very optimistic. In any case, it was not maintained. At that time an automobile was taking perhaps four pounds of tin (most in solder). By the 1960s the amount was down to perhaps 1½ to 2½ pounds, and the process of thinning down the tin content of solder has continued. With six times as many automobiles produced in 1971 as in 1936 the total amount of tin used in the process has fallen to perhaps under 5,000 tons a year.

This is perhaps over-stressing what tin has lost. It is not abnormal for non-ferrous metals to lose particular markets to one another or to an outsider. Copper usage in electricity has been sometimes replaced by aluminium usage. Lead is heavily dependent on the automobile industry and its consumption would be badly affected if that industry shifted to electrical transport motors or, in smaller degree, if the lead battery were replaced by the nickel-cadmium battery. Plastics are a common enemy of most non-ferrous metals. Against its losses over the last forty years tin must take into account the arithmetic of tinplate (a rate of increase in the aggregate production of tinplate which, even on the thinner coating, has meant some increase in the total tonnage of tin metal consumed); the remarkably successful fight in the use of solder in the field of electronic micro-miniaturisation and elsewhere (which has meant that between 1960 and 1970 the tin used in solder in the more important industrially developed countries, including Japan, rose by one-half although total tin consumption in those countries remained almost unchanged); the modern recovery in the use of tin oxide and tin chemicals in the U.S.A.; and the general potentialities opened up by research work into the use of tin compounds and organo-tins and of tin in cast iron.

A comparison with nickel may be rewarding. Nickel, of the other non-ferous metals, is nearest to tin in size, (roughly twice the annual production of tin). It is the nearest in price, and both are regarded by many users as expensive. Both have a curious parallelism in usage, in their heavy concentration, first, on a small number of large uses (nickel with about 60 per cent of its total in stainless steel, plating and high metal alloys, tin with perhaps 65 per cent in tinplate and solders) and then their scattering on a very large number of minor uses (nickel in iron and steel castings, constructional steel alloys, copper and brass products, nickel alloys, nickel-cadmium batteries and chemicals, and tin in white metals, tin chemicals, tin oxide, brass and bronze). Yet statistics show that the consumption of nickel was ten times as high in 1971 as it had been in 1925.

It is difficult not to say that the world tin mining industry has suffered from a lack of sense of urgency as to the importance of expanding tin consumption. The producers' own research organisation (the Tin Research Institute) has, in its forty years of existence, not been as successful as the other non-ferrous organisations in translating its very interesting and advanced scientific research work into development expressed in actual terms of more metal consumed.

II

The Pattern of Tin Production

AT all times over the past century world mine production of tin has been highly concentrated within a relatively small number of producing countries. In the period there have been seen some changes in the order of importance of those producers, the entry of new countries and the weakening of the status of some older members, but the industry has been dominated geographically by a handful of countries.

In the 1870s the largest single producer in the world was still the United Kingdom. Close behind it ran Australia, where the industry was relatively new, and the Netherlands East Indies (Indonesia) and Malaya, both with old-established mines. These four countries produced some 90 per cent of the then 37,000 long tons of tin being mined in the whole world outside China. The two south-east Asian countries – the Netherlands Indies and Malaya – held combined about 40 per cent and the United Kingdom and Australia combined 50 per cent. Production elsewhere – except Siam (Thailand) – was of no importance.

By the 1890s the role in tin of the south-east Asian countries – Malaya, Indonesia and now Siam – had become dominant. Malaya alone was producing well over one-half of the world total of 77,000 tons and Indonesia nearly one-fifth. If Siam were included, south-east Asia was mining approaching 80 per cent of the world's tin. This marked the peak of the south-east Asian dominance, but a reduction in the order of magnitude of importance still left it the decisive force. In 1921, during a world slump, the three south-east Asian countries were still responsible for 60 per cent of world production. In 1941, when the world was producing tin not only at twice the 1921 level, but also at the highest tonnage figure ever known, Malaya, the Netherlands East Indies and Thailand yielded 62 per cent of the output; if Burma and Laos were included in the south-east Asian figure the proportion became 64 per cent. Within this pattern of general control by south-east Asia there were, of course, shifts in the

FIGURE 1

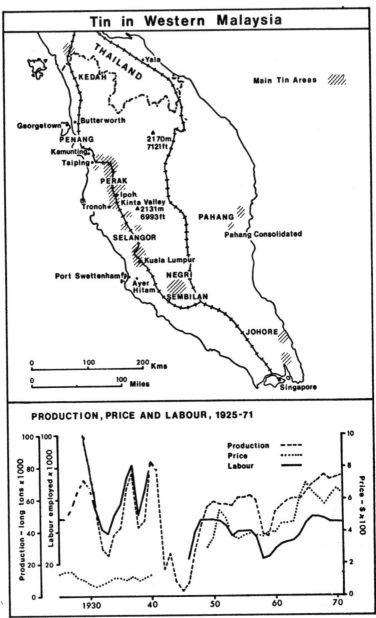

Tin in Western Malaysia

Main Tin Areas

PRODUCTION, PRICE AND LABOUR, 1925-71

Production ----
Price ....
Labour ——

relative importance of the three main partners, but Malaya was dominant at all times and could normally be expected to produce two-thirds of the south-east Asian total and over one-third of the world total.

This hegemony by south-east Asia was threatened only twice. Bolivia did not enter the tin mining field on any substantial scale till after 1900, but then increased its share of the world market so dramatically that by the 1920s it could claim about one quarter of the whole. Its competitive position then suffered from the degradation of its tin-mining ground and, apart from the immediate years of the Second World War, it slid backwards in the international league.

The second threat came from the temporary occupation of south-east Asia by the Japanese during 1942-45 when, for the moment, the share of the area in world production fell to about only one-fifth. But recovery was rapid and by 1951 the proportion was back again to 60 per cent. Over the last twenty years Malaysia and Thailand, if not Indonesia, have more than kept pace with the movement of world production, and in 1972 south-east Asia produced 64 per cent of the world's tin. This is a remarkable long-term domination of a non-ferrous mining industry by one geographical bloc.

Malaysia

The great bulk of the tin deposits of Malaysia lies within a thin strip, at the most 400 miles long and fifty miles wide, along the western coast in the zones of contact between the sediments and the granite Main Range which forms Malaysia's backbone. In this western strip the two main sources of tin supply are from the states of Perak (especially in the Kinta Valley) and of Selangor (especially around Kuala Lumpur). The Kinta Valley has been unchallenged since the 1890s as the largest and most consistent tin producer in the world and has mined probably well over 1½ million tons of tin. Perak as a whole is responsible normally for about 55 per cent and Selangor for over 30 per cent of the total Malayan output. Elsewhere in western Malaysia there is a small mine production in the states of Negri Sembilan (not expanding) and Johore (rising slightly). All production in the west is from alluvial deposits.

In the eastern strip of Malaya, separated from Perak and Selangor by 100 to 150 miles of mountains, there is mined in the states of Pahang and Trengganu only 5 per cent of the Malayan output. This includes output from the underground lode mine of Pahang Consolidated, the only underground tin mine of any importance currently operating in the whole of south-east Asia.

There is as yet no underseas mining along the Malaysian coast. Under-sea mineralisation is known to exist along the western coastline, but very prolonged discussions between state and federal governments and between them and prospecting companies have not yet (by 1972) brought up any tin.

For an industry with such long pre-eminence in the world tin economy Malaysia has always shown a surprising reluctance to count its reserves. The government has never fully understood the political importance of keeping world consumers aware of the potential life of the tin field and its likely availability for consumption. It is highly probable that estimates of the reserves at perhaps 1·5 million tons of tin made in 1938 before independence, 1·0 million tons made in 1962 after independence, and of 0·9 to 1·0 million tons made in 1972 – all excluding underseas deposits – are all under-estimates.

Two great technical advances in the first quarter of the twentieth century formed the foundations for Malaysian mining success. The first was the intensive application by the Chinese miners after 1906 of a European idea – the shift to mining by gravel pumps using monitor jets. The second was the introduction into Malaysia in 1912 of the tin dredge, a step which brought in a flood of European (mainly British) capital to the industry.

[1] "European" mines is a loose but convenient term to cover mines owned by companies originating in the United Kingdom, Australia, France and the U.S.A.

[2] On this question see Yip Yat Hoong: *The Development of the Tin Mining Industry of Malaya* (1969) for very full information.

[3] On behalf of the United Kingdom Ministry of Supply, A. D. Storke undertook an enquiry on the spot in 1945 into the rehabilitation of the Malayan tin industry (especially on the dredge side). His report to the Ministry considered drastic rationalisation, including a very much smaller number of larger new or reconditioned dredges. These proposals did not become effective.

These advances ultimately produced an industry which – by the standards of tin mining, if not of mining elsewhere in other non-ferrous metals – has become relatively highly mechanised. The peak labour force in the twentieth century was reached in 1907 with 231,000 workers, each producing on the average about one-fifth of a ton of tin per annum. By 1920 the labour force was down to 90,000 (producing two-fifths of a ton per annum), and by 1951 to 46,000 (each producing 1·3 tons per annum). The labour force was then fairly stabilised (except during the export control period of 1958-60) and by 1972 with an annual output of about 1·7 tons of tin a head the industry seemed stable with an approximately 45,000 work force.

That labour force was basically Chinese (initially immigrants, later Malaysian-born); the agricultural-based Malaysians took little part in what was to them an alien activity, which took little land but ruined it for agriculture.

In 1920 the Chinese miners in Malaya were responsible for two-thirds of the total Malayan output and the European mines (including the new dredges) for only one-third.[1] But over the next twenty years the influence of the European mines was growing. In the mining boom of 1920-27 the total issued capital of the European mining companies, whether registered inside or outside Malaysia, rose from £5·4 millions in 1920 to £12·1 millions in 1927. By the end of 1930 the 79 dredging companies alone (all European) showed a capital of over £13·0 millions; in addition, the holding of European companies in non-dredging concerns was perhaps another £4·0 millions.[2] The Storke enquiry listed the total issued share capital of 100 European tin mining companies in Malaysia in 1940-41 (dredging and gravel pumping) at approximately £17·0 millions.[3]

It was this investment, especially in the 1920s, which changed the pattern of production within Malaysia. By 1930 the respective positions of the European and Chinese groups had become almost exactly reversed, the European interests now holding 63 per cent of the total output and the Chinese interests only 36 per cent. Through the 1930s the European share (subject to the qualification that it is over-stated since it embraced the production of Chinese sub-contractors) was normally two-thirds of the whole. It reached a peak of 71 per cent in the boom of 1940-41.

This proportion has, however, been substantially weakened over the last twenty years. Through the 1950s the European share was slightly over 60 per cent. A short-lived recovery in the early sixties was followed by another decline and in the later 1960s the share was usually under 60 per cent.

The strength of the European interests had been in dredging. After the Second World War production from the dredges showed a surprising degree of stagnation, given that the dredge as an operating unit is more highly mechanised, more capital intensive and able to treat much poorer grade of tin-bearing ground than the other units – gravel pumps and hydraulic mines – in the industry.

The number of dredges and the size of the average dredge had risen consistently. In 1921 there had been 30 tin dredges with a total output of about 5,000 tons of tin (an average of 165 tons of tin a dredge); in 1929 there were 105 dredges producing over 27,000 tons (or 260 tons a dredge); in 1940, 104 dredges with an average production of 400 tons a year.

Storke's proposals in 1945-46 for the re-organisation of the dredge industry into a smaller number of larger units were not accepted. But at no time in the last twenty years has the number of dredges at work reached anything like the 104 reported for 1940. Rehabilitation after the war brought the number back to 83 in 1951, but these suffered from age and particularly from the export control measures of 1958-60. Through the 1960s they dropped in number and by 1972 were only 61; their average output fell to around 400 tons of tin a year; and their share of the total Malaysian output to around only one-third.

It is not easy to see why the dredging section of the industry, with every economic advantage, has failed in recent years to

---

[4] Between 1929 and 1952 five out of 14 dredges in the Kinta Valley showed a fall in the tin-containing value of their ground by one-third or more, six fell by less than one-third but three actually showed a rise. F. T. Ingham and E. F. Bradford: *The Geology and Mineral Resources of the Kinta Valley,* Federation of Malaya Geological Survey.

[5] It is to be noted that the dredge requires what may not be available, namely a greater body of reserves in sight in order to amortise its larger capital investment.

maintain its position. It is true that the tin content of ground excavated by the tin dredges has generally been reduced, but this is a factor which became evident early on in dredge history,[4] may be expected to continue and presumably runs with a similar, if not exactly parallel, movement in the grade for the Chinese-owned gravel pump mines. It is true that a dredge company requires a greater operating area for mining than does a gravel pump and, therefore, may be more affected in its development by a shortage of land declared open to mining or by inadequate exploration of unmined land.[5] It may be that the dredge industry has in recent years found it increasingly difficult to raise capital.

The main source of financing of the European-owned industry (not merely in dredging) has been the London capital market and the Anglo-Oriental group and the accumulated, undistributed profits of the companies themselves. The London capital market dried up after 1941; it has not been persuaded to re-enter an industry which is linked in the market's mind with nationalisation in Bolivia, with inefficiency in Indonesia, with a long post-war state of political emergency in Malaysia, and with the shadow of communism in neighbouring Vietnam. In other non-ferrous metals the place of the London capital market has been taken very largely by the international mining finance houses, but these houses have not been attracted in recent years into the Malaysian tin industry in any substantial degree. An exception was the investment of about £3·0 millions in 1965 by Charter Consolidated in Tronoh Mines Ltd., but this had some of the characteristics of a rescue operation for an enterprise which had run dangerously short of mining reserves rather than an example of the expansion of productive power. The interest of the international Conzinc Riotinto (now withdrawn) and Billiton groups on the western coast has been concerned with the possibilities and not actualities for dredge mining underseas.

It may be that the technical and financial impetus which in the 1930s was indicating the ultimate control of Malayan tin by a European-owned dredge industry was largely destroyed by the rough time coincidence of the political independence of Malaysia and the ending of the political and financial power in the Bolivian tin industry of Patiño, one of the basic supporters of Anglo-Oriental, the greatest dredge group in Malaya.

The fear, strongly held especially in Malaysia and the United States before 1939, that the whole Malaysian tin industry was being dominated by European companies and capital and that those European companies were being dominated, in turn, by two or three small groups has not been realised. In 1963, 60 per cent of the total Malaysian output was produced by 76 European-owned companies, who owned a total of 100 mines.[6] Three groups – the most important being Anglo-Oriental – covered 47 companies in ownership or in control or in buying agencies; this represented 45 per cent or about 27,000 tons of the then total Malaysian output of 60,000 tons. Within that group Anglo-Oriental represented about 14,000 tons of tin. These figures bear little relation to a picture of the monopoly or quasi-monopoly control to be seen in other non-ferrous metals in other countries. To the outsider, at least, the striking feature of European-owned mining in Malaysia is still the persistence of independent and relatively small companies. Twenty-nine companies (outside the three groups referred to above) still shared 55 per cent or 33,000 tons of the total Malayan output in 1963. The 35 companies (all or nearly all European) listed by *Tin Interna-*

[6] Yip Yat Hoong, *op. cit.,* pp. 21 *et seq.*
[7] *Tin International,* July, 1972.
[8] There are exceptions. As early as 1940 the Chinese-owned Hong Fatt Sungei Besi Ltd. had an issued capital of M.$3·0 millions. Ten other Chinese companies listed in the Storke report had issued capital totalling M.$2·0 million.
[9] F. T. Ingham and E. F. Bradford, *op. cit.,* gave the division of gravel pump and hydraulic mines in the Kinta Valley in 1950-52 as follows:

Monthly production, long tons of tin, approximately	No. of mines
0 to 1·1	50
1·1 to 2·2	86
2·2 to 3·3	75
3·3 to 4·4	47
4·4 to 8·8	58
Over 8·8	23
Total	339

The Malaysian average for gravel pump and hydraulic mines was then about 39 tons of tin a year.

tional in 1972[7] had a total output of around 30,000 tons of tin or 40 per cent of the then Malaysian output. Only eleven of those companies had a production of over 1,000 tons of tin a year; and the largest mine – Berjuntai – had a production of little over 3,500 tons or less than one-twentieth of the Malaysian total.

The strength of the Chinese has always lain in the gravel pump, hydraulic and open-cast mines in Malaysia, mines which call for relatively small capital expenditure,[8] for smaller areas of mining ground, for less intensive mechanisation and for the skilful handling of the labour force. The capital, whose total seems to be quite unknown, is provided by local Chinese entrepreneurs and banks and the industry has owed its survival and its modern revival in part to its direct access to the capital supplies of the prosperous Chinese population in Malaysia and in Singapore.

In the face of the expansion of the European mines in the 1930s and particularly of the dredges, the proportion of the total Malayan tin production held by the Chinese mines fell as low as 28 per cent in 1941. Their position then recovered sharply. They have proved in the last 20 years to be much more resilient than the European dredges, and by 1972 they had so far restored their status that they were producing more than half the national total.

This revival was reflected in the number of mines operating in the field of gravel pumping alone. In the depth of the depression in 1932, the number of these mines had been halved to only 231, but in the boom of 1940 they had trebled again to 733. Through the early 1950s the number ranged between 482 and 634.[9] The recovery after the intensive export control of 1958-60 was even more marked and by 1972 there were over 950 mines in the category. It was a remarkable tribute to the Chinese that the gravel pump mine, even as it became on the average progressively smaller (an average production of about 70 tons of tin in 1929, but of little over 40 tons a year in 1972) should so successfully meet the challenge of tin dredges whose average annual output in the same period had risen from about 260 tons to around 400 tons.

The European position has become weaker perhaps than the figures suggest. Domestic capital (which almost certainly means

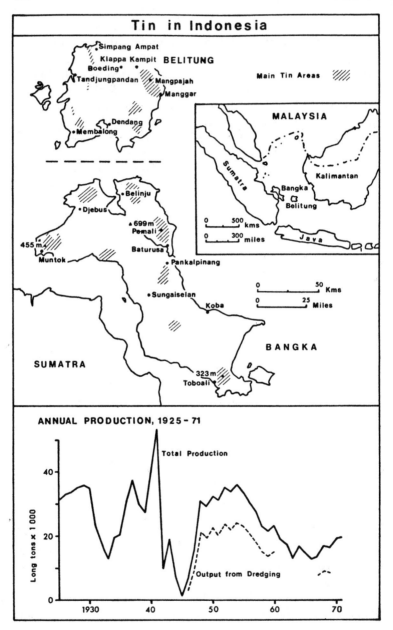

Tin in Indonesia

ANNUAL PRODUCTION, 1925-71

Chinese and not Malaysian capital) has been steadily buying its way into the European concerns. In 1954, 22·3 per cent of the capital of 11 European dredging companies was in local owner-ship (mainly Chinese); by 1964 the local ownership had risen to 64·1 per cent while in eight out of the 12 companies then listed the local ownership was in a majority; and in one of these – the newly established Selangor Dredging Ltd. – local ownership was almost complete. The Chinese share in the local capital in 1964 was even higher than it had been 10 years earlier; the Malaysian share had moved merely from 1·8 to 3·9 per cent and was of any importance only in the Selangor Dredging Ltd.[10]

The European companies have generally accepted this evolu-tion. A significant number (11 of the 35 major mining enter-prises by 1972) has shifted the official address to Malaysia. The habit of appointing Malaysian directors to the boards of Euro-pean companies is common enough. It is possible that the last twenty years has seen an actual outflow of European capital from the Malaysian tin industry. Economic power in the industry has obviously shifted to the Chinese miners and shareholders who have won a clear victory for small-scale locally-based capitalism against a European larger-scale capitalism which may be remembering too well that even such a docile body as the Malayan Trades Union Congress has advocated nationalisation of the industry.

Indonesia

Inside Indonesia production of tin is geographically even more concentrated than in Malaysia. The tin deposits were and are in three island areas – Bangka, Belitung (Billiton) and Sing-kep. These lie in the Sunda Sea, about half way between Singa-pore and Djakarta, the capital of Indonesia. They represent the tail-end of the great tin belt that has run southwards from Burma through southern Thailand and western Malaysia.

Of the three tin islands, Bangka is the oldest tin producer (the records of output date back to 1718) and the most important with 60 per cent of the Indonesian total in 1937 and 66 per cent in

[10]Yip Yat Hoong, *op. cit.*, p. 350.

1971; Belitung, which has worked since 1852, has been declining over the last 30 years (35 per cent of the total in 1937 and 27 per cent in 1971); Singkep, although mining as early as 1890, has always been of little importance (6 per cent in 1937 and 7 per cent in 1971). Small tin deposits exist in the Riau islands nearer to Singapore, off south-west Kalimantan and in central Sumatra, but for all practical purposes tin mining in Indonesia has always meant mining in the three tin islands.

Indonesian mining is almost entirely mining of alluvial deposits, whether they lie on land or under the sea. Most of these deposits are in the *kaksa* layer (a conglomerate of sandy clay, coarse sand, gravel, quartz pebbles and tin ore). An exceptional eluvial deposit at Pemali on Bangka is mined by open-cast methods. The only deep primary deposit, worked by the underground mine at Kelapa Kampit on Belitung island, was giving an annual production of around 2,000 tons of tin a year by 1940, but the mine was flooded during the Japanese occupation of 1942-45 and the very many plans to re-open it have produced no concrete results as yet.

The Indonesian industry was affected before and after the 1920s by technical revolutions similar to those in Malaysia – first, the extension of the use of the gravel pump and then the coming of the dredge. By 1925 Bankatinwinning (the operative company for the whole of Bangka island) had 79 per cent of its output coming from mines mechanised with gravel pumps and monitors and only 21 per cent from hand labour; it had, as yet, no dredges. The first on-land dredge came to Belitung in 1920 and to Bangka in 1926. The principle of the on-land dredge was then modified to take account of the earlier and successful pioneer work of the tin mining companies dredging off the coasts of Thailand; Indonesia took the lead in sea-going dredging. By 1939 Indonesia had 22 dredges (on-land and seagoing) in operation producing about 40 per cent of the total Indonesian tin output.

[11]The grade of ground of reserves in 1966 (which represents roughly the grade of ground being mined) was 236 grammes a cubic metre for dredging reserves and 461 grammes for other mining reserves. The marginal limit for working dredge reserves seemed to be about 130 grammes and for gravel pump and hydraulic reserves 250 grammes.

These dredges indicated one of their economic merits – the ability to step-up output in an emergency – when in 1940 under the pressure of sudden intense demand for tin they almost doubled their production within one year while the Indonesian gravel pump and hydraulic mines could increase by only one-half. By 1969-70 Indonesia had 39 dredges (nine operating at sea and the balance on land sites).

Most were standard bucket dredges, some (but not many) able to work on land and at sea. In addition, the Indonesians had developed the cutter dredge, first as auxiliary to the suction dredge to remove overburden (not containing any tin) so that the suction dredge could tackle the tin-bearing ground; then in 1967 dismountable cutter dredges were put into productive operation on Bangka, particularly to work old tin-containing tailings.

But the expansion of dredge production has had much the same qualifications recently in Indonesia as in Malaya. The Indonesian dredge handles a much greater volume of earth than does the gravel pumping unit (in 1969 the average dredge dealt with 1·1 million cubic metres of ground and the average gravel pump unit with only 110,000 cubic metres), but it also handles ground of far lower tin content, so that the average dredge was producing less than three times as much tin as the average gravel pump mine.[11] The dredge is open to a whole series of limitations (especially the limitation of age) from which the gravel pump mine is free. The dredge fleet was built mainly before 1939; the post-war rehabilitation was limited to the Dutch efforts in 1947-48 and it was not until 1966 that a substantial new ship was added to the sea-going dredges. Even so, that new dredge had a capacity of only 3·7 million cubic metres of ground as compared with nearly 3·0 million cubic metres for a land dredge built for Belitung as long ago as 1937. It was perhaps not surprising that the 28 older dredges in Indonesia were actually moving ground in 1968 at only 60 per cent of capacity.

The gravel pump mines have not lost ground; in fact, as in Malaysia, they gained ground in some recent years. The 66 gravel pump and hydraulic mines of 1939 became 117 in 1969 and their proportion of the Indonesian output remained much the same at around 45 per cent.

The strength of the gravel pumps lies in the quality and quantity of their reserves which totalled in the mid-1960s around 275,000 long tons (by far the greater part in Bangka); the dredges have smaller reserves (about 240,000 long tons with more on Belitung than on Bangka) and poorer ground.[12] But the dredges have one ultimate and perhaps overwhelming advantage. About two-thirds of the potential tin belt in Indonesia is in the drowned river valleys lying on the peneplain under the shallow seas surrounding the tin islands, and 40 per cent of the dredge reserves of Bangka and 70 per cent of Belitung are in near-shore sea areas (within five miles of the land).[13] In its drilling and prospecting work the United Nations Development Programme seems already to have added perhaps 40,000 to 70,000 tons of tin to reserves at sea in up to 100 feet.

The structure of the Indonesian tin industry has always been far simpler than in any other important tin mining country. The Bangka mines had been owned by the Netherlands East Indies Government since 1816 and were operated as a single unit by Bankatinwinning. The mines on Belitung were owned by a joint enterprise (G.M. Billiton) in which the Netherlands East Indies government had a majority and the N.V. Billiton of The Hague a minority shareholding; management of the mines lay, however, with the G.M. Billiton.[14] The general policy on tin (especially when tin entered international commodity agreements in the 1930s) seems to have been directed by the N.V. Billiton interests. After the re-occupation of Indonesia in 1945, when it was fairly clear that, except in the case of the Kampit mine, the damage of the Japanese occupation had been the damage of neglect rather than of destruction, a rapid return to normal with the rehabilita-

[12]This compares with a Dutch estimate of alluvial reserves for the tin islands in 1941 of 497,000 tons of tin.

[13]*Prospects of tin mining in Indonesia: Indications for bids*. Ministry of Mines, Indonesia, 1967.

[14]N.V. Billiton Maatschappij of The Hague had a minority interest in the N.V. Gemeenschappelijke Mijnbouwmaatschappij Billiton (G.M.B.). The Singkep Tin Exploitatie (Sitem) was a subsidiary of the G.M.B. N.V. Billiton owns the Kamativi Tin Mines Ltd. in Rhodesia and owned the Tin Processing Corporation which ran the Texas City smelter for many years on the account of the U.S. government. The Billiton smelter at Arnhem in the Netherlands (N.V. Nederlandse Indonesische Bedrijven) closed in 1971.

tion of the dredges and a simplification of the administration was envisaged, and the Dutch were optimistic enough to estimate for the International Tin Study in early 1948 an output by 1950 as high as 45,000 tons a year – a figure almost as high as the boom of 1941.

In 1947-48 six new dredges, each with a digging capacity of 2·9 million cubic metres a year, were built, and in 1948 all tin mining in Bangka and Belitung was brought under the management of the G.M. Billiton for five years. These steps brought immediate results and by 1954 output was running at 36,000 tons a year and was back to where it had been in the good days of the later 1920s. But tension between Dutch and Indonesian on all matters was growing. The Indonesian concentrates were being shipped very largely to the Arnhem smelter of the Billiton company in the Netherlands and no serious attempt was made to rehabilitate any of the three smelters on Bangka which, before the war, had been treating half the Indonesian mine output. By 1952 the Parti Nasional in Indonesia was pressing for the complete nationalisation of ownership and management in the tin industry. The Indonesian government took over full mining management on Bangka in 1953, and in 1958 the property and rights of the G.M. Billiton on Belitung and Singkep were also transferred. The exclusion of the Dutch from the Indonesian tin industry was regarded as complete and Indonesia became the first major tin producer where the industry was fully nationalised.

By 1954, the impetus of post-war rehabilitation, sustained by the massive long-term contracts for the U.S.A. strategic stockpile, had brought Indonesian production up to one-fifth of the world total. The ending of Dutch managerial efficiency and the cessation of the U.S.A. contract made the Indonesian tin industry more vulnerable to the effects of the price crisis of 1958 and through the 1960s it suffered from almost every disease that could afflict an industry – from the continued depreciation of ageing dredges, from the continuous shortage of spare parts for mining machinery, from the mis-buying of wrong equipment, from gross inefficiency in management and from the political interference to be expected in an industry which was still earning for Indonesia the equivalent of around £16 millions in hard currency in its worst years. The expedient of handing over the running of the mines to the army in 1963 had no result in greater efficiency, except in the art of tin smuggling. The change-over from Sukarno's rule to Suharto's rule in 1967 also affected the labour supply. The labour force on the tin islands had, from early on, been drawn from south China (in 1925 Bankatinwinning, out of a total of 20,000 workers, reported 17,700 as Chinese on contract and 1,500 as Chinese not on contract). They were never popular in Indonesia and they became more unpopular as communism won in China. The savage massacre of the Chinese generally in Indonesia and their expulsion from the tin islands helped still further to weaken the backbone of the industry. By 1966-67 tin output was down to 13,000 tons a year or a figure lower than it had been even seventy years before.

Serious steps were taken, to a flood of outside international advice, to rescue the industry from 1967. Substantial financial inducements were offered under the Foreign Investment Law No. 1 of 1967 to foreign investors to enter the mining industry. Offers were invited to develop tin mining rights in off-shore areas (but generally not in areas where the nationalised organisation – P.N.T. Timah – was operating). The N.V. Billiton, always hankering to get back into Indonesia, secured in 1968 a prospecting concession for forty-years over large off-shore areas between Bangka and Belitung and off south-west Kalimantan (Borneo).

Long-term planning on the dredge fleet produced the largest sea-going dredge yet built with capacity designed at 1,500 tons of tin a year (built in Scotland on an export credit guarantee from the United Kingdom government). The rehabilitation of the older dredge fleet, based on a seven-year programme (largely by Dutch firms under Dutch foreign aid money), began to show effective results in 1971. By 1972 Indonesian tin production at nearly 22,000 tons a year was moving back towards the level at which it had stood during the beginning of the export control crisis of 1958-60; it was, however, well below half of what it had been in 1940-41.

The re-organisation in the 1960s also saw the fulfilment of another Indonesian dream on tin. Like the Bolivians, the

Indonesians had convinced themselves of the necessity of a domestic smelter to treat their concentrates, both to diminish their outgoings in foreign currency and to weaken their dependence on foreign sources for smelting. Prior to 1941, Indonesia had her domestic smelters (three on Bangka island at Pangkal-balem, Muntok and Belingoe) which were taking between one-third to one-half of the domestic mine production. The balance had been shipped initially to the Straits Trading Co. in Singapore but was diverted during the 1930s to the new Billiton smelter in the Netherlands. In the post-war period very half-hearted attempts to rehabilitate the domestic smelter at Muntok had no result for a very long period. Indonesia shifted all her concentrate production for smelting to countries where – and so long as – her political relations were not unfriendly. Exports through the 1950s were mainly to the Netherlands but also in small part to the U.S.A. They were stopped to the Netherlands and diverted to the U.S.A. and Malaysia during 1959-62; they were cut off from Malaysia and resumed to the Netherlands during 1964-68; they were then shifted back to Malaysia and cut off entirely from the Netherlands (thereby helping to enforce the closing down of Billiton's Arnhem smelter there in 1971). But in 1967 the long-awaited new Pelim smelter (owned and managed by the nationalised state organisation and built by a German company) was opened and by 1972 was treating half the rising mine output.

Thailand

The south-east Asian tin belt enters Thailand from Malaysia and runs northwards through the narrow Thai Peninsula along the frontier with Burma. It then splits in one north-westerly direction to link with the tin-wolfram mines of Burma and in another north-eastern direction to appear ultimately in the tin deposits of southern China. The great bulk of the economic tin deposits of Thailand (94 per cent of the production in 1968) are mined in the southern peninsula of southern Thailand in the 300 mile strip of tin-bearing land between Ranong and Yala; within that strip most comes from the four provinces of hill land which

FIGURE 3

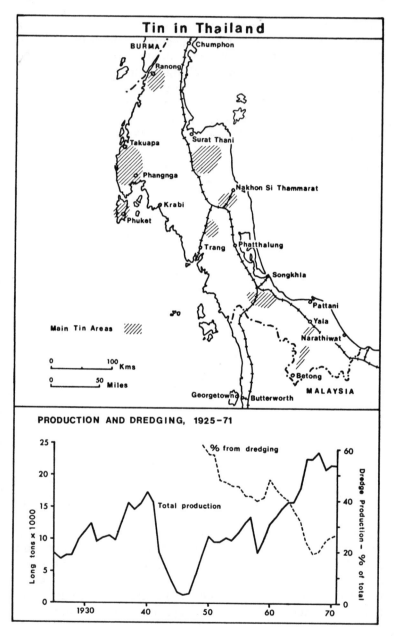

Tin in Thailand

PRODUCTION AND DREDGING, 1925–71

look to Phuket as a centre. A small proportion (only 4 per cent in 1968) comes from the northern section of the peninsula as it joins the main body of Thailand in the approach to Bangkok, and an even smaller amount (only 2 per cent) from the north-western mountain slopes by the Burmese frontier. There is no tin in the flat central belly of Thailand, stretching 300 miles east and north from Bangkok to the frontier with Laos. The tin is almost all eluvial or alluvial; the few lode mines are worked inefficiently on a small scale. By-products and impurities are not important. There are some xenotime, columbium and tantalum ores, and the small mines in the north-west are perhaps more important for wolfram than for tin. Serious trouble has been found with lead impurities at Yala in the extreme south but, in general, the cassiterite is so trouble-free that during the Second World War it could be smelted into reasonably good metal in the equivalent of back garden furnaces.

The type of mining follows the Malaysian practice, except in the case of the off-shore dredges, of which Malaysia has as yet none. About one-quarter of the current output comes from dredges, the remainder from gravel pumping and even smaller mines. The mining expertise has come from old-established Chinese immigrants (who have been accepted within Thai life far more readily than within life in Malaysia and Indonesia) and, in more recent times, from European-financed mining houses, usually working up from Malaysia.

The potentialities of Thailand as a tin producer were shown very clearly during the First World War when it raised its output to one-half of the Indonesian output and to one-fifth of the Malayan level. It could not maintain this position during the 1920s but it was strong enough to impress its importance on the other major tinfields. As a new but essential recruit to the first international tin agreement of 1931 it therefore secured entry on very privileged terms – a guaranteed export tonnage of 10,000 tons a year, beneath which its quota would not fall. Thailand was therefore sheltered from most of the effects of the tin depression which that first agreement had to face. In 1933, almost the worst year for tin production yet seen in the twentieth century, Thailand was actually increasing production at a time when the

Netherlands East Indies, Malaya and Bolivia had reduced their output to one-third of their 1929 tonnages. In the later 1930s (still within the international agreement) Thailand continued to expand until it was moving towards one-tenth of the world total.

Much of that expansion came from dredges. The first off-shore tin dredge in the world was brought by Captain E. T. Miles in 1907 to mine in Tongkah Bay, Phuket island, for the Tongkah Harbour Tin Dredging Co. The dredge was built by Simons Lobnitz of Renfrew, it cost £8,850 and was profitable, probably because of the rich grade of ground (1·52 lb. of cassiterite per cubic yard). The same company had four more dredges built by 1911. The history of the Tongkah dredges is worth a note. The first dredge sank in Tongkah harbour in 1914, the second worked for 22 years, first in Tongkah and then at Bidor in Malaya, the third worked at Tongkah for 20 years and was then sold to Anupas and Sons, the fourth worked 19 years at Tongkah and was then sold to the Penang Tin Mining Corporation, the fifth worked at Tongkah Harbour for 28 years, the sixth worked for 15 years at Tongkah and Chalong Bay, and the seventh for 29 years at Tongkah before it was sold to Lam Nga.[15] On-land dredges were built for the relatively large mining concessions including the Siamese Tin Syndicate, and by 1941 Thailand had a total of 40 dredges (on-land and off-shore), most on the western

[15] Rachan Kanjana-Vanit and others; *Off-shore mining of tin deposits in South Thailand,* in *A Second Technical Conference on Tin, Bangkok, 1969, Vol. 2.*

[16] The very illuminating table below on the importance of the small mine in Thailand is adapted from Poolsawasdi Suwarnarangse: *Tin mining problems in Thailand* in W. Fox: *A Second Technical Conference on Tin, Bangkok, 1969, Vol. 2:*

	Production in 1968		
Method of mining	No. of mines	Total production		Production per unit, ltons (approx.)
		per cent	ltons	
Gravel pumps	396	64·7	15,300	38·6
Inland dredging	20	12·9	3,050	152·5
Off-shore dredging	4	7·4	1,750	437·5
Ground sluicing	245	6·1	1,400	5·7
Dulang washing	6,013*	3·4	800	0·13
Open-cast	32	2·5	590	18·5
Gophering	23	1·6	380	16·5
Hydraulic elevator	10	1·2	280	28·0
Tunnelling	2	0·2	47	23·5

*Number of dulang washers

side of the Thai Peninsula, which were producing about two-thirds of the Thai output.

The post-1945 recovery of production in Thailand was slow. The Thai government had to provide substantial compensation to the non-Thai companies who had suffered war damage. By 1950 the number of dredges at work had crept back to 31. When the International Tin Council imposed export control in 1958-60 Thailand as a member had no guaranteed minimum export tonnage upon which to fall back, but the country was again resilient. By 1963 Thailand overtook Indonesia – not a very difficult feat at that time – to become the world's third largest producer and from 1965 onwards she was producing 20,000 tons or more of tin yearly.

There had been a change in the character of Thai tin production. The dredges, which had been responsible for around 60 per cent of the national output in the 1950s, steadily lost their domination. By 1960 their proportion had fallen to under half and by 1971 – when their number was only 22 – to only one-quarter. The smaller units in Thai tin had moved swiftly towards mechanisation, to the small power unit and to the jig in place of the old-fashioned palong. In the twenty years between 1950 and 1972 the number of non-dredging mines – mainly in the form of gravel pumping – had doubled, their average output per unit more than doubled, and their total production multiplied more than fourfold. Their total horse-power used had increased seven-fold; in the dredges it had merely doubled. The small mines were showing the same vitality and power to compete successfully against the dredge as they were showing over the frontier in Malaysia.[16]

Thai mining production had for very many years turned automatically to the Malayan smelters at Penang and Singapore for metal conversion; none of it went to western Europe. But the Thai government had long hankered for a domestic smelter which would employ local labour. Its offer of special taxation privileges and of a monopoly market with the prohibition of the export of tin ore attracted U.S.A. capital in the form of an American conglomerate firm – the Union Carbide Co. In 1965 Union Carbide built the Thaisarco smelter in Phuket, geared to absorb all the domestic production of tin concentrates. The smelter was successful in smelting in that field and had the distinction of being the first tin smelter in the world outside the U.S.A. to be financed by American money, but it was less successful in maintaining the full interest of Union Carbide (which had perhaps been more concerned with the possible valuable titanium by-products of tin smelting and exploration rather than with tin itself). By 1971, a half share in the ownership of Thaisarco had been taken by the Billiton Company of The Netherlands (now itself a subsidiary of the Royal Dutch Shell oil group) which thereby resumed the interests in smelting which it had recently lost in the Netherlands with the closing down of its shipments from Indonesia.

The contrast between the recent tin histories of Indonesia and Thailand is worth a moment's stress. The Thais had an industry strongly influenced but certainly not dominated by European (that is, mainly British and Australian) capital based on Malaysia; its mining population was largely of Thai-Chinese origin. The Indonesian industry had been dominated entirely until the 1960s by Dutch capital and Dutch expertise in management; its basic initial labour force had been largely immigrant Chinese. But in Thailand there had been no complete external political control or a colonial status and therefore no passionate need to divert post-war energies into securing and asserting its political independence; Indonesia had been a very capably run colonial territory, which had secured a very reluctant consent to independence. It had inherited no substantial bureaucracy and, when it turned in independence to re-organise its tin industry in the 1960s, possibly merely created more inefficiency. The Chinese in Thailand were a full part of Thai social life and a source of mining capital; the Chinese in Indonesia had been within living memory labour imported on contract, were later suspect as both Chinese and communist and had no scope within the commercial mining of Indonesia, except sometimes as tributing subcontractors or *particuliere leverantie*, to improve the quality of the industry or to find capital. In the technical and administrative field the Thais had in the long run gone ahead faster and more publicly. Thailand had established its own

Department of Mines and Geology as early as 1891, although its work was primarily concerned merely with the granting of mine concessions in southern Thailand. In 1942 the geological section became a division whose duties covered the examination of economic mineral deposits. Later a mining technology division was added. The first complete large-scale geological map of Thailand was published in 1951. Outside help – especially from the U.S.A. and Canada – on geochemical and geophysical exploration was welcomed. The desire to keep control of exploitation was reflected in 1952 when aliens were in general excluded from mining concessions north of Chumporn (in effect, from all of Thailand except the Peninsula), but the Thais did not share either the xenophobia of the Indonesians in the early 1960s in keeping all foreigners out of all tin or their sudden desire in the late 1960s to open their un-nationalised mineral resources to the beneficial influence of foreign capital.

Nigeria

The first boom in Nigerian tin in 1909-13 was based in part on the glowing accounts of the potentialities of the Jemaa lode in the south-central field. The boom was carried forward on a new concept as to the life of the tinfield. The earlier belief was that most of the payable deposits were confined to the existing river beds, that these deposits – although valuable – were limited, that the life of the field therefore would be short and that the only reasonable method of exploitation was by fairly primitive and hand-labour mining. The new concept accepted that substantial deposits existed in the original river beds in the Plateau (that is, beds whose course bore no necessary relationship to the course of existing rivers) and that these deposits would justify mechanisation and long-term capital investment. By 1912 Jos Tin Areas Ltd. was operating a small bucket dredge at Jos; another dredge was working, if not very successfully, at Bawa during 1913-15. By 1920 the Bisichi Tin Co. (Nigeria) was using seven monitors and four elevators. But mechanisation was slow. In 1926 the horse-power of machinery in use in the Nigerian tin mines was only 6,300, in 1921 11,000; and in 1938, after a severe decline in

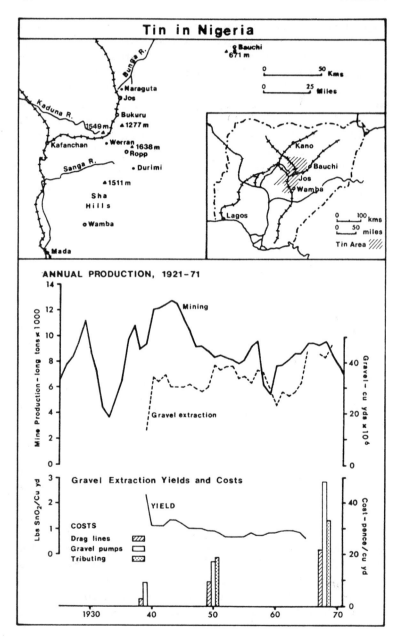

Tin in Nigeria

the hungry thirties, was 10,100. Thereafter, mechanisation in the larger units through elevators, monitors, steam shovels and gravel pumps was fairly rapid. In the two years 1938-39 the two main groupings (Associated Tin Mines of Nigeria and the London Nigerian Tin Mines) spent over £0·5 million on equipment, including a Monighan dragline and a dredge which was completed in 1938.

The problem of communications was solved between 1915 and 1927. The Plateau tinfield lies about 400 miles in a direct line and about 550 miles by rail from the ports of shipment, first Lagos and later Port Harcourt, through which the tin concentrates from the Plateau mines were moved on their way to smelting in the United Kingdom. In 1915 a light railway from the tinfield to Zaria gave a lengthy link with the railway system to Lagos. In 1927 the opening of the Kafanchan-Jos line gave a shorter and more direct connection with Port Harcourt and the Udi coalfield in the south and the rail freight on tin concentrates to the southern coast was reduced by over a quarter.

In the field of power the first hydro-electric plant was installed by the Northern Nigeria (Bauchi) Tin Mines Ltd. at the north end of the Plateau. In 1926 Nesco (the Nigerian Electric Supply Corporation Ltd.) was formed to develop the Kurra hydro-electric scheme at the southern end of the Plateau. By 1949-50 the output generated by its Kurra, Jekko and Kwali hydro-electric plants was over 68 million units, nearly all of it absorbed by the tin miners.

The initial cost of prospecting work on the Plateau after 1900 had been carried by the Niger Co. and the capital for the exploitation of the discoveries after 1909 came primarily from the London capital market. The normal pattern of mining finance was followed. The prospector or prospecting company would prove the ground and would take out a mining lease. These leases he would operate on his own account or would sell (against a payment in cash or in part in shares) to operating companies formed for the purpose. The operating company would raise its capital in the London market or through mining finance houses.

The Niger Company often in the earlier stages held blocks of shares in the new mining companies and occasionally acquired control of some; after 1910 Consolidated Gold Fields was an important financing house. By 1913, 120 companies were reported (not all at the mining stage) with a total authorised capital of £6·8 millions.

The limitations on the degree to which bigger mining units in Nigeria could be developed were important. In general, the tin deposits were scattered over a wide area; in the 1920s, the volume of reserves carried by any individual company was not large. Dredges, which were so important in expanding production in south-east Asia, were tried in Nigeria but had relatively little scope. Draglines and power-driven excavators required capital beyond the resources of the very many small companies. As late as 1952 the tin mining leases in Nigeria were spread over 260,000 acres of land for a total tin production of 8,500 tons of tin; in Malaya at about the same time tin mining concessions covered about 500,000 acres for an output of around 60,000 tons of tin. There was, however, a degree of consolidation in the structure of the industry after 1920. The Anglo-Oriental began to group companies in Nigeria as it was grouping them in Malaya and Thailand. By 1929, three groups (the Associated, the London-Nigerian and the Naraguta) covered 41 out of the 65 companies operating in Nigeria producing over 80 per cent of the total Nigerian tin output. In 1938 the Associated and London Nigerian groups placed the management of their mines under a central organisation and in 1939 merged themselves into a single unit – the Amalgamated Tin Mines of Nigeria Ltd. within the Anglo-Oriental group. This brought a substantial proportion of the overall Nigerian output within the Anglo-Oriental influence and therefore closer to the common international policy on tin to which Anglo-Oriental was committed.

The process of consolidation did not go further. Rather surprisingly, A.T.M.N. has made no serious move in the last thirty years to extend its field of ownership. The smaller unit in Nigerian mining has shown very consistent vitality. Admittedly, numbers suffered badly in the economic slump of the early 1930s and in the Anglo-Oriental groupings. The 83 companies and 70 private operators of 1928 fell to 34 companies and 56 private operators in 1947, but there has been little fundamental change since that latter date. The companies incorporated in the United Kingdom numbered 24 in 1947 but only 11 in 1969; the 10 incorporated in Nigeria had become 56; the three companies registered under business names in Nigeria had become 20 and the 53 private operators of 1947 had become 25. The dominating position of A.T.M.N. and the other European companies remained unchanged. The European-incorporated companies in 1969 held 68 per cent of the total output of tin against 80 per cent in 1947 (when that total output was much the same); the companies incorporated in Nigeria 23 per cent in 1969 against 8 per cent in 1947; the firms registered under business names held 6 per cent against less than 0·5 per cent in 1947; and the private operators held 4 per cent against 12 per cent in 1947. But much of this apparent movement was nominal and followed from a transfer in official residence or a formal transfer of category. The units of production (with, of course, the exception of A.T.M.N.) remained small even by the standards of tin mining. The average annual output of the European companies, apart from the A.T.M.N., in 1969 was under 300 tons of tin, of the Nigerian-registered companies only 40 tons, of the firms with registered business names 30 tons and of the private operators under 15 tons. Allowing for the changes in category this basic smallness was not far different from what it had been in 1937 or 1947.

In its earlier days, the Nigerian tinfield started, as did tinfields everywhere, with the exploitation of high-grade tin-containing ground. As late as 1939, the grade of ground being mined was as high as 1·64 lb of tin contained per cubic yard of ground worked, so that the total Nigerian output of 9,400 tons of tin required the excavation and treatment of only 13 million cubic yards of ground.

From 1940 to 1945, the stimulus of maximum production for patriotic or other reasons operated in Nigeria as it operated elsewhere. Maximum production involved two main factors. The first was the working of the maximum amount of tin-containing ground. This involved excavating ground which was known to contain very little tin (perhaps as little as 0·1 lb of tin per cubic yard) and even the re-working of ground which had been previously worked. This was uneconomic in the sense that the cost of extracting each ton of tin was abnormally high. For the Nigerian industry during the war this was not in itself catastrophic. During 1941-49 the United Kingdom Ministry of Supply was the sole buyer of tin concentrates produced in Nigeria and delivered to the Nigerian port, the Ministry being responsible for all costs beyond that point. The miner was paid on a "costed" basis. His pre-war cost and profit (per ton of tin produced) were arrived at, the Ministry paid his current costs (including increased costs) and the profit per ton so photographed. The result was, of course, a transfer to the Ministry of the increased costs from mining low-grade ground which might never otherwise have been mined. The future of the mine was not generally imperilled. The second aspect of maximum production had more serious consequences. This aspect was the concentration specifically on mining high-grade ground. Operations are shifted on to higher-grade ground in leases known to contain both higher-grade and lower-grade ground. This shift-over is contrary to the normal good mining practice of mixing (where possible) the higher-grade and lower-grade ground so as to secure in the mining programme an economic average of ground. High-grading (sometimes referred to rather dramatically as "picking out the eyes of the mine") increases sharply the tonnage of tin mined. It is at the expense of the future life of the mine. In some cases lower-grade ground which has been deliberately by-passed in the practice of high-grading may not be worked in the future for technical reasons such as the difficulty of moving back to the low-grade ground the necessary working machinery or for economic reasons since working low-grade ground by itself may be unprofitable.

By 1947 the average grade of ground worked in Nigeria had been more than halved to 0·70 lb of tin per cubic yard, so that to produce roughly the same amount of tin required the removal of 29·5 million yards of ground as compared with 13 million cubic

[17]It was not perhaps encouraged by the suggestion in 1972 in the National Development Plan for Nigeria for greater government participation in mining and that "a rational exploration of Nigeria's mineral resources . . . demands that mining is not carried to the point where the over-riding objective is the immediate and maximum return to private investors". Quoted in *Tin International*, August, 1972.

yards in 1939. In more recent years the grade of ground has tended to get still lower and by the later 1960s it was down to around 0·50 lb of tin a cubic yard.

The tin area in the Plateau was geographically well away from the Nigerian civil war of 1967-70. The industry was affected, however, by the loss of semi-skilled labour (much of which was Ibo and willingly or unwillingly abandoned the area) and by restriction on mining supplies (most of which are imported), on capital movements and on repatriation of profits. Production by 1972 at 7,000 long tons of tin was lower even than in 1936-39.

There has been little done to attract in recent years European capital into Nigerian tin.[17] In 1921 Nigeria was the fifth largest producer of tin in the world outside China – a fact perhaps not as impressive as it sounds since that position gave her only 5 per cent of the world total. She trailed sixth, behind the Congo, until 1961, then fifth again until 1968 and sixth thereafter (behind a revitalised Australia). Her share of the world in 1971 was only 4 per cent. The Nigerian government is taking steps, perhaps for social and political rather than economic reasons, to help small mining, but by their nature the addition to production is bound to be small.

The substantial reserves of tin-bearing ground lying under the basalt cap on the Plateau which have been known for many years present mining problems of underground water and soft roofs which cannot be solved without heavy capital outgoings, and it will certainly be some considerable time before the mass mining of low-grade deposits north of the Plateau becomes even a possibility.

The better Nigerian tin mines are not inefficient. The mechanical horse power used in Nigerian mines is, in relation to the size of the industry, on a par with the horse power used in Thailand. But there is lack of specialisation in the forms of mining. Thus for 1949 A.T.M.N. got 9 per cent of its output from a dredge paddock, 19 per cent from six draglines with washing plants, 43 per cent from 34 gravel pump plants, 25 per cent from elevators and hand paddocks and 5 per cent from mill tailings treatment. Another and very much smaller European company – Gold and Base Metals – in the same year drew 56 per cent from

FIGURE 5

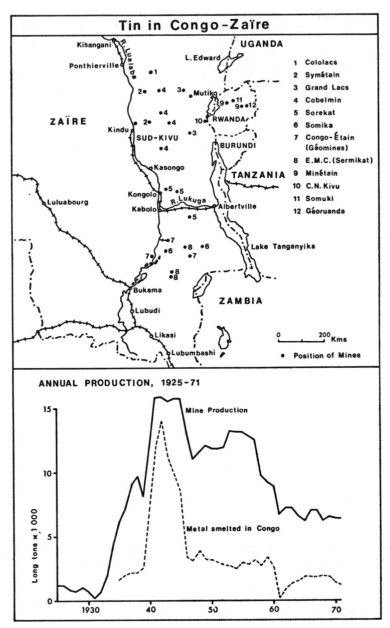

Tin in Congo-Zaïre

1 Cololacs
2 Symétain
3 Grand Lacs
4 Cobelmin
5 Sorekat
6 Somika
7 Congo-Étain (Géomines)
8 E.M.C.(Sermikat)
9 Minétain
10 C.N. Kivu
11 Somuki
12 Géoruanda

0   200 Kms

• Position of Mines

ANNUAL PRODUCTION, 1925-71

Mine Production

Metal smelted in Congo

Long tons x 1,000

tributors and contractors. These characteristics have not changed very much since 1949.

There is a socially desirable but perhaps not very economic dependence on hand labour and on seasonal labour. Water is an important factor in mining. Rainfall on the Plateau is confined to the wet season of May to September; many of the streams (and therefore the water supply for gravel pumps and washing plants) dry up completely in the dry season. The direct hand labour methods are losing importance in Nigerian mining but are by no means dead. In 1947 the direct labour used in hand paddocks and the hand labour used in tributing moved nearly 50 per cent of all the ground moved in the industry; in 1967 the proportion was still 31 per cent; and in 1972 hand labour mines were still producing 30 per cent of the output and employing over 33,000 workers.[18]

Zaïre (The Congo)

The Congo was a new and unimportant tin field in 1921. The first production of tin had not been important enough to be recorded in official statistics until 1913, and eight years later the 600 tons of Congolese production was only 0·5 per cent of the world total. The opening-up of the field on a substantial scale by money from Belgium began in 1933-34 and took not much more than the six years between 1934 and 1940. By that latter date the Congo (including Ruanda-Urundi) was the world's fifth largest producer, providing almost 6 per cent of world output.

The tinfield itself lies in a long belt of irregular deposits running for about 600 miles between the Lualaba (Upper Congo) River and the great lakes, south from the equator. The tin belt is at its widest in its northern section in what was south Kivu province; there it spills over into the new republics of Rwanda and Burundi which were, for all practical purposes, parts of

[18]Many Nigerian tin producers have an interest in columbite which appears as a necessary by-product in mining cassiterite and which is essential in toughening high-grade steel. The production of columbite as an industrial mineral and not as a waste by-product began in 1933, and during and after the Second World War the tin-columbite producers had a price bonanza. The Nigerian share of world columbium production is now down to about one-tenth.

Belgian Congo until the Congo became independent in 1960. The belt is narrower in its southern section – the old area of the Upper Katanga. Thereafter, tin gives way to the gigantic copper interests of what was the Union Minière du Haut Katanga.

On the transport side, the two Zaïre (Congo) tin districts in Katanga Maniema and Kivu have communications with markets which are amongst the longest and most difficult in the whole world of tin. For political reasons the movement of tin concentrates for smelting in Belgium or of metal smelted within the Congo has always been back through the Congo to the Atlantic ports of that state and not through east Africa to the Indian Ocean ports. From the southern tin mines around Manono to the Atlantic coast at Matadi has meant 500 miles by rail and 1,400 miles by river boat (with six intermediate handlings) and from the northern mines of the Symétain company in Kivu to Matadi an average of 200 miles by lorry, 300 miles by rail and 1,300 miles by river boat (with four intermediate handlings).

Production has come from eluvial and alluvial deposits; there is very little lode mining and there are no dredges. Most of the placers are small in size; many deposits contain a total of little over 200 tons, so that companies find themselves with a scattering of effort over a large number of working places. At Manono there are, however, large open-cast mines scooped-out to a depth of 70 metres. From the beginning of mining, the necessary infra-structure was a responsibility of the mining companies. Labour had to be recruited elsewhere and villages built. Power plants had to be erected from scratch; labour with the most primitive background had to be trained in mechanical methods of mining. All medical and educational functions fell to the companies. The Symétain company, through the Fondation Symétain, provided all the social services which the Congo state could not provide and provided them in very generous fashion.

The entry of Belgian tin mining enterprises into the Congo followed classic African lines from the initial grant of vast concessions. Rights over territory were ceded by the state to railway companies. To fulfil their obligations and to attract traffic, the

[19]These names have been changed, sometimes twice, to meet the requirements of the Congo (Zaïre) government since 1960.

railway companies opened up part of their concessions to mining prospecting; the prospectors explored mining areas and then transferred the deposits so discovered to mining enterprises; the mining companies drew on capital from Belgium (sometimes from the same sources which were financing the railway companies). All the capital came from Belgium. There was no Congolese capital available except on a small scale after the independence of the Congo in 1960. There was a very limited production from European settlers (*colons*) who were Belgian, Greek or Portuguese. The Belgians (19 companies of any importance in 1950) were often tightly interlocked in shareholdings or in general policy.

Four companies have always dominated the earlier Congo tin position – the Géomines (about one-third of the whole production) in the southern field around Manono, Symétain (another one-third) around Punia and Kalima in the north, Grands Lacs near Lake Kivu and the Comité National du Kivu south of the lake.[19] The main units in the old Ruanda-Urundi are linked with these groups.

The first effective discovery of cassiterite in the Congo was by Robijns in 1903 on the eastern bank of the Lualaba. Katanga was opened for prospecting in 1910. The main mining company—Géomines—was created in that year by a block of engineers and industrialists, mainly from the Liège district of Belgium. It took over a mining concession which had been granted by the Comité Spécial du Katanga. By 1914 Géomines prospecting missions had discovered the tin deposits of Manono and Mwanza and the coal deposits of the Lukuga; the first lots of cassiterite were shipped out in 1916 to the United Kingdom for smelting. By 1921 the Géomines output, realised by the most primitive pick and shovel methods, had reached an annual rate approaching 150 tons of tin. Real expansion of Géomines did not come until after the great tin slump of 1931-32. A wide plan of mechanisation was then put into effect. A new 15,000 hp hydro-electric plant was built on the river Luvua, a water reserve to cover dry season needs was established on the river Lukushi, and a smelter and refinery—the biggest in Africa and exceeded only by the smelters in Western Europe and south-east Asia—was established at Manono in 1935. The earlier methods of mining shallow deposits were abandoned after 1934 and operations were shifted to larger and deeper open cast mines. By the beginning of the Second World War Géomines was producing at the rate of 3,000 tons of tin a year.

In the post-war period Géomines decided to expand its reserve position. Its initial reserves, which had been calculated in 1929 at 100,000 tons of cassiterite, were almost exhausted by 1950. Intensive prospecting had opened up at greater depth one of the biggest deposits of tin in the world – deposits whose tonnage was estimated at hundreds of thousands of tons of tin. These deposits, in stony pegmatites, had mining problems and, in particular, problems of capital. The company's capital was lifted from 200 to 700 million Belgian francs; 50 million francs were raised on a 10-year loan and a credit obtained from the Economic Co-operation Administration of the U.S. government. The extended power plant and the new mine equipment meant that the proportion of output from the new stony pegmatitic deposits rose from 8 per cent in 1951 to two-thirds by 1961. An indirect effect of the expansion of tin production was an increase in the by-product tantalite-columbite where in 1950 Géomines found itself the world's largest producer. The smelter treated during the war all the tin concentrates of the Congo which were not shipped to the United Kingdom or to the Texas City smelter.[20]

The Symétain (now Cométain) enterprise in Kivu arose from among the great land concessions of the Chemin du Fer Grands Lacs. The Chemin du Fer (CFL) had secured an agreement in 1902 with the government of the Congo Free State to open up a railway system based on Stanleyville (Kisangani) in the northeast Congo. This agreement gave C.F.L., financed by the Baron Edouard Empain group in Brussels, prospecting rights over an enormous area. The retention of a monopoly of prospecting rights on such a vast scale could not be justified, and in 1925 118,000 sq. kms of the C.F.L. domain along the River Lualaba were declared open to public prospecting. There was little result

---

[20]The Longhorn smelter in Texas City was built for, owned by and operated on the account of the U.S. government during the Second World War and thereafter until 1957

until the Symaf group (Syndicat Minier Africain) began to prospect. Symaf was set up in 1926 in Belgium (its initial promoters including the Banque de Bruxelles and the banking Nagelmakers and Allards) to prospect for mines in a vast territory along the Congo valley from Léopoldville (Kinshasa) back to Stanleyville and then south to Nyangwe. Operating in territory between the Lualaba River and the areas in Maniema and Kivu in the old C.F.L. concession, Symaf discovered the Punia and Kalima deposits in 1930. To exploit the large number of claims which it took out, Symaf formed in 1932 the Symétain company (with an initial capital of 12 million Belgian francs).

Symétain's early work was exhausting. The country was mountainous and very densely wooded; it had long been removed from any outside African, let alone European, influences; the inhabitants were on a particularly low standard of living; the area was picturesquely presented as a land of gorillas and pygmies.

Between 1928 and 1931 the Symétain company took out no fewer than 415 prospecting titles. By 1935 its yearly output had reached 1,500 tons of tin; by 1938 the figure was 2,500 tons and the company was employing some 7,000 native workers. During the war-time boom of 1940-45 output was pushed as high as 4,500 tons (in 1945) and the labour force to 12,000. Symétain was committed to expansion. Its capital, already raised to 40 million Belgian francs in 1939, was raised again to 85 million francs in 1947. It looked to the future development of water power for gravel pump mining and for electric power in its other mechanised operations. Its first central hydro-electric plant was created in 1944, its Lutshurukuru plant on the Kalima side was opened in 1952 and its prospecting work was accentuated when in 1949 the C.F.L. domain was re-opened to prospecting after war-time closure. By 1951 the accumulated investment of the company in tin had reached 331 million francs.

The Grands-Lacs mining group lay east of the Symétain mines towards Lake Kivu. From 1903 prospecting missions had covered much of the C.F.L. domain lying between 1° N and Nyangwe on the Upper Congo, looking initially for gold rather than for tin. In 1923 the C.F.L. created a mining subsidiary – the

Minière Grands-Lacs (M.G.L.) – to whom it ceded all its mining activity. By 1929 M.G.L. had become the second largest producer of gold in the Congo; after 1936 it began to work its tin deposits and by the end of the Second World War it was producing well over 1,000 tons of tin a year.

The Comité National du Kivu (C.N. Kivu) began to mine tin after 1926 in the Costermansville district by Lake Kivu, and a group of concerns (Cololacs, Belgikaor, Belgikaétain, Kinorétain, Miluba and Minerga) opened up mines east of the Lualaba river in Maniema and Kivu.

The Congo industry, even as early as 1939, was on the road to high prosperity. As a new field the quality of its tin-containing ground was high. In 1932-35 the Symétain ground under exploitation held 2·5 kg of tin per cubic metre and in 1939 still 1·50 kg. High-grading during the Second World War pushed the level up to no less than 2·35 kg in 1941. The Congo production (with a peak labour force of perhaps 70,000 in 1943) rose (excluding Ruanda) to 15,700 tons of tin in 1944-45 or double the 1939 level. In the absence of south-east Asia from the world market, the Congo became responsible for nearly one-fifth of the available world production. The effort at maximum production was intense (in the six years 1940-45 the Congo produced much more tin than in the previous 30 years) but it was not crippling on a rich and relatively young field. The quality of the Symétain ground was down to 1·10 kg in 1951 and to 0·98 kg of tin in 1969, but this still left it well above the standard normal to Nigeria. The industry was mechanised intelligently, so far as the physical conditions of the Congo allowed. With Symétain the volume of ground handled per worker per day had risen from 1·66 cubic metres in 1938 to 6·6 in 1957.

In mid-1960 the Congo was hurried into independence as the Democratic Republic of the Congo (Kinshasa), later the Repub-

[21]Names have been changed. Géomines became Congo-Etain in 1968, owned jointly by the Zaïre government and Géomines with the latter providing management, recruitment and personnel and marketing in Belgium. Sermikat was taken over by a Congolese citizen in 1964; some Kivu concessions were turned over to Congolese. Cobelmin (operating for Kinorétain, Belgikamines, Minerga and Miluba) was registered as a Congolese company in 1962. Symétain became Cométair.

lic of Zaïre. The immediate moment of independence was unfortunate for the tin industry, which had not begun to recover from
the export restrictions imposed since the end of 1957 by the
International Tin Council. In September, 1960 the Géomines'
administrative centre at Manono was set on fire and the Congo
tin industry entered into a long period of plundering, maltreat-
ment and disorder from both official and unofficial armies. Tin
output fell in 1961 to 6,600 tons (or half its pre-independence
level) and was to remain little above this dreary figure for the rest
of the 1960s. Capital movements and the repatriation of profits,
salaries and savings were subject to restriction; more important
were the delays and restrictions on import of the essential mining supplies. The new Congo (Zaïre) government drew natural
attention to the substantial shareholding rights which it had
inherited from the previous administration and the Congo Free
State. The Belgian companies transferred, where necessary,
their official seats to the Congo with appropriate name
changes.[21] A new and elaborate mining code was drawn up by the
Congo (Zaïre) government to emphasise that the policy of the
government was to ensure its own effective control and to point
the possible approved lines of development (essentially, the linking of foreign and national private interests with the latter pre-
dominating). The existing tin companies were, however, as yet
neither fully nationalised nor replaced, and the entry (during the
1960s) of other foreign capital (for example, Phibraki, formed by
Sobaki and Philipp Bros. of New York) or indeed of indigenous
Zaïre capital was not very important. The Zaïre government had
fatter fish to fry in taking over the huge Belgian copper company
Union Minière du Haut Katanga and in opening up the Congo
copper and cobalt deposits to prospecting and exploitation by
other non-Belgian international capital.

Bolivia

There is no contrast in the world of tin more glaring than that
between mining in Bolivia and mining elsewhere. The contrast
exists in every field – physical, geological, political, economic
and psychological – and in every field the elements are all loaded

FIGURE 6

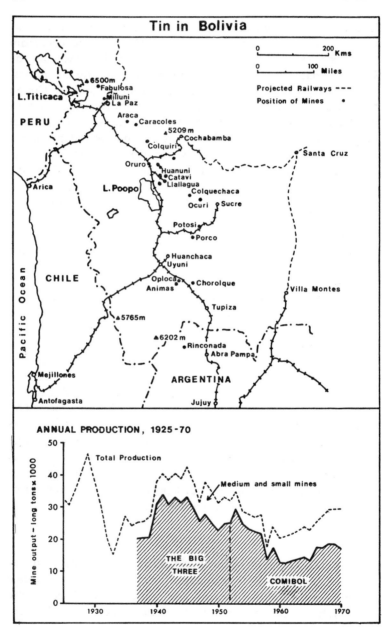

Tin in Bolivia

ANNUAL PRODUCTION, 1925-70

against the Bolivians. The Bolivian's approach to his tin problems has therefore an intensity and perhaps even a justified unreasonability which keeps him constantly out of step with the other tin fields of the world.

Geographically, the Bolivian tinfield lies in a high belt about 500 miles long and 100 to 200 miles deep running along the mountain backbone of the country – the Altiplano – from Lake Titicaca in the north past La Paz and through Oruri and Potosí to the Argentine border on the south. Within this belt there is a parallel north-south railway system with connections westwards to ports outside the country – to Mejillones, Antofagasta and Arica in Chile. The tinfield, in spite of its length, covers perhaps one-fifth of the total area of the highly mineralised country. Outside the mountain backbone there are vast stretches of low lying, tropical and almost unoccupied territories looking to tributaries of the Amazon and Paraguay rivers. There is no tin (at least, none yet discovered) outside the mountains. Bolivia has eight times the area of West Malaysia but only one-half of the population; it is one-quarter bigger than Nigeria but has only one-tenth of the people. Something like 50,000 are employed in the Bolivian tin industry. This number is small in itself but is statistically important (it is more than one-third of the number employed in all industrial manufacture). The miners are organised and, on occasion, they have been politically decisive.

The first problem of Bolivian tin is the simplest but perhaps the most difficult to convey to the outsider. The mines in Bolivia (in tin and other non-ferrous metals) are in high mountain ranges or on high mountain plateaus. They are at heights sometimes above the snow line in a country not far from the equator. Twenty-five of the major mines in Bolivia are listed as being more than 11,500 feet above sea level, and, of these, one (Chorolqúe) is at 16,500 feet, one (Tasna) is at 16,000 feet and three (Fabulosa, Avicaya and Atoroma) are over 14,000 feet. The biggest tin mine in Bolivia – the Llallagua-Catavi complex – has its administrative headquarters at 12,700 feet above sea level; but the headframe of the Salvadora mine at Llallagua is at 15,300 feet. At Potosí, the oldest mining centre in Bolivia, the town shelters – if ever such a word is appropriate to a Bolivian mining centre – at the foot of the tin-rich and once silver-rich pyramid of the Cerro de Potosí. The hummock rises 2,000 feet above the town which is itself at 13,000 feet above sea level. The miner climbs a road with an ascent of one in two to reach the entrance of the Unificada mine at 14,000 feet, and the small mine operators climb beyond that towards the peak of the mountain.

Mountain heights, and all that comes from them, are the norm. The landscape is arid, bleak, cold, sometimes very beautiful and always without humanity. The air is thin; heavy physical effort is difficult. The hundreds of mines are disconnected; roads are poor, sometimes little more than tracks. The railway system looks outside Bolivia. The possibilities for decent agriculture are limited; there is almost no industrial manufacturing development within reach of the mining townships. Hard drinking is the miner's industry.

The geographical problem is made worse by geology. Almost all the tin in Bolivia is produced from underground mining in hard rock – a form of tin mining almost unknown in south-east Asia or in Africa. The mining is from adits driven into the steeply sloping mountain faces or by shafts. At Cerro de Potosí the basic geology has produced a mushroom-shaped igneous intrusion to form the core of the hill, around 1,500 yards wide at the top and narrowing to 100 yards at the bottom. Within this mushroom the ore veins may reach 1,000 yards long. The veins run at all angles and thicknesses. There are hundreds of parallel veinlets running from 1 cm to 5 cms in thickness; the average tin vein being worked is only 20 cms or 8 inches. At the Unificada mine there are sixteen different underground levels worked at about 30 yard intervals. At Caracoles, where the highest level is at 14,500 feet, there is also exploratory work going down to 13,000 feet. The ores are often complex in metal content. At the Cerro de Potosí, for example, the upper veins are rich in silver and tin, but in the middle veins the tin and silver are flanked by veins richer in zinc blende and galena (lead). The tin industry also produces lead, antimony, wolfram, zinc and copper.

[22]For details on the tin losses and on other matters of mining and processing see D. J. Fox: *Tin and the Bolivian Economy*, 1970.
[23]The theoretical maximum is about 78 per cent.

Underground work is sometimes in the poorest conditions. Blasting, often very reckless, is almost continuous. Temperature and humidity levels in the lower sections of a mine may be almost beyond toleration; the temperature of a working face within a level may be 40° to 50° C (say, 70 to 90° F) higher than the temperature of the mountain air outside from which the miner has come at the beginning of his shift.

The Bolivian miner, whether in a large unit of the nationalised Corporación Minera de Bolivia (Comibol) or in a mining co-operative of a dozen workers, is skilled. He has, after all, under one form or another of harsh compulsion, been working on silver mines for 400 years and on tin mines for a century. But, outside the mine, the techniques for treating the ore, particularly mixed ores, are sometimes lamentable. Much of the labour employed in handling broken ore is female and it is possible that in some cases a quarter of the tin won by the men with such labour is at once thrown away. Even where treatment, crushing and flotation techniques are quite advanced, there may still be another loss of one-third of the original ore mined. In the case of Comibol, covering over half the total Bolivian production, loss of tin[22] in 1964-67 was no less than 49 per cent.

The concentrates actually produced in Bolivia have a tin content of between 20 and 65 per cent[23]; the commercial concentrate of 70 per cent or higher tin content which is standard in Africa and south-east Asia is almost unknown in Bolivia. The freight and smelting charges on this lower-grade material from Bolivia are disproportionately high.

Mining costs are generally higher in Bolivia than in the alluvial mining areas of the world. The nationalised corporation (Comibol) lost about $U.S.42 millions in the four years 1961-64. Some of its mines were producing at a cost far beyond the then world price of tin, but some losses arose from the government's currency policies and some private enterprises still left in private hands (and not all of them small) have been profitable enough. The importance of the tin price is not so much in relation to the profitability or otherwise of the mining units inside Bolivia (after all, many highly-advanced countries are still prepared to subsidise their coal mining industries) as in relation to the total export earnings of the whole of Bolivia. Exports of tin from Bolivia are responsible for about 70 per cent of the country's export earnings. When the price of tin drops, the whole of Bolivia may have to eat less. The country is therefore tied to an avid and intense public concentration at all times on the tin price and on the factors affecting that price.

The final problem for Bolivia has been the acute shrinkage in the value of the tin-containing ground which she has been mining. In its early development in the 1890s as a new tin field where valuable tin ores had been untouched because the miner was looking for silver, the grade of tin ground mined was sometimes astonishingly high. When Simon Patiño's manager struck a new lode at La Salvadora mine in 1900 some of the ore extracted assayed no less than 47 per cent tin content – a strike which formed one of the bases of Patiño's fortune. This was an incredible bonanza, but grades of 8 to 12 per cent tin were not uncommon. At Llallagua in 1908 the grade was still 12 to 15 per cent and even as late as 1924 9·2 per cent. By 1938 Llallagua was down to 2·45 per cent but the average for the larger mines as a whole was still 3 per cent. On this very high quality ground, mined at Bolivian wages, the profits could not avoid being enormous.

By 1921 Bolivia's tin output had been pushed up to 28,500 tons or over one-quarter of the world's production. She was the second largest producer; she threatened to overtake Malaya, whose output so far in the twentieth century had been declining while that of Bolivia had doubled.

The initial tin mining capital of Bolivia had come in important degree from Chile and Europe. But the industry became dominated for nearly fifty years by a purely Bolivian figure – Simon Patiño from near Cochabamba, well outside the mining areas. Patiño, after training in the buying of ore and supplies in Oruro, with (so far as we know) no geological, mining or engineering training but with a meticulous head for business details and a fierce driving force on all employees, bought full ownership of the La Salvadora mine (in the Llallagua-Catavi area) in 1897 and struck his phenomenally rich find about 1900. The never-empty pot of gold at Salvadora was producing 10,000 tons of tin a year by 1910. Working through the merchant house of Duncan Fox and the Anglo-South American Bank (later the Bank of London and South America) he pulled off by 1924 the biggest deal of all – the taking over from its Chilean owners of the Companía Estañifera de Llallagua. The group was now turning out in the neighbourhood of 15,000 tons of tin a year – that is, half the Bolivian production, the equal of the joint production of Thailand and Nigeria and about 11 per cent of world production. Most of the Patiño Bolivian interests were brought together in 1924 in Patiño Mines and Enterprises Consolidated. This was, significantly, a dollar corporation, located in Delaware, U.S.A.

It also marked the move of Patiño (or perhaps more accurately of his corporation) into wider international fields and towards the formation of the first – and the only – international group in tin smelting. He acquired control of the Williams, Harvey smelter in the United Kingdom which, both before and after the collapse of the Patiño mining interests within Bolivia, smelted the bulk of Bolivian tin concentrate output; control of the Eastern Smelting Co. in Malaya, smelting about half the Malayan concentrate output; links with the Anglo-Oriental/London Tin Corporation mining groups in Nigeria and Malaya; and an important – but not dominating – voice in the work of the International Tin Committee during the 1930s.

A long way behind Patiño were two other groups. in Bolivia. Luis Soux held by 1921 the lion's share of the deeper tin workings at Potosi through the Compania Minera de Potosi; but in a merger with silver companies in 1929, control passed to Mauricio Hochschild, who became the second of the "tin barons" of Bolivia, whose background was not generally told, whose personality was unpleasant and whose capital connections were sometimes not clear. The third group was the Aramayo group, based in Switzerland, with interests both in tin and silver, whose head during the 1940-50 period, C. V. Aramayo, was the only one of the barons with any sense of obligation within him to Bolivia. By 1952, these three tin barons, with a total capital of about $U.S.33 ·0 millions, were responsible for three-quarters of all Bolivian tin exports and a labour force of perhaps 20-25,000.

Their ownership of the industry justified itself in terms of production and technical advance. In 1921 the Patiño group brought a rail link to Catavi (before 1921 all concentrates had been moved out of Catavi on the back of llamas and took five days to reach the railhead at Machacamarca sixty miles away). By 1937 Catavi had been changed to shrinkage stoping and had turned to re-working old tailings; by 1948 it had introduced block caving to treat low-grade ore reserves. The Patiño social policy was less progressive. In the great slump of 1930-33, when the labour force at Catavi was reduced by 70 per cent, Patiño Enterprises paid rates so low that it began to employ women underground, and in 1942 it was involved in a massacre at Catavi of workers.

Great temporary opportunities came to Bolivia with the cutting-off in 1942 of the south-east Asian tin supplies. Bolivia found herself the nearest tin producer to the U.S.A. and with unlimited scope for sales against dollars and gold to the U.S.A. and the United Kingdom. During 1942-45 the Bolivian production averaged 40,000 tons of tin a year or almost half the available world output.

The profits of the tin barons were at the expense of the long-term health of the industry. Deliberately or unwillingly the grade of tin-containing ground mined was being pushed down rapidly.[24] At Llallagua-Catavi the grade of ore mined, which had been nearly halved between 1925 and 1935, fell by one-fifth in 1935-40, again by one-fifth in 1940-45 and again by one-half in 1945-50; and at the last of these dates, 1950, it stood at less than one-fifth of what it had been twenty-five years earlier.

In 1952, after a social and political revolution in Bolivia in which the miners' union played an important and perhaps determining part, the mines of the "big three" (Patiño, Hochschild and Aramayo) were nationalised under the Corporación Minera de Bolivia (Comibol). Some, but not much, compensation was provided. The medium mines outside the ownership of the tin barons and the small mines were regulated but not nationalised.

[24]See J. M. Molloy and R. S. Thorn: *Beyond the Revolution: Bolivia since 1952* and D. J. Fox: *Tin and the Bolivian Economy.*

Comibol had a most unhappy start in life. The price paid to the trade unions for their support was feather-bedding and an increase between 1952 and 1956 of nearly half in the total labour force (an increase heavily biased in favour of those working on the mine surface). At the same time the total tin production of Comibol fell by 4,000 tons a year. The worsening in the quality of ground being mined continued and by 1960 Catavi was down to a tin content of only 0·73 per cent. Machinery was in short supply; the import of mining machinery under the old regime throughout 1940-52 had been no higher than in 1925-29. Engineers and mine managers were in even shorter supply; 170 of the 200 foreign mining engineers left the country after nationalisation. The governmental control of the foreign exchange earned by the mining industry meant that the earnings of tin exports were used to subsidise agricultural and other imports and not to buy mining equipment and supplies essential to modernisation. The whole of the Bolivian tin industry was affected first by the cessation of tin purchases by the U.S. strategic stockpile and then by the imposition of export control by the International Tin Council during 1958-60. Between 1957 and 1960 Comibol lost $U.S.31 millions.

Comibol and Bolivia (perhaps in that order since, in 1961, Comibol's gross expenditure was twice the expenditure of the central government) were saved by a curious alliance. The U.S. accepted the principle of economic aid to a policy led by a Paz Estenssoro government which it had thought to be revolutionary and anti-American; the Bolivian government accepted the principle (but not any extensive practice) of compensation for tin nationalisation; both disliked the control of the miners' union over the industry; both agreed that if the tin industry foundered Bolivia would die. During 1961-64 U.S. economic aid to Bolivia totalled no less than $U.S.213 millions. The Triangular Plan of 1961 between the Bolivian government, the U.S.A. and the Inter-American Development Bank provided for loans of around $U.S.50 millions (mainly from the U.S.A.) first to float and then to rehabilitate Comibol. The labour force was curtailed; by 1964 it was back again to where it had been just before nationalisation. The resistance of the miners' union to wholesale dismissals was broken in 1964-65. Comibol, whose costs during the early sixties had been usually about one-quarter higher than the price it received for its tin, began to recover financially.

The haul back was a long one. Total Bolivian tin production (Comibol and others) had been forced down in 1958 to 18,000 tons, the lowest figure for twenty-five years. But through the 1960s, output rose almost unchecked until by 1970-71 it had moved back to 30,000 tons or one-sixth of world production. That recovery was due only in part, and indeed not in a very substantial part, to the money poured into Comibol. Nationalisation had grouped within Comibol some 17 companies (*empresas*) of very varying size, quality and profitability. Of these empresas in 1955 three (Catavi-Llallagua, Huanuni and Colquiri) represented 70 per cent of the total Comibol output of 23,500 tons of tin; the others ranged in size from 50 tons to 1,960 tons. Comibol production remained in the first half of the 1960s only around 13,000 tons of tin a year and in the years 1961-64 Comibol lost another $U.S.46·7 millions. Real recovery did not come until 1966 and was then largely limited to Llallagua-Catavi, although still enough to provide a profit of $U.S.23·8 millions over 1965-70. The 1971 Comibol tonnage was still below its 1953 level.

The increase in recent years in the Bolivian total has come disproportionately from the medium, unnationalised mines (the *mineria mediana*) and the small mines (the *mineria chica*). The medium mines are accepted as the mines with minor capitalisation, partially mechanised and with technical services. The small mines, sometimes consisting of little more than one family, work with little capital, with no or little mechanisation, and sell their output through a state organisation – the Banco Minero de Bolivia.

The medium mines were not necessarily Bolivian in character. Out of 19 medium mines producing tin in 1966 the capital for 11 (with 40 per cent of the output of this category) was Bolivian, of three (with 17 per cent of the output) was Bolivian-Chilean, of one (with 20 per cent) was Bolivian-North American, of one (with 15 per cent) was Bolivian-British and of three (with 8 per cent) was North American. They were not necessarily small; the average output per company in 1966 was about 220 tons of tin a year (that is, some five times the size of the average gravel pump in Malaysia); but four of them (International Mining, Fabulosa, Cerro Grande and Tihua) were over 400 tons a year and therefore in the class of average Malaysian dredge mine output. In the year of nationalisation, 1952, these medium mines produced 4,100 tons of tin out of a Bolivian total of 32,000 tons and in 1961 only 2,300 tons out of 20,400 tons. By 1968 their revival was so sharp that they were producing 6,400 tons out of 29,000 tons and by 1971 perhaps 6,000 tons out of 30,300 tons.

The medium mines seem to hold one great advantage for the future. Their reserves are adequate; they showed in 1967 a total reserve (proved, probable and possible) of 135,000 long tons of tin or over thirty years of working life. These reserves are of a much higher grade of ground than are the Comibol reserves. The labour force of the medium mines has a productivity per head of around a ton of tin per annum, substantially higher than productivity at Comibol.

The small mines are almost legion. They work on a variety of rights. At Cerro de Potosí small mines use traditional rights to work on the mountain above 14,500 feet; in 1968 there were no fewer than 575 registered *bocaminas* worked by private companies, cooperative societies, dependants of the Banco Minero and individuals. In the Frias province around Potosí the Banco Minero listed 253 producers (with a labour force of perhaps 2,000) whose output in 1964 was averaging about 2 tons a year per unit. Catavi is typical of the anomalies and differences in Bolivian mining. Close to the largest and one of the most highly mechanised tin mines in the world a small unit of four men drives its own shaft down to about 100 feet, instals a windlass, sends down one man to hew out the ore in the immediate underground standing space, winds up the broken ore in a bucket, prays to God and lights candles to the Devil for no cave-in. Catavi has an excellent flotation and dressing plant; outside, *veneristas* wash material in the alluvium and valleys; *lameristas* work the fine slimes; women (*pallyris*) re-pick the discarded ore.

The waste slime from the mill has accumulated in a settling lake, it was being treated by an American-owned suction dredge, but the dredge was seized by the government in 1970 and, during the discussions on compensation, sank by accident.

The small miners (*mineria chica*, those who sold to the Banco Minero, as opposed to those who sold through Comibol) were producing perhaps 3,000 tons of tin in 1952 or about one-tenth of the Bolivian total. They reached a peak of over 6,000 tons in 1965 when the tin price touched over 178 U.S. cents a pound, the highest annual price yet reported in tin, but they fell with the tin price to 4,000 tons by 1968.

In one field, Bolivia has recently brought herself into line with other tin producing countries. The country had long resented her position purely as a producer of raw material – tin concentrate – which was then shipped out for processing into tin metal in the U.S.A. or in the United Kingdom (in the latter case, in a smelter of the Patiño group). Long years of self-persuasion as to the economic benefits of building a large smelter within Bolivia made the plan a matter of political necessity.[25] In 1971, with aid

[25]The plan had greater financial benefits than most outside critics accepted. In the year 1971-72, with a larger tonnage of tin metal exported, Bolivian earnings from all tin exports were \$U.S.116·6 millions against \$97·3 millions in 1970-71.

[26]It is axiomatic in Bolivian thinking that the tin barons made colossal profits from the exploitation of the tin mines. Looking at the various elements of the grade of ground being mined, the low rate of Bolivian mine wages (even if they were high by Bolivian standards), the sharp increase in total Bolivian production until the mid-1940s, the low rate of export duty (especially before 1920) and the activities of Simon Patiño outside Bolivia on the basis of his income from Bolivia, it is very difficult not to accept the Bolivian viewpoint. The Patiño Mines and Enterprises Incorporated (based on the Llallagua-Catavi complex) made a profit of \$U.S.74·4 millions during 1924-52 on an initial paid-up capital of \$27·6 millions. C. V. Aramayo: *Memorandum sur les problèmes de l'industrie minière de Bolivie*, 1947 (translated from the original Spanish) gave the following average annual figures for the period 1930-44:

	% of profit to capital	% of dividend to capital
Unificada de Potosí (Hochschild)	(loss) −24·6	—
Minero de Oruro	8·39	5·6
Aramayo Co.	10·8	5·8
Patiño Mines	7·32	5·79

But Aramayo forgot to point out that over 1921-38 the Aramayo company paid out dividends which seem to have totalled 275 per cent (plus capitalised share bonuses).

from Federal Germany, the first part of the plan to build a smelter to treat the whole of Bolivian tin production became a reality. Bolivia therefore joined the other tin producers – Nigeria, Thailand and Indonesia – who in the last dozen years have turned to process for themselves all or a substantial proportion of their domestic mine output.[26]

Brazil

Tin mining in Brazil was of negligible importance prior to the 1950s when what little output there was came from the São João del Rei area of the Minas Gerais in the east of the country. But from 1962 onwards a new and distant virgin field was being opened up. This new field lay far away in the territory of Rondônia in the south of the upper Amazon valley. The first discovery of tin had been made there in 1950. Aerophotographic coverage of Rondônia in 1963-65 and geological prospecting in the most difficult conditions to be expected there showed a series of tin deposits running in a belt along the younger granites for about 200 miles from the Brazilian-Bolivian frontier on the river Madeira south-east towards the Matto Grosso. The deposits are eluvial or alluvial in character and have no relation to the mountain lodes of Bolivia in the High Andes over 700 miles away. The centre of Rondonian activity in transport and mining supplies at Porto Velho is little more than 100 miles from the Bolivian frontier. The cassiterite found at Abuna is actually on the Bolivian frontier and the Bolivian geologists hope for an extension into the low-lying tropical areas of the Beni area of Bolivia.

The Rondonian deposits are apparently rich in character and quality. Some of the early discoveries ran as high as 200 kgs of cassiterite per cubic metre of ground – a phenomenal richness. Early mining in 1958-60 relied on crude manual working methods, but by 1964, when production had reached 700 tons of tin a year, more mechanical means, including bulldozers, mechanical shovels, jig concentrators and diesel engines were being brought in. Mining materials were being flown into and cassiterite out of Porto Velho. The country is inhospitable but climatically not disastrous. Initial mining has been under a thin

FIGURE 7

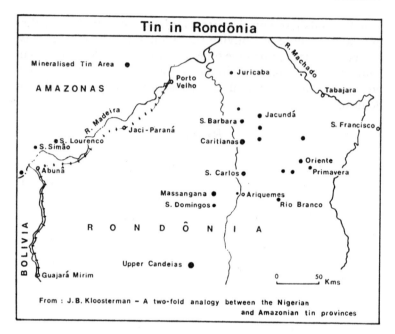

Tin in Rondônia

Mineralised Tin Area

AMAZONAS

Porto Velho

• Juricaba

R. Machado

Tabajara

R. Madeira

Jaci-Paraná

S. Lourenço
S. Simão

• Jacundá

S. Barbara •

Caritianas •

S. Francisco

Abunã

S. Carlos •

• Oriente
• Primavera

BOLIVIA

Massangana •
S. Domingos •

• Ariquemes

Rio Branco

R   O   N   D   Ô   N   I   A

Upper Candeias •

Guajará Mirim

0   50
Kms

From : J.B. Kloosterman – A two-fold analogy between the Nigerian
and Amazonian tin provinces

overburden of perhaps 6 to 10 feet into tin-containing gravel resting on decomposed bedrock with a tin content usually between 2 and 10 kg of cassiterite per cubic metre.

The local labour is of Indian origin, has no concept of wage bargaining, is illiterate, is cheap and is wide open to exploitation; it is naturally inefficient and at one time it was reported that up to 40 per cent of the labour force could be expected to have debilitating malaria.

By the later 1960s, the field was beginning to attract international attention and capital, within the limits of the insistence of the Brazilian government on the necessary participation of Brazilian citizens. The territory is particularly attractive to American capital which perhaps felt itself safer in a tin field so much nearer geographically to the United States and still willing enough to provide safeguards for investment. By 1972 there was

[27] J. B. Kloosterman: *A two-fold analogy between the Nigerian and Amazonian tin provinces* in W. Fox: *A second technical conference on tin, Bangkok, 1969.*

even a dredge in Rondônia and the interests involved in tin mining included the U.S. firms of W. R. Grace and Co. (which was also involved among the non-nationalised tin enterprises in Bolivia) and N. L. Industries, formerly the National Lead Co. (which had been interested forty years before in the formation of Patiño Enterprises in Bolivia) as well as the Billiton Co. of the Netherlands. Total mine production by 1972 had become perhaps 4,000 tons of tin or over 2 per cent of world tin production.

Small as it was, the new Brazilian field had still established itself within ten years as the eighth tin producer in the world. The parallel with the rapid development of the Nigerian field after 1911 or of the Belgian Congo field after 1921 could not be ignored. At the United Nations Tin Conference at Geneva in 1970 which established the fourth international tin agreement, the other producers made strong efforts to persuade Brazil to enter the agreement, but the Brazilian government was not prepared to handicap the future of the field by exposing it to the risk of early export control.

If the estimates of the future Rondonian position, made mainly by Brazilian politicians, have any basis in fact there can be no doubt as to the future weight of Rondônia in tin. The estimates of reserves are very variable, as might be expected from the extent of the field and the sketchiness of detailed geological investigation. The Brazilian government quotes figures too high to be repeated here; a more reliable observer[27] speaks of hundreds of thousands of tons of cassiterite on the assumption of minimum workable ground of 0·6 kg/cu. m of cassiterite; the geologists of the U.S. Geological Survey refer to a range of likely reserves between 97,000 and 537,000 tons of tin and possible additional reserves of over a million tons; and others elsewhere thought in terms of a minimum of 300,000 tons of tin in reserves.

The Amazon valley has created in the past its own visions of a land rich in gold and rich in rubber; and it is necessary not to be dazzled by the prospects in tin. But, if any of the estimates were to be realised, Rondônia would move into the category of a potential major world tinfield.

FIGURE 8

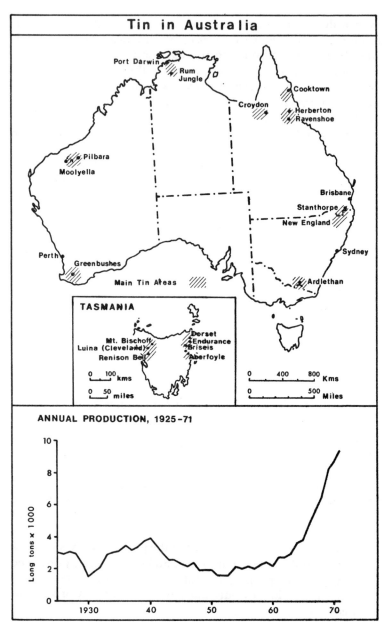

Tin in Australia

ANNUAL PRODUCTION, 1925-71

Australia

The nineteenth century development of Australian tin resources had been sudden and dramatic. In a short five years after 1871 the country shot to a production of over 11,000 tons of tin, becoming the foremost producer of tin in the world, greater even than the United Kingdom. She did not keep that proud position for very long. Her output began to decline after 1886 and in the thirty years before 1921 moved rather spasmodically between a minimum of 3,000 tons and a maximum of 8,000 tons a year. The world slump of 1922 halved output to 2,600 tons and gave Australian tin mining a blow from which it was not to recover for 40 years. By 1946 Australia had ceased to be an exporter of tin. By 1959 production seemed to be disappearing in such old-established areas as New South Wales and Western Australia and the industry was settling down to a production rate of only about 2,000 tons a year. It had one stroke of luck. As a consumer member of the first International Tin Agreement Australia stood outside the restrictions of export control in 1958-60 and her tin industry may therefore have been saved from destruction.

A most remarkable transformation in the fortunes of the industry took place after 1964. The industry accepted the principle that in tin (as in other non-ferrous metals) the future would lie in the treatment on a massive scale of low-grade deposits, in the application for that purpose of great capital investments (including investment from mining houses outside Australia) and in the necessary increase in output per man hour through mechanisation of highly expensive labour. It was also fortunate that the fever of mining prospecting which swept Australia at the time and which so successfully scoured every corner of the country was not confined to uranium, bauxite, iron ore and nickel but also spread very generously into tin. The industry also became convinced that the international tin agreements had the will and the power to guarantee a world floor price for tin. This attitude was confirmed when the Australians in 1968-69, although officially consumers under the third tin agreement, volunteered to accept some control over the exports of their producing industry and in 1971 when they entered the fourth agreement officially as producers.

The concentration of mining effort through the 1960s had results both quick and substantial. The number of tin mines in Australia (some of them on the smallest scale) almost doubled during the period, and mechanisation in the larger more than quadrupled the output per miner. From 1965, Australia was again an exporter of tin. She overhauled the Congo-Zaïre in 1969 and Nigeria in 1970 to become the fifth ranking producer in the world, excluding the U.S.S.R. and China. By 1972 she was producing about 12,000 tons of tin a year and was shipping well over half of this on to the world market, thus contributing to the Tin Council's decision to limit export quotas again.

The revival was primarily in New South Wales and northern Tasmania and came mainly from massive underground or open-cast hard rock operations. In Tasmania, the mine at Renison Bell, which had been working rather fitfully since 1890 first on alluvial and then on deep mining, was transformed after 1964 by a development programme financed from the Consolidated Gold Fields group,[28] a programme which drove a slanting adit down to 800 feet below ground. By 1971, Renison was one of the great tin mines of the world producing itself 3,800 tons of tin or twice as much as the whole of Australia had been producing twenty years earlier. The Aberfoyle group (which was largely responsible for the revival of faith in Australian tin mining) opened up the Mount Cleveland mine at Luina in Tasmania as an entirely new mine in an old field, driving a horizontal adit into the mountain side.[29] These two Tasmanian concerns, along with the hard rock open-cast Ardlethan in New South Wales, covered by 1972 about two-thirds of total Australian output. Dredging had become unimportant, and the two remaining bucket dredges – Ravenshoe and Tableland – were responsible for

[28]Consolidated Gold Fields had created a small international hard rock grouping in tin. In 1971 it controlled Renison Bell in Tasmania, Wheal Jane in Cornwall, Union Tin and Rooiberg in South Africa. The total annual output of these companies was over 7,000 tons a year or about 4 per cent of the world total.

[29]The Aberfoyle interests were taken over by Cominco, the Canadian zinc-lead producer.

only 6 per cent. The prospects for off-shore sea dredging for tin in the Bass Strait off Tasmania have not as yet matured. In spite of the domination of the industry by the revitalised large units of production, the small miner is still vigorous and, particularly in Queensland, he still rummages for small pockets as assiduously as does his fellow miner in the Bolivian mountains.

The new mining methods have brought out more tin from Australia than the most optimistic observer believed to be possible in the 1950s. But the massive working of ground not in itself of any high grade has serious technical problems and the Australians have found themselves with milling and flotation difficulties, especially on fine tin, not unlike those in Bolivia. Renison has had a grade of ground which in recent years has varied between 0.8 and 1.4 per cent tin content, that is, substantially higher than the grade in the same period at the Llallagua-Catavi mine in Bolivia. The tin recovery at Renison on treatment was as low as 41 per cent in 1967-68 and only 52 per cent in 1970; in the latter year the high-grade concentrates produced were only of 56 to 63 per cent quality while one-fifth of the total production was in the form of low-grade concentrates of 19 to 22 per cent tin. The result is no better (if for different reasons) than the Bolivian results.

Burma

In Burma the tin-wolfram deposits are widely distributed in a narrow belt about 750 miles long running north from Victoria Point in the southern tip into the Southern Shan States in the north. Within that zone the main mining areas are Mergui, Tavoy and Karenni. In all areas the cassiterite and wolfram are closely associated with the biotite granite; in general, in the north the proportion of wolfram is greater and the cassiterite may be regarded as a by-product (so that the industry there represented in the past by the Mawchi mines was more dependent on a world wolfram market more highly erratic even than tin); in the south cassiterite is the main product of the tin-wolfram mines, although in some of the alluvial deposits there the wolfram disappears. Mining is mainly alluvial (including at one time dredging in the Heinze basin in Tavoy and at Thabawleik and Theindaw in Mergui) but Mawchi in Karenni is a lode mine and was the greatest tin-wolfram producer in the country, if not in the world, prior to 1941.

As a tin mining country Burma had a long and unimportant history before 1921, when its total tin output (in tin or in mixed concentrates) had reached 1,400 tons. The stimulus to production came from British capital following the tin price boom of 1924-27. The boom in production was unchecked through the 1930s and was, in fact, accentuated by the recovery of the wolfram price. Tin output quadrupled in the twenty years after 1921 and had reached 6,000 tons or approaching 4 per cent of the world output by 1939.

At that stage the Burmese industry had all the variety of disorganisation which characterises tinfields in their earlier years. European capital was strong. Most of the total Burmese tin output came from a group of larger mines (in 1939 75 per cent of the tin production and 62 per cent of the wolfram output came from only 4 per cent of the total number of mines, including Mawchi, Tavoy Tin Dredging, High Speed Steel Alloys Mining, Consolidated Tin Mines of Burma and Thabawleik Tin Dredging). In many mines apart from the dredging plants, winning of ore was on the tribute system where the workers (or the contractors who supplied the workers) were paid a price for each viss of concentrates handed in. The price paid by the management depended on the degree of freedom of the concentrates from impurities, the overheads incurred by the mine owners on the provision of plant and power, the price paid by neighbouring mines, the richness of the deposit and finally the current market price of tin.

These tributing methods were guaranteed to produce more tin, to employ the maximum number of part-time workers, to fit in with the agricultural life of a tin-containing area and to weaken the general future of the industry.

[30]This melancholy report of pre-1939 conditions requires some discounting. The more hopeful Colonial officials, especially in the old British and French overseas empires, often expected or anticipated from the areas they governed an unreasonable amount of intelligent response to the economic stimulus.

Most of the small mines were hill sluicing. The underground mines worked mineralised adits in the dry season and closed or reverted to hill sluicing in the wet season. The small mines in 1939 suffered from lack of capital and mechanical equipment, shortage of labour, seasonal operations in the wet season, bad communications, and high fuel and transport costs and even the water wheel was regarded in some areas as relatively advanced mechanisation.[30]

The Burmese collapsed under the Japanese military occupation during 1942-45. Mining (not only in tin) met every obstacle against its revival under an independent Burmese government. The Chinese directly employed by the major companies in Tavoy, Mergui and Karemni and the Ghurkas in the underground mines were replaced largely by local Burmese labour. The government was reluctant to permit the full return of the foreign capital which had dominated the pre-war industry. The internal local rebellions were looting the stocks of tin, explosives and stores as a matter of routine. Initial joint companies with government participation were set up for a number of enterprises (including Mawchi) but this merely preceded full government ownership. The mining industry was subject to nationalisation and to re-organisation under different names and ended under the Burma Minerals Development Corporation, later the Myanwa Tin and Tungsten Corporation. No proposal for the restoration of the industry became effective.

Through the 1950s tin production ran at only around 1,000 tons a year or about one-sixth of the peak 1941 level. Through the 1960s the output (so far as it can be traced) seems to have withered down to around 500 tons a year. A proposal in 1970 from the U.S.S.R. to help to re-open the Mawchi mine is not yet effective.

The minor tinfields

Outside the major tinfields which we have already examined there is left perhaps less than 5 per cent of world tin production. Many of these minor fields have more history than production.

In the United Kingdom the tin industry could still list, as late as 1919, 24 main mining companies – many of them with names like a roll of honour (Dolcoath, East Pool and Agar, Geevor, Grenville United, South Crofty and Tincroft) – operating or proposing to operate on a production which still remained as high as 3,300 tons a year. In the price crisis of 1920 the industry met disaster. Mine after mine collapsed. The deep hard-rock Cornish mines could not operate on a price that even the open-cast alluvial mines of south-east Asia found unprofitable. Output in 1922 was little over one-tenth of what it had been in 1920. But the industry was tough. It recovered its spirit and its output as prices moved up again and in the boom of 1929 it was able to report an annual output of over 3,000 tons. The industry was, however, largely a marginal one. A sharp break in the world tin price had almost as serious an effect on Cornwall in 1931 as it had had in 1921, but the recovery from the slump of 1931 was by no means as resilient and at the beginning of the Second World War Cornish output was only at the rate of 1,600 tons a year. There was no war boom in production; in fact, the contrary. During that war the historic mine of East Pool and Agar was finished and after the war labour shortages kept the output, now effectively limited to two mines (Geevor and South Crofty), at little over 1,000 tons a year.

Cornwall saw a revival in the middle 1960s. New management and fresh capital were pumped into South Crofty from the Siamese Tin Syndicate (now renamed the St. Piran Mining Co.). Geevor expanded and the output of Cornwall had risen again by 1970 to 1,700 tons a year. Outside capital was by now persuaded that the International Tin Agreement would shore up the tin price for a period reasonably enough far ahead; on the other hand, the United Kingdom government was inducing invest-ment in Cornwall as a development area. In 1971 the deep Wheal Jane mine was opened with Consolidated Gold Fields capital. It was the first new mine on a massive scale seen in Cornwall for very many years.

The tin belt of the Iberian Peninsula runs for about 300 miles from Badajoz towards La Coruña on either side of the frontier between Spain and Portugal. It was worked in Roman times but then disappeared as an effective industry for centuries. The tin on the Portuguese side is linked with wolfram and this conjunction helped in the minor re-emergence of Iberia during the First World War. By the later 1920s the combined Spanish-Portuguese output was running under 1,000 tons a year. During the Second World War the area, and especially Portugal, leaped into prominence as the only external tin producing area left with direct land communication to Germany and a savage and not very intelligent pre-emptive buying war took place between Germany on the one hand, and the United States and the United Kingdom on the other hand. This windfall pushed production in Portugal to over 3,000 tons in 1943 and in Spain to 1,000 tons in 1945. Neither country maintained its position in the peace. Even in the Korean war in 1950-51 and the consequent scramble for raw materials the Iberian mines could stage only a limited recovery to a joint production of only about 2,000 tons a year. Spanish production shrank in the 1960s; Portuguese production followed it downhill; and a very intensive enquiry on the industry in the two countries by D. J. Fox in 1970 produced – except for the large Beralt wolfram mine in Portugal – a dismal story of closed and derelict mines, of a lack of capital and of the steady drifting away of a long underpaid, part-time mining population.

In the old French Indochina, the presence of tin in the Mekong valley had been revealed as early as 1867 when villages in the Nam Pathène valley were paying taxes to the Siamese authorities in locally-produced tin ingots. Fench operations began with the Société des Mines d'Etain du Haut Tonkin in 1901 and the ore bodies in the Pia Ouac mountains and the Tinh Tuc valley were mined for tin and wolfram. These deposits lay in Cao Bang province on the frontier of China in modern North Vietnam.

The second of the Indochinese tin districts lay in the Nam Pathène basin, a sub-tributary of the river Mekong, in middle Laos. Open-cast mining under French ownership began effectively here after 1923. By 1940 Indochina as a whole was producing 1,600 tons of tin a year. Most of the domestic concentrates were shipped to Malaya for smelting; the local smelter at Haiphong handled and purified only crude tin metal from China for re-export to France and the U.S.A.

The French plan to restore the shrunken production of the 1941-45 period proved only paper. The Tonkin deposits fell early on into the sphere of North Vietnam and nothing is known of what has happened to them during the last twenty years. The Laos deposits fell to the new state of Laos with the French companies re-organised but still in ownership. By 1966 tin production in Laos had crept up to around 300 tons a year. The later 1960s saw a stimulus which, in the light of the political conditions of Laos, was most remarkable and by 1972 concentrates from Laos were coming into Malaysia to be smelted at a rate of over 1,000 tons of tin a year.

In the Republic of South Africa, tin mining is based on the deposits in the Potgietersrus-Pietersburg area in the northern Transvaal. Production began effectively in 1906 and within five years had risen to over 2,000 tons of tin annually (that is, double the rate of Nigeria which had started production under European companies at much the same date). South Africa did not fulfil her early promise and in the middle twenties was down to a rate of little over 1,000 tons of tin. Even this rate was not maintained in the next twenty years and it was not till 1953 that the industry saw over 1,000 tons annual production repeated. Since that date a more encouraging story, based on the treatment of massive reserves in lodes of low-grade disseminated ores, especially by the Rooiberg Minerals Development Co., has brought annual production to 2,000 tons, only one per cent of world total output but roughly enough to cover the consumption requirements of the country.

The other minor tin producers in the world need take up little space here. The Argentine tin mines have in recent years pushed tin output towards 2,000 tons a year, mainly from an extension in Jujuy province of the mountain mineralised belt from the Bolivia altiplano. The Belgian companies in Rwanda who had established themselves through subsidiaries from the old Belgian Congo days have managed to maintain production at around 1,300 tons of tin a year, a better relative rate than they have been able to maintain in Congo-Zaïre. In South West Africa, where there is an unusual concentration of ownership (the South West Africa Co. in tin-wolfram at Brandenberg West and the South

African Iron and Steel Industrial Corporation at Uis), mass mining methods have since 1965 doubled the tin output to around 1,000 tons a year. Mexico has, especially in Durango and San Luis Potosí, a very large number of small deposits of poor quality, many irregularly worked.

Since the San Antonio mine closed in 1939, Mexican tin output has normally been around only 500 tons a year, except for a small boom in the high price years of 1963-66. Japan, now the world's second customer for tin metal, has an old mining industry in the north-east and south of Kyushu and in the Akenobe area of Honshu. The mines have a Cornish character of depth, age and mixed minerals (the Akenobe deposits have been mined for over 400 years, initially for copper and silver, later for tin and wolfram). The mines reached a peak production of 2,200 tons in 1937 and in 1941, but in the last twenty years they have been running normally between 800 and 1,000 tons a year, a mere fraction of the Japanese consumption.

Perhaps a word or two is due on the position of the world's largest consumer – the U.S.A. – in the field of production. The United States may feel with some justification that God, who endowed the country so plentifully with copper, lead, zinc and iron ore, was spiteful when He forgot the tin. There is, for all practical purposes, no tin deposit in the United States outside Alaska; and even in Alaska the deposits are small, frozen, difficult of access and highly expensive to mine.

China and the U.S.S.R.

This picture of the movement since 1921 in the tin producing countries of the world has omitted the People's Republic of China and the U.S.S.R. Both are substantial producers; both have great potentialities as consumers. In neither case it is yet possible to say with any official authority what are their tonnages of production or consumption. The U.S.S.R. in 1957-58 had a disastrous effect on the world tin price through its unrestricted exports of tin, likely to have been of Chinese origin; in more recent years the U.S.S.R., in transforming itself into an importer, probably did something to prevent a slide towards a world surplus of production. Direct Chinese exports have scarcely affected the world in the last thirty years.

In China there are very large deposits of tin along two arms of the continuation of the Malaysia-Thailand tin belt. The more important are lode deposits mined, sometimes at great depth underground, in the mountains in a stretch of about 400 square miles between Kokiu and Mengtse in southern Yunnan; the less important (usually alluvial and sometimes tin and tungsten combined) are in the Nam-Ling range from Kiangsi to Kweichow province and from the Fukien-Kwantung province into the island of Hainan. Tin mining has been known in Kokiu for centuries and production was reported on fairly reliable authority to be 6,300 tons in 1921 or approaching 6 per cent of the rest of the world. The industry was primitively run with perhaps as many as 100,000 workers. The tin came on the world market as metal, much of the crude metal smelted in Kokiu being re-smelted to higher marketable standards in Hong Kong. By 1938 production may have been 13,000 tons or 9 per cent of the world total. It was perhaps only the pre-occupation of foreign mining capital with the rapid expansion of tin production at that time in Nigeria, Thailand and Indonesia and the greater ease of working alluvial tin deposits in those countries that prevented a large spill-off of this capital investment into south China.

The investment opportunity then lost did not re-appear. China became involved in the war against Japan and in an internal political struggle. In the mists, first of battle and then of intense secrecy, all true outside knowledge of the tin position of China disappeared. As late as 1949 the non-Communist government was telling the International Tin Study Group of an estimated production of 5,500 tons a year, but thereafter even estimates ceased to have any apparent relevance. At no time in the last 25 years have any official Chinese figures been given.[31]

[31]The International Tin Council in 1956 made an estimate of production at 14,000 tons a year; the Council emphasised that this figure was a nominal one but continued to print figures which went on rising to 26,000 tons in 1965. The Council then accepted that it had no real basis for these figures and very wisely stopped printing them. The U.S. Department of Mines estimated 20,000 tons for 1956, and 30,000 tons for 1961 but then reduced the figure to 20,000 tons for 1967; it seemed to have little real heart in the job. Other sources usually quoted variants on the U.S. estimates.

Two firm facts are, however, available. The first is the receipt of Chinese tin as reported officially by the main importing countries (excluding the socialist centrally-planned ones). In the 13 years 1959-72 the exports so measured reached a total of 65,000 tons or under 5,000 tons a year. The range of the figures was between a minimum of 3,000 and a maximum of 7,000 tons a year. These Chinese exports did not rise or fall with the world price of tin; in fact, they were at their lowest during the tin price boom of the middle 1960s. The volume of trade was not significant as a proportion of world trade in tin. The second firm fact relates to the trade with the U.S.S.R. From – and perhaps before – 1956 to 1961 the movement of tin metal from China to the U.S.S.R. was, quietly and inconspicuously in its initial years, a traffic third only to the movement in tin metal into the U.S.A. or of tin-in-concentrates into the United Kingdom. In those seven years China moved no less than 120,000 tons of metal to the U.S.S.R. These shipments were in accordance with aid agreements between the two countries, and it is very likely that some of the tin came from stocks accumulated but unshipped during the civil war in China. Exports from China to the U.S.S.R. tailed off in 1962-64, then stopped abruptly and have not, as yet, been resumed in any substantial degree.

We cannot say how the production of tin is moving within China (although we may assume that, with the past known quality of mining ground and with stability of government, production should be moving upwards). We do not know how consumption is moving (although we know that consumption does not necessarily move in relation with population and that even thickly populated countries with a growing tinplate industry and growing manufactures like India can use surprisingly little tin). We do not know whether China (like the U.S.A.) is stockpiling tin for strategic, economic or political reasons, and we do not know whether or not it will release more tin into international trade with the capitalist countries.

Whatever is vague amongst economists there is one general agreement amongst geologists – that China has the tin resources to be one of the world's largest producers. One estimate in 1946 put tin reserves in Yunnan alone at 1·25 to 1·50 million tons;

another in 1946 estimated all China at 1·8 million tons; another in 1964 was more modest at 0·7 million tons; and a fourth has given measured, indicated and inferred reserves at 1·5 million tons.

Prior to 1921, and perhaps much later, the production of tin in the U.S.S.R. was of no significance. But the Russian drive for self-sufficiency and her dependence for tin on her allies during the Second World War (when there was shipped to her over 20,000 tons of tin from the United Kingdom and 150,000 tons of tinplate from the U.S.A.) pushed her into a great prospecting drive east of the Urals through and after the 1950s. The results in the number of deposits discovered was most encouraging and the map of Eastern Siberia became dotted with a host of tin deposits of unknown size and quality. These deposits lie mainly in the Onon River valley in Transbaikal, in the Maritime Territory north and east of Vladivostok and in areas around Verkhoyansk in the Yakutsk Republic, along the Upper Kolyma River and east of Chaunskaya Bay. Full and often admirable geological information is made public, but nothing of actual or potential tin output. We know (and can compare with Alaska) the conditions likely to apply to mining in some of the more northern Siberian areas.

Mining in frozen ground or permafrost is extremely difficult, but these conditions are no more unfavourable or insuperable than in Alaska. Communications are deplorable, but this is not an uncommon feature in mining elsewhere. Working costs are likely to be high or very high, but we know nothing of another factor – the grade of ground being mined. Mining conditions in the tin area of the Maritime Territory are by no means excessively difficult and perhaps it is no worse to die of frostbite there than of sunstroke in south-east Asia or lack of oxygen in Bolivia.

[32]Sutulov: *The Soviet Challenge in Base Metals.* U. of Utah, 1971. But V. V. Strishkov (of the U.S. Bureau of Mines) estimates tin metal production at 26,000 tons for 1968 and 28,000 tons for 1971 in the *Mining Annual Review* (of the *Mining Journal*), 1972.

[33]Tinned goods reflect a western conception of advancing standards. In the U.S.S.R. the production of tinplate trebled in 1954-70; in the rest of the world it doubled. But Russian production at 0·5 million tons in 1970 was lower than in any one of the U.S.A., Japan, France, Federal Germany or the United Kingdom.

An unofficial estimate, perhaps a very approximate one, of mine production in 1939 was 3,000 tons of tin a year. Later estimates, all by outside sources and none from the U.S.S.R., have been very widely varying. The U.S. Bureau of Mines, less generous than others, estimated a smelter production rising from 11,800 tons in 1956 to 20,000 tons in 1961 and then to 25,000 tons in 1967. Metallgesellschaft put forward mine production at 13,700 metric tons for 1957, 20,000 tons for 1962, 23,000 tons for 1967 and 27,000 tons for 1971. Neither the International Tin Study Group nor the International Tin Council has ever known enough and they wisely refused to guess. The most recent estimate[32] believes that the Russian output of tin was as little as 1,000 tons in 1938 but had been driven up to 16,000 tons in 1960, to 23,000 tons in 1965 and then to no less than 35,000 tons in 1970. Consumption was estimated at 5,000 tons in 1932 and at 42,000 tons in 1970.

We have already referred to the first and calamitous impact of the U.S.S.R. in the late 1950s when a great volume of tin metal was thrown on to the world market. The second impact of the U.S.S.R. later undid a little of this harm. From 1963 the U.S.S.R. ceased to be an exporter and reversed its position to become an importer from the world market. Between then and 1970 she was taking an average of about 5,000 tons a year, mainly from Malaysia and the United Kingdom but also in the later stages some from Bolivia. This intake by the U.S.S.R., combined with the demand from other socialist centrally-planned countries in Eastern Europe which were no longer drawing tin from the U.S.S.R., involved tonnages important enough to prevent in the late 1960s a serious world surplus of tin arising.

In the absence of any reliable figures of past production and consumption it is impossible to guess whether domestic mine production of tin will meet the consumption requirements of a Russian society believed by the outsider to be itching for consumer goods.[33] I assume that the U.S.S.R. will not allow her domestic reserves to be exploited in part or in whole by outside capital.

III

Surpluses and Shortages, 1901-72

MUCH very influential opinion in the tin industry was converted in the 1930s to a belief that tin held an inbuilt tendency to create conditions of production surplus. The international arrangements of that time, with their emphasis on the limitation of production as the basic cure for tin's long-term problems, were successful enough, at least in their price results, to confirm this attitude.

The viewpoint was carried forward into the International Study Group in 1948 and then into the first international agreement of producing and consuming governments in 1956. It was perhaps in some degree unfortunate for thinking in tin that this agreement secured its first and greatest success in 1958-60 through the operation of export control. Mentalities change slowly; as late as 1968-69, a body of opinion in the International Tin Council turned almost automatically to restriction as the cure for a relative short-term and not very substantial degree of surplus in production. At the United Nations and in the World Bank and the International Monetary Fund the more modern thinking on the absorption of surpluses, even fairly long-term ones, within a buffer stock is helping to change the older negative approach of destroying surpluses by denying production.

Tin production and consumption before the Second World War

World mine production of tin increased almost every year between 1901 and 1913; it then fell back a little to a fairly stable plateau for the next ten years; and it rose again to a period of stability at a higher level around 144,000 tons a year through 1924-26.

In a period before 1914 which was lavish and even extravagant in its use of all raw materials, it was to be expected that the consumption of tin would be both booming and spasmodic. Consumption in 1901-13 rose by about one-half, a rate of increase as fast as the rate of increase in production. From 1914 to 1920, in spite of the First World War, consumption was stable enough. In general, the comparison of production and consumption gave a slight shortage of production before 1913 and a small surplus of production over 1913-20; but, even in the latter period, the maximum surplus in any one year was not of an overwhelming character, only 10 to 12 per cent of annual output.

In 1921 the world shot suddenly into a consumption crisis with a total lower than for at least twenty years; production merely staggered a little and the result was a production surplus of no less than 36,000 tons or one-third of a year's supply. But 1921 was not a long-term deviation; recovery in both spheres was remarkably rapid, particularly in consumption where within one year there was a revival of over one-half in usage, a speed of recovery from a crisis not to be seen at any other time in tin in the next fifty years.

From 1922 to 1927 both consumption and production were rising in rough balance. The consequence was not, curiously enough, stability but a sharply rising tin price which almost doubled in the six years. The annual tin price averaged £242 a ton and it is not unfair to assume that it was this figure to which the International Tin Committee tried to hark back in its export control and buffer stock planning during the 1930s.

The trends in production and consumption, which so far had not been very dissimilar, began, however, to diverge more sharply after 1926. Production was rising spasmodically but very sharply in the next three years. This movement, normally but not necessarily correctly attributed to the flow of capital into the new tin dredging units in Malaya, brought output to a boom level in 1929. Consumption was moving on a similar trend upwards but more than a short head behind. The coincidence of production and consumption booms in 1929 on different levels did not seem too dangerous; the surplus of production for the year was only 12,000 tons or no more than three weeks of world supply; and there was no heavy carry forward of accumulated surplus built up over previous years.

The critical year was 1930. The world depression and U.S.A. depression in particular then hit the tin industry. The initial blow was none too severe – a drop in world consumption of only about one-ninth – but it was followed by another drop of over one-eighth in 1931 and yet again by a catastrophic drop of one-quarter in 1932. Production followed downwards but not in time harmony – a drop of one-eleventh in 1930 and of one-fifth in 1931 but approaching one-third in 1932.

The result statistically was not to produce an actual overwhelming surplus of tin. This was prevented by the export control measures of the producers' first international tin arrangements, and the total statistical surplus of production permitted over the three years 1930-32 was only 14,000 tons, or much the same as it had been in the single boom year 1929. But the chart of the market price followed in highly exaggerated form the surplus production trend. In 1930 with a surplus of 9 per cent of production the annual tin price fell by 30 per cent; in 1931 with a surplus of 2 per cent it fell by 17 per cent; and in 1932 with a production shortage of 5 per cent the price rose by 13 per cent.

The recovery, first in consumption and only later in permitted production, was faster than might have been expected. Within five years consumption had moved rather irregularly upwards – by 19 per cent in 1935, by 7 per cent in 1936 and by no less than 21 per cent in 1937. World controlled production moved even faster from 88,000 tons in 1933 to 206,000 tons by 1937, that is, by 37 per cent in 1934, by 15 per cent in 1935, by 31 per cent in 1936 and by 14 per cent in 1937. The tin price shared in the recovery and in the optimism and rose in total, if irregularly, by 78 per cent between 1932 and 1937 (all sterling prices).

The position after 1939

The outbreak of the Second World War in 1939 probably saved the tin producers from a repetition of the 1930-32 crisis. The production recovery had come fundamentally in 1936, but the consumption recovery not till 1937; there had accumulated in those two years (in spite of the rising price) a statistical surplus in production of 42,000 tons, a figure as high as the surplus over the four years 1928-31. Production and consumption fell in 1938 at a speed exceeded only in 1930-32 and in 1921; and the pattern of surplus production of around 11 per cent which had been seen for 1936-37 was continued at a similar level in 1938-39.

For the next ten years statistical shortages and surpluses in tin were almost without meaning. Through 1940-41 the production of tin was without limitation and the industry surprised the world and even its own dreams of capacity by mining on an annual basis half as much metal again as it had mined annually over the previous five years. The stimulus was not in consumption (except in the U.S.A.) and not in price (which in 1941 in the major buying market was little above the 1935-39 average), but in a guaranteed price for unlimited tonnage. The production surplus of 1940-41 (which was very substantial indeed) went in large degree into the U.S.A. strategic stockpile and the same buyer of last resort appeared, irregularly enough until 1954, when any surplus arose. Consumption was restricted on military or political grounds in almost every country through most of the 1940s and again in the U.S.A. in the early 1950s.

When the picture began to clear after the ending of U.S.A. buying around 1954 the tin industry was slow to realise the very important changes – indeed, almost fundamental changes – in consumption trends to which it had been subject. The productive capacity of the mines in south-east Asia had not been destroyed during the Japanese occupation and by 1949-51 the world mining industry had brought itself back to the relatively high levels maintained throughout the later 1930s. It was a recovery which had no relation to the demands of commercial consumers, who were now taking a fifth less than they had been taking in 1935-39. Most important for consumption was the technological change fostered by the war on the development of the thinly-coated electrolytic tinplate in the U.S.A. By the end of the war, one-third of the tinplate in the U.S.A. was electrolytic, by 1951 two-thirds and by 1961, 95 per cent. Other countries followed inevitably, if more slowly. More than half of the tinplate production of Belgium was electrolytic by 1953, of the United Kingdom by 1958, of Japan by 1959, of France by 1960 and of the Netherlands and Federal Germany by 1961. The effect on immediate consumption was partly disguised by the greater volume of tinplate being produced; but the new technology in tinplate was a cardinal factor in the consumption picture of the 1950s and 1960s. It was true that up to 1939 the rate of increased consumption of tin had been lagging behind the rate of increased consumption in most of the other major non-ferrous metals but the lag of tin behind copper and zinc and even behind steel was not insuperably wide. It was the limitation on the major use of tin in tinplate which was to stamp tin with a future expansion rate far behind that of all the other metals.

For the 1950s, the average annual rate of aggregate tin consumption was about the same as for the 1920s but somewhat below the average for the 1930s. The only consolation was a certain smoothing out of fluctuations. Consumption between 1950 and 1959 showed a swing of only 17 per cent between the highest and lowest annual figures; in the 1930s the swing had been as high as 44 per cent and in the 1920s higher still at 57 per cent.

Sales by the U.S.S.R. and the U.S.A.

The producing side of the industry was very unwillingly broken loose from the support of the U.S. strategic buying in 1954 and the mining countries found themselves naked after a period of six or seven years in which they were cushioned with almost all surpluses mopped up by the U.S.A. They could not and did not adjust production and perhaps deceived themselves because of the short-lived upward movement in consumption in the middle 1950s. But even the relatively high consumption of 1956 was still one-tenth below production and in 1957 the gap was wider still. The world market also deceived itself. The concealed surplus of 1954 (which the Americans were taking) and the actual substantial surpluses thereafter did not prevent the price rising into 1956. The coincidence of yet another production surplus in 1957 which could not be disguised and of heavy sales of tin from a new quarter, the U.S.S.R., from which the worst might be expected, brought a price collapse from which the producing industry ultimately rescued itself by savage export control. The crisis was not a crisis of consumption. This fell in 1958 by only 5 per cent; it more than recovered in 1959; and within another year was higher than it had been for twenty years.

Through the first half of the 1960s the statistical relation of production and consumption remained unreal. Consumption had a long and almost continuous rise – a run that was to go on into 1971-72 to give the longest period of increase the industry had known in the twentieth century. It was also the least spasmodic increase, if not the most striking. Between 1961 and 1965 consumption rose by 5 per cent and between 1965 and 1972 by about 10 per cent.

Production shared some common sense of growth, but was increased from a far lower basis and not geared in time with consumption. Recovery in the mines from the export control crisis of 1958-60 was slow. The reasons were not entirely economic. By 1962 the market price was moving back towards where it had been ten years earlier. If the speed of restoration of production to pre-control levels shown in Thailand and Malaysia had been followed elsewhere, it is probable that most of the growing consumption requirements of the early 1960s could have been met easily enough from normal mining channels. What did happen, namely, the shortages of 1960-65, were due primarily to the failure of the newly independent Republic of the Congo and the worse failure of Indonesia to meet obligations to produce even at prices which through 1961-3 were relatively high and steady or through 1964-5 were rising again and were exceptionally high.

There was, therefore, a shortage of total traditional supply in every one of the five years from 1960 to 1965, a shortage almost as bad towards the end of the period as at the beginning.

The absence of traditional supplies was an open invitation to the entry of abnormal supplies. The U.S.A. had overbought earlier in building up the stockpile. In doing so, it had given the producing countries long-term support to the tin price; it now decided to unfreeze a substantial portion of its holdings and, by doing so, prevented the tin price rocketing against consumers (including that one-third of world consumption represented by itself). In effect, the total world deficit in supplies during

FIGURE 9

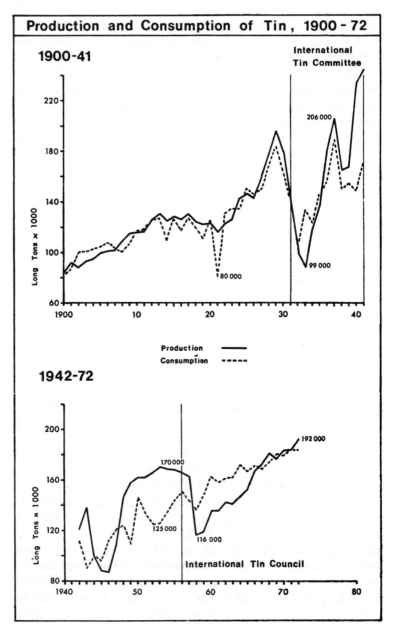

1963-67 was met by U.S. disposals. Neither timing nor tonnage of disposals could practically be related to particular shortages in particular periods but the result of American intervention was, after an initial price rise and with very wide daily fluctuations, to keep the price level on an annual basis over 1964-66 between 6 per cent below and 7 per cent above an average of £1,316 a ton.

Tin returned to a degree of normality in production and consumption only when disposals withered away in 1967. The circumstances were propitious for producers. The recovery in total production had come in 1966; an increase of 10 per cent in that year was followed by another 3 per cent in 1967 and another 5 per cent in 1968. In 1966 the shortage of production and in 1967 the shortage of consumption were negligible; and in those happy balanced circumstances the price over the two years settled round £1,260 a ton average or 159 cents a pound.

The breathing space of freedom from external factors did not last long. Consumption, already high, became slightly higher still in 1968. But the impetus of production recovery could not be stopped and the year saw a surplus which, even if of only 4 per cent, was still the first production surplus of any importance for ten years. The surplus pushed down the dollar price by over 5 cents a pound or by about 3 per cent. The medicine of the International Tin Council – a very small dose of export control which reduced production by about 2 per cent and sales from the buffer stock to meet part of the resultant shortage in supplies – resulted in a price rise of around 10 per cent. Through 1970–71 both production and consumption stood at almost record levels – production at its highest since 1941, consumption since 1937.

In the seventy years between 1901 and 1972 both consumption and production had more than doubled – the first from 87,000 tons to 184,000 tons, the second from 92,000 tons to 192,000 tons. The tin price in 1901 (completely free and without any external control or support) stood at £119 a ton in London and 27 cents a pound in New York; in 1972 (a year with no external limitation on production or consumption but with some degree of price support from buffer stock buying) the sterling price was twelve times as high at £1,506 a metric ton and the dollar price was six times as high at 177 cents a pound.

As already shown, there was an earlier belief in many sections of the tin industry in its inbuilt surplus character. This view, itself possibly no more than a rationalisation of the experiences of the 1930s, may not seem so tenable in the light of the fact that over the fifteen years between 1956 and 1971, even with the existence of permanent and approved machinery to apply export control, it has been thought desirable to exercise that power only for two periods and these for a total of less than four years.

In the seventy years 1901–72 there was surplus production in 38 years and surplus consumption in 33 years[1]. There is some difference in the pattern of the first and the second halves of that long period. In the first half (1901–35) the position, as set out in Table I, almost balances out with 17 production excesses and 18 consumption excesses. In the second half (1936–72) there were 21 years of production excess and 15 years of consumption excess. The initial impression is that production excess has appeared more commonly in the later than in the earlier period. If we take as the dividing line in the whole seventy years the year

TABLE 1

Relation of production and commercial consumption:
in number of years, 1901-72

Period (inclusive)	Production excess	Consumption excess
1901-12	3	9
1913-20	7	1
1921-31	7	4
1932-35	0	4
Total, 1901-35	17	18
1936-41	6	0
1942-47	2	4
1948-57	10	0
1958-66	0	9
1967-72	3	2
Total, 1936-72	21	15
Total, 1901-72	38	33

[1] That is, the difference between world mine production and commercial metal consumption.

1931, in which for the first time an international policy of attacking surpluses by a restriction of exports was accepted, the thirty years back to 1901 show 17 production excesses and 14 consumption excesses; the forty years after 1931 show 21 excesses in production and 19 excesses in consumption. On the face of it this shows a slight balance in favour of over-production.

It may be argued that this calculation is loaded since years included in the table as years of consumption excess may be in that category only because of the artificial reduction of production in those particular years as a result of export control. It is very probable, if not certain, that without such control the years 1932–35, 1958–60 and 1969 would have swung from the consumption excess to the production excess listing. The consequence would be a total since 1901 of 46 production excess years against 25 consumption excess years, with the ratio of 21:14 in the first half of the period and 25:11 in the second half.

This conclusion supports the thesis of inherent production surplus in tin. But, if there are excluded the periods 1940–41 and 1950–54, when completely extraneous factors of unlimited production and guaranteed buying were operative, the balance between excess production and excess consumption (in numbers of years) is almost equal.

Table 2 shows an attempt to assess, over ten yearly periods, the degree of fluctuations on the producing and consuming sides of the industry. In general, according to that table, production excesses, when they occurred, were growing larger in tonnage

TABLE 2

Production and consumption excesses, 1901-72

| Period (inclusive) | excess 000 tons | Production excesses | | Consumption excesses | | |
		Range of excess 000 tons	Average excess per cent	excess 000 tons	Range of excess 000 tons	Average excess per cent
1900-09	3 to 9	6	4 to 8	1 to 12	7	1 to 14
1910-19	2 to 15	8	2 to 12	1	1	1
1920-29	7 to 36	14	4 to 31	2 to 8	4	2 to 6
1930-39	4 to 26	16	3 to 14	3 to 44	16	3 to 50
1950-59	16 to 44	28	10 to 24	20 to 29	24	18 to 25
1960-72	2 to 8	5	2 to 4	1 to 26	15	2 to 16

Note: Years 1940-49 not relevant.

terms between 1900 and 1959; the same was true of consumption excesses. In the 1960s the average production excess was very much smaller in tonnage and percentage; the consumption excess was still large. I do not think we can draw much conclusion from the figures of the 1960s since this was largely a period of failure of production.

TABLE 3

Production and consumption cycles showing:
Number of consecutive years of excess

Period	Production excess years	Consumption excess years
1902-07	1	5
1913-21	6	—
1922-26	—	2
1927-31	4	—
1932-35	—	3
1936-41	5	—
1944-47	—	3
1948-57	9	—
1958-66	—	8
1967-71	1	—

We may, however, draw some guidance from Table 3 as to the periodicity of economic cycles (measured by production and consumption, and not necessarily by price) over the seventy years. Prior to the Second World War production excesses ranged in swings of between 4 and 6 years, consumption excesses in swings of 2 to 5 years. After that war the first swing of excess production and the succeeding swing of excess consumption were both much longer – nine years in production and eight years in consumption.

IV

The Price of Tin

THE price of tin was notoriously volatile even long ago during the times when Cornwall held a tight-fisted monopoly of supply. This volatility continued even with the widening of the supply basis in the nineteenth century to include south-east Asia and Australia and in the twentieth century to include Bolivia, Nigeria and Congo-Zaïre. To the fluctuations to be expected from the geographical distortion of the industry between producers and consumers there were added, as tin remained dependent for its consumption on industrialised countries with relative high standards of living and technology, the reactions on the tin price of slump or boom in these countries. Further, within the mining industry itself, there were contradictory movements – the steady deterioration in the tin grade of the ground being mined which was pushing up the mining costs and the spasmodic improvements in mining techniques (in gravel pumps, in dredging, in mass treatment and in the recovery processes) which were pushing them down. There was to be added, especially after the 1930s, the challenge to tin from substitutes or from economy in usage. Finally, it had to be taken into account that the political stability of many of the major sources of tin supply was or was felt to be seriously threatened after the Second World War.

Tin lived on world industrialism. In general the price was rising through the 1880s when the south-east Asian countries were appearing and the United Kingdom was continuing as major producers, and falling through the 1890s, when Malaya was blossoming fully as a producer and the massive expansion of U.S. consumption in the tinplate trade had yet to come. In those twenty years the range of annual prices was nearly 100 per cent. In the next twenty years 1900-19 the industralised demand for raw materials – the demand which the next generation was to tie so closely to its concept of imperialism – was intense in every direction. The main user of tin – the tinplate industry – doubled its world output but, most significantly, in relation to the growing world population, tin was showing a rise of one-sixth per head in consumption[1]. The buoyant tin price between 1900 and 1919 at no time fell below a daily lowest figure of £100 a ton; it was even above a daily price maximum of £200 in eight of the 20 years. The annual average price over 1900-14 was £153 a ton; in the exceptional war circumstances of 1915-19 it rose to £234. But, in general, price fluctuations between 1900 and 1914 were less sharp than in the previous thirty years[2].

Tin also established a rough pattern in its price relationship to the other major non-ferrous metals. Over 1904-13 the dollar price of tin was more than twice the price of copper, seven times the price of zinc and eight times the price of lead – always the poor relation. Twenty years later, over the ten years 1929-38, the U.S. dollar market price of tin was four times the price of copper and nine times the price of lead and zinc.

The years of the first World War and its immediate aftermath (with a minimum annual average price of £151 in 1914 and a maximum of £330 in 1918) were no more typical for tin than they were for other raw materials regarded as essential to a war-time economy or dependent on a depleted shipping; but the temporary bonanza was all the sweeter because of its almost immediate collapse.

The price position was perhaps showing a change. The rosy circumstances of 1900-14, when nine of the fifteen years had shown an upward movement in price and only six a downward movement, were not being repeated. Over the next twenty years 1920-39 the line became flatter; ten of those twenty years had upward and ten downward movements in price. In the outcome, however, the general price level over the twenty years 1920-39 proved to be, at least in sterling, substantially higher at £213 a ton than it had been before the war. Fluctuations, measured in annual prices within each decade, remained high. Fluctuations within each year, measured as the distance between the lowest and highest daily prices, which showed a curious and accidental closeness of about £60 a ton during the

---

[1] 0.130 lb per annum in 1900-04 and 0.156 lb in 1920-24.
[2] Margin between lowest and highest annual prices 77 per cent in 1900-14, 89 per cent in 1890-99, 45 per cent in 1880-89 and 133 per cent in 1870-79.

twenties, became more irregular in the thirties; the abnormally wide movements were both in years of rapidly rising prices (1924 and 1937).

The price history between 1939 and 1949 may be ignored here. Governments controlled prices. The London Metal Exchange was closed and the governmental prices in force, although not unchangeable, bore no necessary relation to factors of production, consumption or costs. Reopening of the Exchange for tin dealings in 1949 gave the tin market a complete freedom from outside control which it had not enjoyed since the establishment of the producers' international tin agreement in 1931. The market fell straight away into the price cyclone of the Korean War of 1950-51 when the price of tin rose for a moment to a figure four times as high as the highest daily price quoted at any time between 1900 and 1939 and, even when it fell, still rode at eight times the lowest figure in that period. But in fundamental trends the market showed little change as compared with the pre-1939 position. In the period 1950-55 the range between the highest and lowest annual averages (between £1,077 and £719 a ton) was very wide in absolute cash terms but in relative terms was not greatly out of line with the quinquennial periods from 1915 onwards; and similarly the range between the highest daily and the lowest daily prices in the period (£1,620 and £566) was relatively not abnormal.

From the beginning of the First International Tin Agreement in July, 1956 –or perhaps, more accurately, from the point of time in 1957 when the Council's buffer stock Manager began to buy tin in support of the price – the market was subject to a degree of influence which sometimes became control. For roughly one-half of the fifteen years between 1956 and 1971 the International Tin Council controlled the price (or, rather, the variations in the price) by limiting the tonnage of tin entering the market or by acting through the buffer stock or by both. For less than another one-third of the period the price and its variations were determined by actual or anticipated disposals from the U.S. surplus strategic stockpile, and for less than one-third of the period neither the Tin Council nor the stockpile could claim effective control over the price.

Price movements after 1956

Under the three international agreements operative in 1956 to 1972 the price of tin, measured in a sterling that was towards the end becoming unsteady, averaged £1,120 a long ton[3]. This high price, which compares with only £830 a ton over the six years 1950-55 immediately preceding the first agreement, may perhaps give a misleading impression. In fact, under the first agreement the price in annual terms was not only relatively stable in relation to the pre-agreement years but was also one-fourteenth below the level of those years. In the second agreement, which was quite ineffective in price control during almost all its life, the price explosion did not come till half way through its five years. The price jump in 1963-64 of £330 a ton (annual averages) or of 36 per cent was amongst the sharpest movements proportionately in the tin price for any one year in the twentieth century. The rise in the next year 1964-65, although substantial in cash, had proportionately been exceeded in many previous years. The significance of 1965 was rather that it represented the peak point in an uninterrupted and unprecedented price rise continuous for seven years during which the tin price had doubled. The third agreement could not maintain this record. The sterling price cracked badly in 1966-67 and was saved only by the devaluation of the pound sterling towards the end of 1967, a devaluation which made the average sterling price of £1,369 a long ton over the third agreement artificially high.

Over the seventeen years 1956-72 the annual tin price moved roughly through a range of over 100 per cent, from a low of £735 in 1958 to a high of £1,554 (long ton basis) in 1970; and the daily price through a range of nearly 200 per cent between a minimum of £640 in 1958 and a maximum of £1,715 in 1964. Relatively, those ranges were not very different from what they had been before the agreements. In 1950-55 the range in annual prices between lowest at £719 and highest at £1,077 had been around 50 per cent and in the 1930s the range from the lowest at £118 to the

---

[3] This average is for the seventeen years 1956-72 inclusive. In the following paragraphs averages for the first agreement are for the years 1956-60 inclusive, for the second agreement 1961-65 and for the third agreement 1966-70.

highest at £242 had been over over 100 per cent. The range between the lowest daily and the highest daily in 1950-55 had been between £566 and £1,620 or about 200 per cent and in the 1930s between £100 and £311 or over 200 per cent.

The sterling devaluation of 1967 emphasised the importance of the U.S. dollar price for tin. Nigeria and Malaysia usually remitted substantial profits in sterling and bought some mining supplies in sterling. Nigeria sold much and Malaysia relatively little metal to sterling countries; Malaysia and Thailand sold solid tonnages of metal in the U.S.A.; Indonesia sold outside the sterling area; and Bolivia lived in terms of dollars. The dollar price had for many years followed the general trend of the sterling price and it moved down in 1966-67 with the sterling price. But after devaluation of the pound in November, 1967 the dollar price rose by only 16 per cent during 1967-72, the sterling price by 24 per cent.

This divergence in the trend of sterling and dollar prices in recent years merely emphasised what had been happening over a much longer period. Over the seventy years after 1901 the London price, expressed in sterling, had risen thirteenfold, the Malaysian price, expressed in Malaysian dollars, had risen nearly ninefold and the New York price, expressed in dollars, had risen only between fivefold and sixfold.

This examination of the tin price over seventy years indicates the degree to which that price has been subject to fluctuations (both inside and outside periods of international control) and the extent of those fluctuations. It shows that the price trend has been upwards, but in two distorted movements – a doubling of the price (in sterling and in dollars) in the first and longer section of the period up to 1939 but no less than a sixfold rise in sterling and over a threefold rise in the dollar price in the second and shorter period. But the basic reasons for the long-term price movement still remain to be answered.

Tin is regarded by economists as a commodity inelastic to price. Users, in general, are not persuaded by substantial price reductions to use more and are not dissuaded by substantial price increases to use less. It is not for a non-econonist to challenge this truth. But he may sometimes find it difficult to swallow its

FIGURE 10

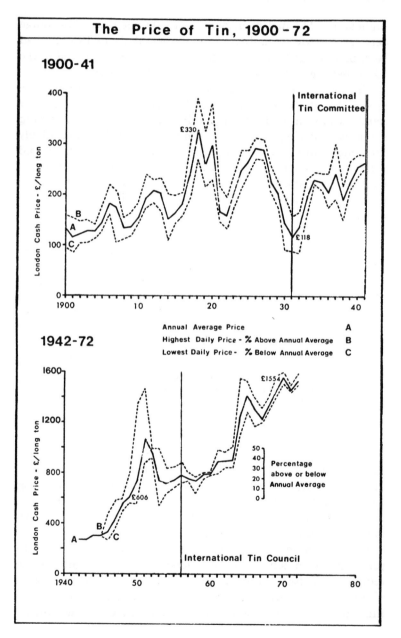

The Price of Tin, 1900-72

application in practice. Over seventy years the sixfold rise in the U.S. dollar price of the metal has seen a rough doubling of the consumption in tonnage terms; but in the first half of those seventy years a doubling in the dollar price went hand in hand with an increase in the tonnage consumed of about 80 to 90 per cent; in the second half the trebling in the dollar price saw the tonnage consumed rise by only about one-quarter. In general, over seventy years the price of tin has risen more than the price of other non-ferrous metals and the consumption of tin has not risen in anything like the same degree. The growth of tin consumption in the last thirty years may have been handicapped by the relatively sharp rise in the tin price.

Price changes and the levels of consumption

Is it possible to support any generalisation on the price of tin by looking at particular cases over particular shorter-term periods? The longest run of rising prices in the present century was in 1959-65 for sterling and in 1960-65 for dollar quotations. The rise was very sharp – a rise of three-quarters or more. The run was similar in movement for a shorter period in 1931-34 when the sterling price almost doubled and the dollar price more than doubled. During the 1960-66 price run consumption of tin in the U.S.A. rose by 17 per cent but consumption in other countries remained almost unchanged. If we adopt a time-lag theory – that high prices would affect actual usage of tin about three years after these prices had been in effect – and look at consumption in 1968, we find that U.S.A. consumption was the same as it has been in 1965 but consumption in the rest of the world was up by about 5 per cent: if we extend the time-lag to five years the result is a drop in 1970 as compared with 1966 of one-ninth in U.S. consumption, but a rise of one-sixth in the rest of the world. In the earlier run of general price rises over 1931-34 a doubling of the price saw consumption go down although within three years of the prolonged price rise world consumption was up by half.

On the other side of the picture the longest run of falling prices over 1926-31, when prices were much more than halved, saw world consumption start at 146,000 tons, then shoot up to 184,000 tons and fall back to 141,000 tons; but three years later the price had doubled and consumption had fallen again.

It may be that, in some particular cases, price has determined adversely the consumption of tin, particularly in fields where tin at that time was open to substitution for technical reasons. One example is in the use of tin foil (tin metal in very thin sheets) as a wrapping material for packaging. In 1928 the U.S.A. was using over 5,000 tons of tin a year in tinfoil; this was cut to 1,600 tons by 1935. In those years the price of tin showed almost no change but tinfoil was killed by a cheaper substitute in aluminium. In the same period the use of tin in bronze in the U.S.A. was almost halved but the use of tin in collapsible tubes (not yet threatened by lead or aluminium) went up by nearly one-third. The price of tin more than doubled between 1927-36 and 1948-50; in that period (during which there had been strong governmental pressure to economise in tin) the use of tin in collapsible tubes in the world was reduced to almost one-third.

In the United Kingdom during 1946 to 1952 when the price of tin trebled, the use of tin in solder, bronze and white metal rose but in no particular relation to the rise in the tin price; the use of tin in cable sheathing, in collapsible tubes and in type metal fell sharply. In Canada between 1954 and 1965 the dollar tin price almost doubled; the use of tin in the manufacture of solder showed no change, but the tonnage of solder in a similar period 1954-69 doubled on an almost unchanged use of tin (in other words, the amount of tin in each ton of tin-lead solder was roughly halved). In the United Kingdom between 1960 and 1971, with a two-thirds rise in the tin price, the total use of tin fell by one-fifth but more in foil, sheets and collapsible tubes and less in white metal and bronze. In the U.S.A. over the same period total consumption showed little change but within that stability the use of tin in solder, in chemicals and in tin oxide doubled or more than doubled, the use of tin in collapsible tubes, foil and babbitt metal remained unchanged and the use of tin in bronze fell slightly.

How far has the tin price been a factor in the tinplate industry? It is difficult not to believe that the price of tin had a substantial long-term effect on the total consumption of tin in that industry. It is of course well known, especially in the tin producing countries, that the cost element represented by tin in the ultimate item manufactured from tinplate, that is, the packaging can, is a small, and perhaps very small, proportion of what the ultimate consumer pays. But this thinking has persuaded many producers into believing that an increase in the tin price does not squeeze the ultimate consumer and therefore does not affect consumption. This belief is not relevant to the subject. The ultimate consumer has no control over and very little interest in the price of the tin on his canned goods. But the consumer of the tin producer is not the ultimate consumer but the fabricator and the tinplate maker. In tinplate the current cost of the tin in a fairly thinly-coated sheet of steel is about $U.S.21 a ton; in 1937 it was on a very thickly-coated sheet about $18 a ton. For the world tinplating industry as a whole the bill for the tin used in 1937 was about $U.S.80 millions and in 1971 about $300 millions, although the thickness of the tin coating in 1971 was only one-third of what it had been in 1937. Price therefore is important to the tinplaters' purchasing agencies, even though normally price fluctuations may be passed on to the tinplate customers. Price will be of no importance to the tinplate trade only in the event of there being no substitute available for a tin coating on steel in terms of health and technical application. If, however, technicians devise a new and more economical process for thinning the coating on steel, this technical development need not coincide with a period of high prices or unstable prices (the development of electrolytic tinplate was initially in the first half of the 1940s when the tin price was fixed by the U.S. government at a relatively low level); it may coincide (as it did in the U.S.A.) with economy measures enforced by a government. Once the technical process has been approved, all tinplate producers with varying time lags will apply the electrolytic process in practice. In 1971, if there had been no shift over to the electrolytic process, the cost of tin would have been at prevailing tin prices no less than $U.S.60 a ton of tinplate.

How far has the consumption of tin been affected by the element of fluctuation in supplies and prices? Over the last fifty years (excluding the unrepresentative war years) tin has suffered almost as much from statistical shortages as from statistical surpluses[4].

In the period 1920-39 (see Table 4 below) years of excess production were years in which normally but not invariably the price went down and years of excess consumption were years in which the price normally but not invariably went up. The years 1950-71 showed a slightly different pattern and the years of surplus production statistically were divided roughly half with the price going down and half with the price going up.

TABLE 4

Excess production, excess consumption and price

Surplus of production:	1928-39	1950-71
no. of years:	11	11
in which, price down	8	5
in which, price up	3	6
Excess of consumption:		
no. of years:	9	11
in which, price down	2	3
in which, price up	7	8

There is very little pattern even if we look at the upward or downward movement (that is, movement in relation to the immediately preceding year). During 1921-39 consumption was rising in 12 years, but the price went up in only six years and down in six years; consumption was falling in seven years but the price was down only in four years and actually up in three years.

Over the period 1950-71, when mere statistics of production and consumption were by no means the only factors to be taken into account, consumption was rising in 14 years but the price was up in only nine of those years; consumption was falling in 8 years, half of them with a falling price and half with a rising price.

The reference is to shortages and surpluses, but it is to be noted that the statistical surpluses of production in the first half of the 1950s did not in general come on the market and that the statistical shortages of the 1960s were met largely by releases from the U.S. stockpile which did not come into the statistics of production. To compare merely arithmetic shortages and surpluses, without reference to other factors (stock changes, abnormal supplies, etc.), may not mean very much.

Stocks and the tin price

Can we find a determining influence on the tin price from the accumulation of stocks built up over a period of depressed years or released over a period of prosperous years? During the 1930s, a very considerable impression was made on tin thinking by the direct inverse relationship between stocks and the price, an impression which lead ultimately to the fairly general acceptance of the conception of the buffer stock to influence the price through stock withholding or stock selling. But in the last twenty years it has been far more difficult to see the position and therefore the influence of normal commercial stocks as distinct from other associated factors. These associated factors have included, first, the release of hitherto unknown stocks on the market by the Russians in the late fifties, secondly, the release over five years from the U.S. strategic stockpile of tonnages negotiated from time to time and in any case all outside normally declared commercial stock and thirdly, in 1957-60, in 1968 and in 1971-72, the acquisition and holding in the buffer stock of a tonnage of tin representing a dominating or influential share of the available world stock. These factors have to be taken into account in a comparison of world stocks and world prices during 1950-71.

In that period the price had first an immediate peak during the Korean crisis but had dropped by 1953; it was then for seven years on a fairly straight line between £700 and £800 a ton; it then rose, at first slowly but after 1963 very sharply, to over £1,450 a ton for 1971. Against this price movement I have compared (see Figure 11) four lines of metal stocks: (a) all total declared stocks of tin metal (including the major part, but not necessarily the whole, of the Tin Council's buffer stock), (b) the U.S. consumers' stock alone, (c) the world metal stock *less* the consumers' commercial stocks in the U.S.A., (d) the end-year buffer stock and (e) the nominal free stock outside the U.S. consumers and the Tin Council's buffer stock. These figures do not include the U.S. surplus strategic stockpile. The first category of total world metal stocks swung down and then up during 1950-60 while the price was fairly quiescent; it swung down and up again in 1960-68 while the price was generally rising. The second

FIGURE 11

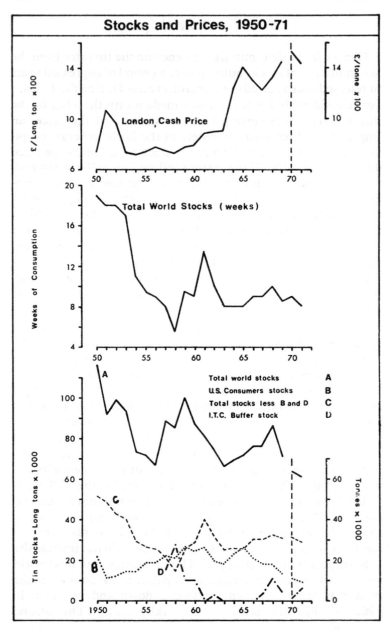

Stocks and Prices, 1950-71

category of U.S. commercial stocks followed a sharply different road. The third category of stocks – the world without the U.S.A. – followed a roughly similar pattern. In general, these U.S. stocks formed between a minimum of one-fifth and a maximum of one-half of total world stocks. They were rising steadily through the 1950s, reaching 26,000 tons or almost 40 per cent of the world total in 1959-61; they were rising in 1963-65 (as prices were shooting upwards); but from 1965 to 1972, when prices were fluctuating but fairly high in dollar terms, the stocks were in a continuous slide until they stood by 1972 at well under three months of U.S. consumption instead of nearly six as in 1960.

TABLE 5

Tin stocks, 1950-71

(In thousand long tons[1])

End year	All metal stocks, including afloats (a)	U.S. consumers (b)	Others, excluding (b) (c)	I.T.C. buffer stocks (d)	Net other (e)	Year price £
1950	73·7	21·9	51·8	—	51·8	744·6
1951	59·4	11·0	48·4	—	48·4	1077·3
1952	54·7	12·2	42·5	—	42·5	964·4
1953	54·7	14·5	40·2	—	40·2	731·7
1954	43·0	14·6	28·4	—	28·4	719·4
1955	45·6	18·8	26·8	—	26·8	740·1
1956	44·6	19·0	25·6	—	25·6	787·7
1957	59·4	22·4	37·0	15·3	21·7	754·8
1958	59·0	21·4	37·6	23·3	14·3	734·9
1959	63·5	27·0	36·5	10·0	26·5	785·4
1960	63·5	24·8	38·7	10·0	28·7	796·6
1961	56·5	26·3	40·2	—	40·2	888·6
1962	51·3	19·9	31·4	—	31·4	896·5
1963	46·8	18·5	28·3	3·3	25·0	909·7
1964	48·8	22·8	26·0	—	26·0	1239·4
1965	52·6	26·9	25·7	—	25·7	1412·7
1966	51·4	20·9	30·5	—	30·5	1295·8
1967	53·1	18·2	34·9	4·8	30·1	1228·8
1968	62·1	18·2	43·9	11·3	32·6	1323·3
1969	49·1	14·0	35·1	4·6	30·5	1451·3
1970	42·7	10·1	32·6	1·2	31·4	1529·5
1971	45·6	9·8	35·8	6·6	29·2	1437·4
1972	48·9	11·2	37·6	12·5	24·7	1505·9

Note: Some, but not necessarily all, of the I.T.C. buffer stock is included in reported metal stocks.

[1] Metric tons 1970 and after.

The buffer stock of the International Tin Council, at its most active, left only about 14,000 tons of tin metal in free world stocks outside the U.S.A. (that is, after being deducted from the net other stocks)[5] in 1958, but even on this scarcity the buffer stock could only hold the price at almost the lowest level for seven years. In 1968 and again in 1971-72 the buffer stock left between 25,000 and 31,000 tons available (outside the U.S.A.), but in the first period its activity was associated with a price that had been falling for three years and in the second period with a price that (in dollar terms) had been rising for three years.

It does not seem possible to draw from the chart in Figure 11 or the tonnages in Table 5 a clear direct statistical relationship between stocks and prices. It does seem that the U.S.A. consumer over the last twenty years has been prepared to see his metal stocks vary between 10,000 tons and 27,000 tons – irrespective of price – and that the consumer outside the U.S.A. has been prepared to see his metal stocks move more narrowly, especially in the last ten years, again irrespective of price.

How far has tin continued to move in the price relationship with other non-ferrous metals which it had established in the earlier part of the century? Prices of non-ferrous metals generally rose rapidly after the Second World War, but the tin price began to rise earlier and rose more. For the decade 1961-71 tin stood at four times the copper price and eleven times the lead and zinc price (see Table 6). It is possible, but unproven, that this slower rise in the copper, lead and zinc prices may have been reflected in the much more rapid rate of increase in the consumption of those metals[6] than in tin. But, of these metals, the most important – that is, potentially the sharpest competitor against tin – was aluminium. Aluminium has remained the cheapest of the non-ferrous metals at a price which through the 1960s was still only

---

[5] At the lowest point in 1959 the amount of free stock in the world (that is, world stocks, less U.S.A. consumers and less I.T.C. buffer and special fund) seems to have been only about 6,000 tons.

[6] Over sevenfold in copper, over sixfold in zinc and over threefold in zinc between 1901-70, according to Metallgesellschaft *Metal Statistics*, while in the same period tin slightly more than doubled.

[7] But 7 per cent of the U.S. can market means the equivalent of about 2,000 tons of tin a year.

TABLE 6

Non-ferrous metal prices in the U.S.A.
(Cents per pound)

Period	Tin	Copper	Lead	Zinc	Aluminium
1904-13	36·48	15·36	4·55	5·64	—
1929-38	40·79	10·26	4·55	4·61	22·0
1961-71	150·2	39·2	13·4	13·7	25·5

Source: *Metal Statistics* (The American Metal Market) and American Bureau of Metal Statistics: *Year Book*.

one-sixth of the price of tin. In the last thirty years, world consumption of aluminium metal has moved up sixfold and of tin by one-third. The loss of individual markets from tin to aluminium has already been referred to, although those losses were not always due to price factors. As a serious competitor aluminium has not yet eaten substantially into the tinplate market (in the U.S.A. aluminium cans held only 7 per cent of the combined aluminium and tinplate packaging market in 1971)[7].

Tin has, therefore, become dearer in relation to the other non-ferrous metals. This may merely mean that advanced industrialised societies wanting tin for advanced industrial products (including packaging) are accepting the fact that a relatively higher price for tin is necessary to maintain the quality of their industrialism. It may merely mean that more competitive industrialism, when it can choose its raw materials, looks first to aluminium and then to other non-ferrous metals before it turns to tin.

V

International Co-operation, 1921-31, The First Steps

THE price bonanza of the First World War for tin producers was late, artificial and temporary. It did not come from any substantial increase in consumption. Tin supplies to Germany were largely cut off by the British blockade; Russia disappeared as a consumer and was not to re-emerge for almost twenty years; the world tinplate industry shifted the bulk of production to the United States but used little more tin. The bonanza did not come from a shortage of mine production since during the war output, although very variable in different fields, had been running at an aggregate rate higher than consumption. But the shipping shortage which piled up stocks in the main producing areas also reduced the tonnages afloat and thinned the stocks available in the main consuming areas. At the end of 1918 stocks in the United Kingdom were less than half what they had been in 1914 and in the U.S.A. only one-sixth. The combination of this stock shortage with a fairly small fall in the mine production of Malaya and exports of China (although total world production still remained in excess) produced in 1918 a maximum daily price of no less than £399 a ton, the highest yet reported on the London Metal Exchange.

The post-war boom anticipated in tin suffered the same fate as in other world commodities and in manufacturing. The high hopes and high prices lasted little over a year. The London capital market re-opened its activities for Malayan tin development and the amount of capital raised there, for that purpose in the two years 1919-20, was higher than even in the three golden years of 1911-13. The immediate revival of world consumption

[1] K. E. Knorr: *Tin under control*, p. 73, quoting E. Baliol Scott, section on tin in *The Mineral Industry during 1920* (ed. G. A. Roush).

[2] Yip Yat Hoong: *The development of the tin mining industry of Malaya*, p. 154, says that representatives of the "foreign" (that is, British) mining firms approached the government.

produced another and higher peak daily price in 1920 (this time, £419 a ton). But tin could not itself hold out against the general and unexpected depression into which the world was plunging. The tinplate industry halved its activity in 1921 and the total world consumption of tin for all purposes fell by over one-third. The world of tin producers could not adjust itself downwards to the speed of this calamitous movement and, in fact, one of the most important fields – the Netherlands East Indies (later Indonesia) – actually increased its output substantially. World stocks of tin metal doubled in the three years to 1921. Equally depressing on the price was the accumulation of stocks of tin concentrates in Malaya, the Netherlands Indies and Bolivia. The three tin markets – Singapore, London and New York – competed from the middle of 1920 in pushing the tin price ever downwards. The New York price was halved to only 34 cents a pound at the end of the year, it fell again in the summer and autumn of 1921 to 26 cents, and the annual average for 1921 was the lowest for a dozen years. In London and Singapore the collapse was as dramatic.

Malaya and Thailand, unlike Bolivia or the Netherlands East Indies (N.E.I.), had been slackening in production for a number of years and therefore could not immediately modify the effects of a falling unit price by a greater tonnage. The total value of tin exports from Malaya in 1921 was almost one-half of what it had been in 1920 ($M 49 millions against $M 88 millions) and labour, already almost halved from 172,000 in 1914 to 90,000 in 1920, fell yet again in 1921.

Under the emergency conditions of the war in 1914-15 the Federated Malay States government had been pushed into governmental buying of tin to save the market. It now found itself faced with the halving of the income which it drew from its royalty on the tin industry and, perhaps under pressure from the European interests there, it proceeded towards a revolutionary peace-time intervention. In the spring of 1919 it bought tin concentrates to support the Malayan market price and sold them at a slightly higher price in the autumn.[1]

In 1920 some members of the Malayan tin industry[2] proposed the governmental buying on a minimum price of all the tin produced in Malaya. This stock should be sold when the market recovered. Such stock buying would support not only the Singapore price but, since Malaya represented about one-third of world production, the world price. The proposal was strengthened as the price continued to fall. By the end of November, 1920 the Singapore tin price stood at only $M 90 a picul (£176 a ton). This was not a disastrous price (it was, in fact, above the annual averages for the three war years 1914-16) but it was far below the expectations and actual levels of the immediate post-war boom.

In December, 1920 the F.M.S. government fixed a minimum price at which it would buy and stock all the tin produced in Malaya, and between that date and February, 1921 there was bought and stocked a substantial tonnage (but less than 10,000 tons) at an average price of $M 114 a picul (£223 a ton)[3].

The effect of this buying on the tin price in Malaya was clear; the effect on tin prices outside Malaya was not encouraging. Neither the London nor the New York price responded. The F.M.S. government must have been made painfully aware of its minority position in world production.

The Netherlands East Indies had similar price problems in tin but had a simpler internal organisation in production. Although the recent production history in the N.E.I. had been happier than in Malaya, the N.E.I. had been accumulating stocks as fast as had Malaya,[4] and its centralised management was perhaps more open to the concepts of price control through stock holdings. In

[3] The government was not the only interested party in Malaya. W. F. Nutt, the managing director of the Straits Trading Co. (smelters), had withheld tin metal from the market, profitably. By September, 1921, the company had accumulated a quantity of tin worth some $M 7·5 millions. The result was that for the first time in its history the company failed to pay a half-yearly dividend and Nutt resigned. See K. G. Tregonning: *Straits Tin: A brief account of the first seventy five years of the Straits Trading Company Limited*, pp.40-1.

[4] And the prospect of more since its production rose by one-sixth in 1920-21.

[5] The initial total of 19,138 tons was made up as to 9,508 tons from the F.M.S. and Johore governments, 4,000 tons from the Netherlands East Indies government, 2,800 tons from Billiton, 150 tons from the Singkep Company and 2,680 tons from the Straits Trading Co. Knorr, *op. cit.* p. 74.

[6] The initial tonnage was reduced to 17,188 tons in April, 1921 when the Billiton and Singkep contributions were released from the terms of the arrangement.

combination, the two countries held in 1920 nearly half the world's production of tin, a proportion which would give them a more reasonable prospect of success in pushing up the price. In February, 1921 the two governments, with other local interests, agreed to establish the Bandoeng pool[5].

The Bandoeng Pool

This pool was the first intergovernmental arrangement in the tin industry. It contained part of the essential thinking of a buffer stock. It was careful to limit its objectives. The initial pool was, in effect, the accumulated stocks held by the two governments and the smelter at Singapore. The pool was not concerned with taking new supplies off the market and thus adding to its opening stock. It was not an agreement to control production; on the contrary, the Netherlands Indies output was to be sharply higher in 1921 and again in 1922. The object of the pool was simply to hold the accumulated stocks off the market until the price recovered to an acceptable level. That acceptable level recovery price was envisaged at £240 a ton; the average market price in London for 1920 had been £296 and for 1921 was to be £165.

The tonnage involved in the pool for about two-thirds of its life was 17,188 tons[6]. This was about 28 per cent of the combined N.E.I.—Malaya production and about 15 per cent of total world production for 1921. Although the pool had only part of the mechanism of later tin buffer stocks (it acted in a one-way selling direction only) it was comparable at least in tonnage and in effects on the price. The first international tin pool of 1931-34 accumulated a total of 21,000 tons of tin, or 19 per cent of the average annual world output. The buffer stock of the International Tin Committee in 1935 raised only 8,300 tons of tin, or 6 per cent of world production. The contribution to the buffer stock of 1938-39 was 15,512 tons, or 9 per cent of world production. The buffer stock of the first post-war international tin agreement reached in 1958 a maximum of 23,000 tons (excluding the special fund), or 20 per cent of production; the buffer stock of the second agreement in 1962 a maximum of 3,270 tons, or only 2 per cent of world production; and the buffer stock of the third agreement a maximum of 11,290 tons in 1968, or only 6 per cent of world production.

The freezing of the Bandoeng stocks did not affect the flow of new tin production on to the market or the heavy visible metal stocks outside the pool, but it certainly affected prices. The London price began to show some steadiness from the pool's inception. From the end of the first quarter of 1922 (when the price was at its lowest) until the beginning of releases from the pool in April, 1923 the price was rising almost continuously. The rise, checked then by the pool releases, was resumed again, and by the spring of 1924 the price reached £280 a ton. For 1923 as a whole the tin price was a quarter higher than it had been for 1922 or 1921.

Releases from the pool began in April, 1923 at 5 per cent of the total per month (that is, 880 tons monthly) in daily dealings. By December, 1923 the pool was down to 11,500 tons and by December, 1924 to 3,000 tons. This remnant was disposed of in 1925 and the pool then ceased to influence the market.

The degree of fluctuation within the price range was also altered. The range between the lowest and highest monthly averages over 1918-20 had been between 70 and 115 per cent of the lowest. For 1921 the range was down to 42 per cent, in 1922 it was 35 per cent, 1923 36 per cent and 1924 49 per cent.

The Bandoeng pool was not, of course, the only factor in the price rise. The overall tin supply and consumption position had changed. The price collapse of 1921-22 had followed from the consumption collapse of 1921, but the very sharp consumption recovery in 1922 (which created an apparent statistical shortage) did not reflect itself at all in the market price of that year and it was not until 1923 (also a year with a statistical shortage) that the price rose by almost one-third from year-opening to year-end.

The price recovery which resulted from or coincided with the later months of the Bandoeng pool had very beneficial effects on the export earnings of all producing countries. The value of tin exports from Malaya, for example, had fallen in 1922 to $M 48 millions, the lowest level for twenty years. It was brought back to $M 64 millions for 1923 and $M 92 millions for 1924.

[7] Yip Yat Hoong: *Development*, p. 156.

In the eyes of its proponents, the pool had proved conclusively that a degree of control over stocks meant a degree of control over price. That control was proved profitable. The F.M.S. government alone had made a profit on its holdings in the pool of at least $M 500,000[7].

The boom years, 1924-28

The five years 1924-28 were boom years for tin with, in general, rising production and rising consumption (both accelerating towards the end) and with a high, but not always rising, price (breaking towards the end). The market was free from interferences by any stock pool, and commercial stocks were relatively low.

The production boom came in two steps – one at the beginning in 1923-24, the other and more important at the end in 1926-28 – with a rise in world tin production of almost one-quarter.

World consumption of tin also rose in two uneven steps – the first in 1924-25, the second also in 1927-28 – but at a slightly higher rate than did production. This rising consumption was based in part on the tinplate trade (which added half a million tons to its annual output) and in part on the automobile industry, which was moving towards a peak production of 1929-30 and which was extravagent in its use of tin in solder. Behind everything was the optimistic business outlook that was to be shattered on Wall Street in 1929.

The price rise, which had started as early as the end of 1922, continued for four years until 1927. Thereafter, even though the excess of production was slight, the price was sliding down through 1928 and 1929. Over the five years 1924-28 the London price averaged £264 a ton. The highest London price in the period was £321 a ton, the lowest £201; the range between the highest and lowest daily prices was around £60 a ton in four out of the five years.

Over the five years aggregate world production of tin was only about 3 per cent above aggregate world consumption. These were the good days!

The price collapse, 1929-31

The tin price broke, first but not disastrously, in October-November, 1929 when the vast share collapse in Wall Street so dramatically marked the beginning of the great world depression. It broke again in a series of steps in May, 1930, in October, 1930 and in May, 1931. At that last point the lowest daily price of £100 a ton was not only less than half the lowest of 1928 but was also the lowest price reported for 30 years.

The fundamental reason for the tin crisis was the collapse of world tin consumption and, particularly, of consumption in the U.S.A. Total world consumption had climbed for 1929 to 184,000 tons. This boom level was not to be exceeded for the next forty years. Consumption had been stimulated in the tin-plate industry, which had been growing almost without interruption since 1922, which by 1929 was nearly three times as large as it had been in the depression of 1921 and which, as yet, has made no very major technical changes in its usage of tin metal.

Tinplate flagged in 1929, creaked ominously in 1930 and again in 1931, and collapsed in 1932. But this was within the United States alone. The world outside the U.S.A., which had produced 1·1 million tons of tinplate in 1924, still produced 1·3 million tons in 1929 and 1·25 million tons in the worst depression year 1932. The U.S. production had risen from 1·3 million tons in 1924 to 1·8 million tons in 1929, but dropped to under 1·0 million tons in 1932.

The total U.S. usage of tin (including tin in tinplate) at 85,000 tons was still, in 1929, 44 per cent of the world total. The slump of 1932 cut U.S. tin consumption by well over half to 35,500 tons, or only 34 per cent of the world whole. The rest of the world took 82,000 tons of tin in 1925, 99,000 tons in the peak year 1929 and still 70,000 tons in 1932. In short, over the eight years 1925-32, U.S. consumption fell by 41,000 tons per annum but the consumption of the rest of the world fell by only 12,000 tons.

---

[8] In 1921-24 on a tin price of £194 capital flotations averaged £0·2 millions; the average tin price for 1925-28 was £267. Knorr, *op. cit.*, p. 83. See also Yip Yat Hoong, *Developments*, pp. 158-160 on Malaysian capital.

The peak year for consumption—1929—was also the peak year for mine production (196,000 tons). This was a rise in production in five years of 54,000 tons; the rise in consumption in the same period was 50,000 tons. If the bottoms of 1921 and the peaks of 1929 are taken for comparison, the rise in production was 80,000 tons and in consumption 104,000 tons. The rush of production came in 1927-29, when 53,000 tons were added to production and only 38,000 tons to consumption.

The production explosion came mainly from Malaya and Bolivia, where there had been a total increase of 35,000 tons in two years. The long run of higher prices in the 1920s had naturally pushed investment activity in tin (as in other mineral and manufacturing industries) into boom levels. New flotations for capital for the Malayan tin industry averaged £1·4 million a year over 1925-28[8]. Much, probably most, of this money went into dredging units with bigger capacity. The 41 dredges operating in Malaya in 1926 became 105 by 1929 and their output of tin nearly trebled. The blame for the uncontrolled expansion of production in tin, and therefore for the subsequent price disaster of 1929-32, has often been placed, very fully and very conveniently, on the shoulders of the big financial groups which were believed to dominate the Malayan and the world tin industry at that time. Admittedly, the dredges, as the most efficient and economical means of mining, gave a very sharp boost to Malayan production. Between 1926 and 1929 they added about 18,000 tons of tin a year. This was especially true for 1929 alone when dredge production rose from an annual rate of 23,000 tons in the first quarter to 30,000 tons in the last quarter. But even that last figure was no more than one-third of what the Malayan Mines Department thought to be the capacity of the industry. Some of the dredge capital may have been diverted from the development of the other European-owned mines. This may help to explain the relatively low rate of expansion (6,000 tons of tin in three years) outside dredging. In any case, the whole dredge output of Malaya in 1929 was below 40 per cent of the Malayan total and was only one-seventh of world total.

Outside Malaya the sharpest increase in production had come from Bolivia. That country reached a peak of 46,000 tons (the highest in its history) in 1929. This was 16,000 tons more than in 1926. The dredge (always negligible in Bolivia) was not responsible; the drive came almost entirely from Patiño's re-organised mining group (Patiño Mines and Enterprises Consolidated).

It is possible that the assessment of the boom of 1929, and therefore of the consequent slump, may be made more cold-bloodedly now. Earlier views may have been doubly slanted. The advocates of a cure through a system of international limitation of production had, both to confirm their own opinions and to convert others, a need to emphasise the peculiar tendency of the industry to a surplus (a tendency which might shelve awkward questions as to whether mechanisation, rationalisation of ownership or an over-supply of capital had any determining hand in the process). So far as the examination of the tin problem was concerned, economic opinion, then created very largely in the atmosphere of the two major consuming countries (the U.S.A. and the United Kingdom), usually disliked strongly the medicine of export limitation and very strongly the existence of international combines. The economists were perhaps willing to find in the tin boom of 1929 a consequence not of the use of more efficient means of production but of the short-sighted policy of over-capitalisation by the tin companies, who were now trying to control a market to save themselves from the consequence of a boom of their own provoking.

There was no abnormal upward swing in tin production (as opposed to other non-ferrous metals) prior to 1929. On the contrary a comparison with steel and the other non-ferrous metals (see Table 7) seems to indicate otherwise. Over the seven years 1923-29 in which the boom was made, the upward surge in tin was less than elsewhere. In those years production in steel rose by 57 per cent, in copper by 59 per cent, in zinc by 54 per cent, in lead by 46 per cent, and in nickel by 80 per cent. The increase in tin metal production was no more than 34 per cent. The League of Nations general index of world metals production rose by 30 per cent during 1925-29, the index of tin metal production by only 27 per cent. Over longer comparable periods – the ten years 1911-20 and the ten years 1921-30 running up to the boom – the production of the other major non-ferrous metals rose in much

TABLE 7

World production in steel and non-ferrous metals, 1923-32
(In 000 metric tons)

	Steel (a)	Copper	Lead	Zinc	Tin	Nickel
1923	79	1,249	1,253	946	142	31
1924	79	1,343	1,382	1,004	149	35
1925	91	1,425	1,568	1,133	150	37
1926	93	1,494	1,660	1,219	155	34
1927	102	1,554	1,771	1,307	164	35
1928	110	1,758	1,762	1,401	185	50
1929	121	1,981	1,823	1,452	190	56
23-29*	57	59	46	54	34	80
1930	95	1,647	1,669	1,394	179	54
1931	69	1,416	1,405	1,000	151	36
1932	50	920	1,150	780	104	20
29-32†	59	54	37	46	45	64

* % Increase † % Decrease smelter production. (a) In million tons.
Source: *League of Nations Statistical Summary*

closer harmony (copper by 20 per cent, lead by 27 per cent and zinc by 26 per cent) but tin was lower with only 17 per cent.

Even in the years 1927-29 immediately prior to the peak of the boom, when the new dredges were most influential in stimulating tin output, the rate of increase in tin metal production, although more than in lead and a little more than in zinc, was still less than in steel and very much less than in copper or nickel.

It is difficult to accept that the break in the boom was a consequence of the reckless over-expansion of the tin industry, unless we also apply the same charge (in even stronger degree) to other non-ferrous metals and to steel.

TABLE 8

World production of metals
in 000 metric tons

	1911-20	1921-30	% increase
tin	128	150	17
copper	1,127	1,351	20
lead	1,118	1,425	27
zinc	876	1,105	26
nickel	36	37	1

All smelter output except nickel (mine) *Metallgesellschaft Metal Statistics.*

The Tin Producers' Association

Opinion in the industry during the boom did not concern itself much with the consequences of a slump in consumption that was not yet obvious. It was shifting to accept that the downward drift in the price from the spring of 1927 was the result of the weight of growing stocks, and it remembered the quite recent success of the Bandoeng pool in handling the stock problem.

The industry could look as a rule to a stock of visible metal which would run between a maximum of 22,000 tons and a minimum of 11,000 tons[9]. In 1926-27 visible stocks, at only five weeks of consumption, were below average, but they began to rise ominously after the middle of 1928. They reached a first record high level by the end of that year; they rose again through 1929; by the end of 1930 they were to be 42,000 tons or three times the 1927 figure.

A reference to Fig. 12 shows very clearly how stocks and prices moved inversely after 1927. The lesson drawn, at least by the groups that had cohered behind the Anglo-Oriental and the Patiño money, was simple and clear. An essential step to the stability and then to the recovery in the price was a degree of control over the level of stocks. The bigger units in the industry were willing to apply the remedy and now had the financial resources to do so. However, they still had to convert governments.

In 1928 the two largest groups (Patiño and the Anglo-Oriental) were reported[10] to have held between 4,000 and 7,000 tons of tin off the market in order to prevent the price continuing to fall. It was true that the drift was stopped in the second half of 1928 and that the price was held fairly stable until May, 1929. But visible stocks were also rising, and in that month the price fell below £200 a ton – a mystical figure to which the tin world, for no logical reason, was beginning to attach sanctity.

[9] The average over the twenty-two years 1907-28 was 16,000 tons.
[10] E. S. May: *The international tin cartel* in W. Y. Elliott: *International control in the non-ferrous metals.*
[11] In February, 1930 a cut to 80 per cent of the 1929 production; later a complete mining stoppage for two months. The N.E.I. and Bolivian producers agreed on a 20 per cent reduction.

In June, 1929 there was formed in London the Tin Producers' Association. The initial membership of this voluntary body was mainly of British capital interests in Malàya, Nigeria and Burma; the Association covered slightly over one-fifth of total world production. Circumstances were forcing many producers to move beyond the concept of stock-holding into the concept of limitation of production. The objectives of the new body, as set out in November, 1929, were to ensure a reduction in the supplies of new tin coming on to the market and to provide that excess supplies of tin concentrates from the mines to the smelter (that is, excess to the tonnages used for smelting) should be stockpiled by the smelters. The level of the smelter output was to be so arranged as to keep the world price within a certain (but unstated) range. The stock of tin concentrates would be financed by a newly formed Anglo-American Tin Corporation.

The proposals of November, 1929 came too late. The tin price, lifted temporarily after, and probably as a consequence of, the formation of the Association, began to fall again in the autumn. Very bravely, and in spite of its limited membership, the Association decided to abandon the stock-withholding policy and agreed on a voluntary limitation of output. These limitations applied only to the individual members of the Association; they did not apply to non-members (who in Malaya included the Chinese miners, highly suspicious of the scheme, and some lower-cost European mining groups). The restrictions were extended in scope and severity in 1930 when the N.E.I. producers and the major Bolivian producers, including Patiño, agreed to join in restriction and thereby very much widened the authority of the Association. As the price and stock positions worsened the Association's measures tightened.[11].

The Association was not strong enough to cope with an accumulated excess of world production of 30,000 tons, while world visible stocks had built up to a figure perhaps 25,000 tons higher than normal. A drastic reduction throughout all producing areas of about one-third in the actual production levels of 1930 would have been necessary to bring about a price recovery. The Association did not realise the magnitude of its task or the limitations of its voluntary membership, even reinforced by the

FIGURE 12

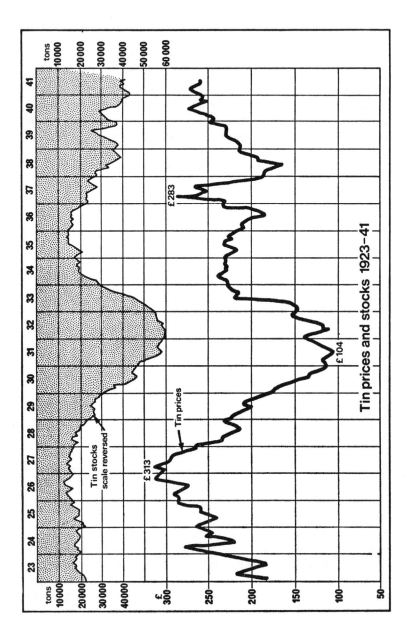

Tin prices and stocks 1923–41

support of the Netherlands East Indies and main Bolivian producers, or the degree to which the industry was dependent on a recovery in the tin-using economy of the U.S.A.

The most successful pressure for and application of restriction came in Nigeria (where the Associated Tin Mines and the London Nigerian Tin Mines dominated production), but even here the fall in total output in 1930 was less than a quarter. In Malaya, the initial stronghold of the Association (if such a word could be applied to a body not supported by the Chinese miners) not even the European dredges dropped output during 1930 by as much as 10 per cent. The main world loss of output came from Bolivia (where it fell by no less than 18 per cent) and the least from the Netherlands East Indies (a drop of 3 per cent). Thailand and Burma actually increased output.

The year 1930 closed with world production down by less than one-tenth. Even all this reduction could not be credited fully to the voluntary restriction. The price had squeezed out of production many uneconomic mines, especially in Australia, Bolivia and the United Kingdom.

The experiences of 1929-30 showed clearly enough:

(i) that a voluntary scheme of restriction, limited to members of the Tin Producers' Association, might merely leave the market for tin in any one country open to those non-Association companies who could weather the rock-bottom 1930 price (for example, in Malaya because they were low-cost European companies or because they had the social structure of Chinese companies able to pass on to the working miner the cost of a long and hungry depression);

(ii) that the voluntary system resulted in great inequalities of sacrifice between different countries;

(iii) that the voluntary system should be replaced by compulsory, intergovernmental restriction, taking in the great bulk of world production on a national basis;

(iv) that restriction of output in itself alone was not a cure for the ever-growing mass of visible stocks which had so depressing an effect on the price; and

(v) that the big financial groupings in the industry, however influential, were not strong enough to control a world industry where the operations of medium and small miners were still important factors.

Signing the first international tin agreement, 1931

Governments in the tin-producing countries had every reason to listen with sympathy to proposals for a compulsory governmental control scheme. Earnings of the producing companies from tin production and, in consequence, the governmental revenues from duties and royalties on export or production, were falling rapidly and even catastrophically. In Malaya national earnings from tin exports in 1930 were down by over one-third as compared with 1929. ($M 117 million in 1929 to $M 76·0 million in 1930). The 1930 figure was the lowest in seven years, and there was no sign that the slide was stopping[12]. The revenue duty collected by the Malayan governments from the tin industry, which was an important contribution to the total Malayan internal revenue, fell in 1930 by over $M 6 million and was no higher than it had been in the decade before 1914, although production now was a quarter higher. Employment was badly affected. In 1929-30 the number of workers employed in tin in Malaya dropped by about a quarter. The number was to drop again in 1931 and in 1932, so that by this latter year it was to be half the 1929 level. Other countries had similar disasters. In Nigeria the value of tin production fell from £2·1 millions in 1929 to £1·1 millions in 1930, in Bolivia from £9·3 millions to £5·9 millions and in Indonesia from 89 million to 59 million guilders. Unemployment and short time working were rampant everywhere.

The main drive for the acceptance of compulsory control seems to have come from the Netherlands East Indies (that is, from the Billiton Company) and from Bolivia (that is, from Patiño). In May, 1930 these threatened to withdraw from their

[12]Earnings in 1931 were to be $M 51 millions (the lowest since 1921-22) and in 1932 $M 31 millions (the lowest since 1898).
[13]According to Houwert of the Billiton Co. in *Draft: Memorandum on proposals for consideration at an international conference on trade and employment*, The Hague, September, 1946 (duplicated). Houwert says that the draft agreement was drawn up by J. van den Broek, managing director of Billiton. But the Colonial Office of the United Kingdom certainly had a hand in it very early on.

convention with the Tin Producers' Association "if it were not re-organised upon efficient lines"[13], that is with binding agreements sanctioned by the governments of the countries concerned.

A draft intergovernmental tin agreement was approved in principle at a meeting of delegates of Malaya, Nigeria, Bolivia and the Netherlands in November, 1930 when the London price was as low as £112 a ton. The scheme could not be agreed to immediately by the United Kingdom, which showed perhaps greater obligation to public opinion in Nigeria and Malaya than did the Billiton Company in the Netherlands East Indies or Patiño in Bolivia. There was some opposition from the Chinese and lower-cost European mines in Malaya. An informal meeting under Sir John Campbell in the Colonial Office with delegates from Nigeria, Malaya, Bolivia and the N.E.I. discussed points at issue. The United Kingdom had already agreed in principle with the idea of restriction but wanted assurance that the proposed system would safeguard adequately the interests of consumers and that Malaya and Nigeria should have a complete plan of the scheme before approval. The meeting did not trouble over points already largely agreed (the basis of 1929 for standard tonnages, the six months' basis for quotas etc.) but concentrated on the statement (which must become public) as to the objectives of the scheme.

The Colonial Office question was: "(Is the scheme) intended to secure a fair equilibrium between production and consumption or . . . (is it)intended to reduce the visible stocks from about 50,000 tons to 25,000 or 20,000 tons or . . (is it) intended to raise the price of tin to £150 or £165 or £200 per ton?". The danger of putting into the agreement specific figures on which public opinion might seize was obvious. The Netherlands delegates wished to bring in a reference to the stabilisation of the price. Sir Philip Cunliffe-Lister made the point clear. "We all mean the same thing", he said. "We know the price has to be higher than it is today. On the other hand, we want increased consumption and none of us mean to stand for a price that would limit consumption. On the contrary, we want consumption to be increased but if you put in some words about a fair price you will at once be asked what price". Mr. Groothof (Netherlands) replied: "No, I do not want that"[14]. The meeting compromised on the more tactful general objective of securing an equilibrium between production and consumption, so preventing rapid and severe price oscillation. In March, 1931 the first international tin agreement came into force for two years. It laid the foundations for tin control for the next ten years.

[14]Informal meeting in the Colonial Office, 27, February, 1931.

VI

Principles of the Tin Agreements, 1931-46

THE first international tin agreement lasted until 1933. Its general principle – the regulation of production by a system of quotas and the enforcement of that regulation by governmental action – was maintained in the second agreement of 1934-36, in the third agreement of 1937-41 and in the nominal fourth agreement of 1942-46. That same principle was also to be contained in the four wider international agreements operative from 1956 onwards.

There were, of course, differences, sometimes important, in detail in the agreements of 1931-46 which arose from the need to strengthen the initial machinery of the first agreement, to tighten its administrative control or to soothe outside public opinion.[1]

There was a development in the statement of the objectives of the agreements. The stated objectives of the agreement of 1931 "to secure a fair and reasonable equilibrium between production and consumption with the view of preventing rapid and severe oscillations of price" had made no reference to the influence of stock changes, although the members of the International Tin Committee were perfectly well aware of this problem from the beginning. Consumption of tin in the world and especially in the United States was liable to changes which might be substantial, sudden, not easily seen and of relatively long-term duration and a factor which the agreement members could scarcely hope to influence.

At any given moment, the level of world stocks might be a determining factor in the tin price. Control over exports was a weapon in meeting long-term over-production; but control over stocks or, at least, over a substantial proportion of those stocks

---

[1] For a comparison of the terms of the four agreements, see *A statement on the position and prospects of the tin industry*, published by the International Tin Study Group in 1950.

could meet short-term fluctuations between supply and market demand.

The second agreement of 1934-36 therefore extended its objectives so as "to ensure the absorption of surplus stocks"; the third and fourth agreements of 1937 and 1942 added "with a view to . . . maintaining reasonable stocks". The third and fourth agreements limited the size of stocks in member countries. The application of these extended objectives was bound to push the International Tin Committee towards the holding of stocks and the use of a buffer stock. The international tin pool of 1931-34 was a private pool, but it acted with the cognisance and approval of the countries signatory to the control agreement. The inter-governmental buffer stock of 1934-35 was still closer to the control agreement. That stock was financed by the member governments of the I.T.C.; it was to be used "as an adjunct to the International Tin Control scheme"; and the Buffer Stock Committee acted on the instructions of the International Tin Committee. The buffer stock of 1938-39 was set up nominally by a separate agreement signed by the member governments of the Committee but it was very tightly tied to that Committee. Contributions to the buffer stock were placed at the disposal of the Committee, upon the understanding that such stock would again be used as an adjunct to the control scheme; the initial size of the buffer stock might be increased by the I.T.C.; the I.T.C. fixed the quota for the purposes of providing the tin for the buffer stock, and the I.T.C. gave general instructions on the operations of the buffer stock executive. The fusion of control agreement and buffer stock was completed when in all the post-1956 agreements both functions were embraced within a single document.

The signatories to the first agreement in 1931 were Malaya, Nigeria, Bolivia and the Netherlands East Indies. These four countries had produced, during the boom of 1929, no less than 85 per cent of the world's tin. They were well aware of the dangers

[2] Japan entered the post-1956 agreements as a consumer and Australia in 1956-71 as a consumer and in 1971-76 as a producer.

[3] In fact, the post-1956 comparison is on the basis of world tonnage without China and is therefore nearer the 1931-41 percentages which are on world production including China.

that the mere fact of restriction of output on themselves would reduce the importance of their own production in the world picture and that, if their own restriction affected the tin price favourably, there would be a strong incentive to increased production in the unrestricted areas. The signatories were eager from the beginning of the control agreements to secure the adhesion of outside producers. They were therefore in a poor bargaining position and, in general, new recruits received terms on joining which were far more lenient than those of the original members. Siam (Thailand) entered in 1931 and remained a party to the agreements until 1941, but on terms of certain guarantees as to the tonnages which it would be permitted to export, guarantees which meant that it was subject to no control or to a far smaller degree of export control. The Belgian Congo joined the second and later agreements but, contrary to the concept of a controlled and reduced export, received a standard tonnage which increased steadily each year. From 1934 to 1941 French Indo-China and during 1934-36 the Cornish tin companies and Portugal were members on terms favourable to themselves but equally to the disadvantage of the other original participants. Some producing countries – China, Burma, Japan and Australia – either would not or could not be bribed into membership.[2]

Over the whole period 1931-41 the membership of the agreements covered an average of 83 per cent of world production. The minimum covered was 75 per cent in 1933 which represented about 25,000 tons a year of "outside" production and the maximum covered was 90 per cent in 1940-41, still leaving about 24,000 tons "outside". The Committee was prepared to make important sacrifices of principle to obtain universality of membership. It was true that over some periods of operation of the agreements "outsiders" maintained production at times when "insiders" were suffering restriction, but there is little evidence that "outsiders" were at any time a serious threat to the effective operations of the agreements. In the post-1956 agreements, the producing members usually accounted for between 88 and 91 per cent of world production, that is, not far beyond the pre-1939 coverage.[3] They had then no serious problem from normal "outside" production but only from abnormal exports from the

U.S.S.R. in 1957-60 and abnormal sales from the surplus strategic U.S. stockpile in 1962-65.

The first agreement pointed out that "research designed to stimulate consumption of tin was a most desirable adjunct" to the control scheme. This objective was repeated in all the later agreements and was embodied in the separate Research Agreement of 1938.

Consumers and the agreements

The first two agreements made no provision for representation of consumers. This left the International Tin Committee open to severe criticism as a producers' cartel from a world where the thinking of economists was still dominated by concepts of the virtues of competition. Under pressure from this opinion, especially strong in the U.S. administration, the third agreement included formally an invitation to a representative of the consumers in each of the two largest tin consuming countries (i.e., the U.S.A. and the United Kingdom) to attend its meetings and to tender advice to the Committee regarding world stocks and consumption. Under this provision, consumer representatives came to Committee meetings during 1937-41. So far as they stated any general line, it was naturally (a) that the consuming interests agreed on the desirability of a stable price for tin, (b) that the price level on which the Committee was basing its operations was always too high and (c) the result of this high level was or would be to push consumers into research work for the use of a smaller amount of tin in products or to lessen the market for tinplate in face of substitutes.

TABLE 9

Percentage of world output covered by countries participating in the pre-1941 international tin agreements

1931	81	1937	86
1932	80	1938	80
1933	75	1939	84
1934	83	1940	90
1935	83	1941	90
1936	86		

In drafting the fourth agreement the Committee was given to understand that the U.S. government was very anxious to maintain the consumer representation of the U.S.A. in as efficient a manner as possible (without, of course, U.S. membership of the agreement). That agreement, therefore, provided for invitations to two persons representing the tin-consuming interests of the U.S.A. (one being appointed by the U.S. government and the other being the direct representative of U.S. consumers) and one from consumers other than the U.S.A. These representatives would tender advice on world stocks and consumption, but would have no vote. In practice during 1942-46, although the consumers' representatives attended some meetings, they took little or no part in the activities of the Committee which itself, as a consequence of the war and maximum production, was inactive both on regulation and price. When international regulation of tin emerged again in 1956 it was on a new basis where representation of consumers was on a governmental basis and where the consuming governments so represented held in totality as much voting power as the producing governments.

Voting rights and standard tonnages

The first and second agreements faced the problem of voting rights of member countries. Fortunately they had a simple problem. At any given moment under these agreements the combined output of Malaya and Bolivia dominated the scene but the interests of low-cost Malaya and high-cost Bolivia were so divergent economically that they were most unlikely to act together to run the Committee. The south-east Asian producing members had a similar combined superiority in tonnage, but the differences in the organisation of production and in the temperaments of the producers in Malaya and the Netherlands Indies were so sharp as to make continual union of forces improbable. Malaya alone never had more than half of the total Committee tonnage.

For the first two agreements it was provided, wisely enough, that no votes would be allocated and that, in order to safeguard all interests, all decisions would require to be unanimous as regards changes in quotas. In the third agreement, largely to conciliate restless opinion amongst the Malayan producers that the interests of Malaya were not being sufficiently taken into account, a system of weighted voting (not entirely in line with production) was introduced. Under this a total of 20 votes was allocated to members, with "favourable votes being necessary to carry any proposal". This system was carried forward in principle into the fourth agreement, but with a total of 17 votes of which 10 in favour were required to carry a proposal. The weighting of these votes was such that the necessary majority of 10 votes could not be obtained by a combination of Malaya and Bolivia or of Malaya and the Netherlands East Indies. The voting rights were not strictly related to standard tonnages and resulted generally in under-weighting of the main producer Malaya. The principle of weighted voting was, with some modifications and on a fairer basis, retained in the post-1956 agreements.

The formula on which the quota system worked was simple and was maintained in all four agreements. Each signatory government in each agreement was given a standard tonnage. The totals of these standard tonnages varied in each agreement (162,168 for the initial four participants in the first agreement; 184,000 to 192,000 tons for the nine participants in the second agreement; 197,850 to 209,970 tons in the third agreement; and 232,900 tons for the fourth agreement). The basic year for standard tonnages was normally 1929 and the original members became firmly and almost unreasonably attached to that year. This meant in practice that almost no change could be made downwards. Changes, if any, were usually upwards, and thus inflated the total apparent production capacity of the industry. In spite of the ability or inability of the individual members to fulfil the quota allocations made from time to time, no serious attempt was made to alter the standard tonnages to reflect real and changing capacity to produce. In the first eight years of the agreements, the only important change was a slightly higher tonnage for Malaya, justified by the change in the method of calculating the

---

[4] Actual production of the five members in the exceptionally favourable conditions of 1941 was 203,000 tons.

tin content of the quotas. In 1938, standard tonnages became a little more realistic, since Bolivia accepted part of the consequences of her gross failure to fulfil her quotas and Malaya and the Netherlands East Indies benefited from what Bolivia lost. But the realism did not last long, and Bolivia was brought back in 1939 almost to the 1929 basis. The standard tonnage allocations and guarantees to the new members recruited in 1931 and in 1934 (Siam, the Congo, French Indo-China, etc.) turned the standard tonnage concept almost into nonsense. In consequence, the five major members (including Siam) who had held standard tonnages totalling 172,168 tons in 1931 were (including the Congo, but excluding Siam) to be allocated in 1941 total standard tonnages of no less than 232,900 tons.[4]

There was no specific provision in any of the agreements, merely the opportunity for discussion in the negotiations for a new agreement, for changes in the standard tonnages and therefore no automatic reward for efficiency or penalty for inefficiency. This left the system of standard tonnages wide open to outside criticism and, very wisely, the post-war agreements incorporated provisions which prevented freezing of the initial tonnages and voting position.

The Committee fixed, from time to time, the percentage or quota to be applied to these standard tonnages. Production was to be controlled by each participating government so that throughout the year it corresponded to the quota. Exports should not exceed the quota for the year; they should be distributed uniformly over the year; and quarterly exports should not exceed the quarterly quotas. These practices of the first agreement suffered no substantial changes in the later agreements. As responsibility for controlling exports was delegated specifically in the agreements to member governments, the Committee saw no need for (and every political risk in trying to apply) penalty clauses in the agreements in relation to under-export or over-export against the quota, and any disputes (and some did occur) were settled by discussion within the Committee. The disputes were remembered after the end of the Committee and the post-1956 international agreements provided an inordinate amount of space on provisions and penalties for over-export, provisions which, in fact, were appealed to only once in the fifteen years after 1956.

Quotas were fixed in the first agreement initially for not less than six months and in later agreements for quarterly periods. In practice the quota (as applied to standard tonnages differing in different agreements) ran between a minimum of 33 per cent over 1933 and a maximum of 110 per cent over much of 1937. The movement of the percentage was irregular. It was moving down through 1931-33, and was fairly steady at a slightly higher level through 1934 until the middle of 1935; it was rising sharply through 1935 into 1936 and 1937; it moved down sharply through 1938 until the middle of 1939. Thereafter, the figure for export control purposes was nominal. With the outbreak of the Second World War, the quota figure was lifted by the Committee until it stood during 1941 at 130 per cent. This would have permitted an output of about 270,000 tons a year from the seven parties to the third agreement and was therefore in no sense a limitation, since that figure for the seven member countries was greater than the actual production of the whole world proved to be for that year. Under the fourth agreement, with two-thirds of the world tin production over-run by the Japanese, the application of any quota figure was of academic interest but was an unnecessary irritant to outside opinion, even though the recommendation of the Committee for a policy of intensive and maximum production was observed rigorously through 1940-41 in all the members.

The first agreement had hoped that alterations in the quotas would be made as seldom as possible since this lack of movement in itself would be a stabilising action. But such a policy required a degree of skill in assessing the future tin position, especially in consumption, which the Committee never really enjoyed. In arriving at the quota to be fixed for each quota period in the first three agreements the Committee had before it a statement from the Statistical Office of ITRAD[5] in The Hague as to current production, consumption and stocks and as to future consumption. The statistics were good; the views on the future tended to

[5] The International Tin Research and Development Council.

reflect those current in the Billiton company. The delegates on the Committee made representations as to a desirable level of production and, therefore by implication, of price. In general, throughout the three agreements Bolivia (that is, Patiño) and the Netherlands East Indies (through the Billiton Company) urged the higher and Malaya the lower degree of restriction.

At its worst period in 1932-33 the Committee kept its rate of quota percentage unchanged at 33·3 per cent for as long a period as eighteen months, but this constancy was perhaps not because the Committee was wedded to the idea of stability but because it feared the strain within its own ranks if it adopted the only other practical possibility, namely, a further restriction. In 1934-35 there was a further period of eighteen months when the rate of restriction varied only between 40 and 50 per cent, and the rate remained fairly constant between 100 and 110 per cent for the full year 1937 and between 35 and 40 per cent in the year from June, 1938. The rate of change was often very sharp (the quota doubling during 1935, halving in 1938 and trebling in 1939). Only in ten out of the 33 quota periods during 1931-39 was the quota unchanged in comparison with the immediately preceding figure.

The quota system, whatever its drawbacks, was generally adhered to in the earlier and economically more depressing days of the Committee's control. Over the first agreement and for the first year of the second agreement the correspondence between permitted exports and actual exports was very close, indeed to an almost suspicious exactitude (456,476 tons of permitted exports and 455,355 tons of actual exports from participants over 1931-35). This exactitude did not continue in the years of recovery 1936-37, when actual exports were in total some 53,000 tons below total permitted exports. On the other hand, the total actual exports again over 1938-39 were very close – only 1 per cent below permitted exports.

The question of a buffer stock

The first agreement in 1931 made no reference to a buffer stock or tin pool, but the Committee was well aware of the international tin pool which operated during 1931-34. This was a private scheme, which accumulated and then released a total of about 21,000 tons of tin. It acted with the cognisance and approval of the countries signatory to the international agreement. Its effect on stabilising prices was real. The international buffer stock of 1935 was for all practical purposes part of the control system. Its initial stock of 8,300 tons was contributed by the four signatory countries to the second agreement. These tonnages were raised as a proportion of the standard tonnage of each contributory, and the stock was to be used as an adjunct to the control scheme. This buffer stock was very short-lived and its tonnage small, and the surprising stability of prices in the middle of 1935 was perhaps almost an accident.

The members of the control agreement were by now convinced on the philosophy of buffer stocks; they were by no means convinced that the circumstances of 1936-37 (which required only a limited degree of export control, which provided a London tin price almost always above £200 a ton and in which stocks were running at a not abnormally high level) needed the application of a buffer stock whose cost would fall entirely on them and whose use through most of 1937 would have been only to prevent the price rising. The sharp downturn in the tin price in the first half of 1938 and pressure from the Dutch interests helped to stimulate the coming into effect of the second buffer stock. This was created in June, 1938 by a formal agreement between the parties to the control agreement, again as an adjunct to that scheme. Control of the buffer stock by the Committee was tightened.

The contributions to the buffer stock by governments were placed at the disposal of the Committee; the initial tonnage in the buffer stock could be increased by decision of the Committee; the Committee raised the contributions by fixing a special quota or quotas within the control scheme; the Committee gave general instructions as to the operation of the buffer stock; and the life of the buffer stock was to be the same as the life of the control scheme. The buffer stock, which operated during 1938-39 on an important scale, was tied to and was successful in maintaining a price ranging between £200 and £230 a ton.

When the fourth agreement was under consideration, the I.T.C. agreed in December, 1941 a draft scheme for the continuation of the buffer stock but reversed its attitude in a very short time and the scheme was dropped. In the discussions at the International Tin Study Group from 1948 onwards it was throughout assumed that the buffer stock and the control of exports should be integral parts of any inter-governmental regulation agreement and both aspects have been covered in the international tin agreements operative since 1956.

VII
Tin Control in Operation, 1931-46

THE functions of the International Tin Committee under the first, and indeed under the succeeding, tin agreements were to reconcile a number of conceptions whose importance varied from time to time. These were its assessment of the tin position in terms of production, consumption and stocks, the desirable tin price, the tonnage of production (and therefore the likely price which would follow from the application of a particular quota on its members), the safeguarding of the position of its members in relation to non-member or "outside" production, the continuation of the control agreement, the relation of the control agreement to any tin pool or buffer stock in existence and the need if not to conciliate at least not to embitter consumer public opinion in the world.

The first agreement, 1931-33

The Committee's method of assessment of likely production and consumption was initially fairly primitive. The Committee had no real statistical services of its own and required time to develop them. It was largely dependent on the figures collected by the Tin Producers' Association and on the personal views of its members, reinforced in the case of the Dutch by the statistics of the Billiton Co. in The Hague. It had agreed, even before the signature of the agreement, that world tin mine production in 1929 was 186,518 long tons.[1] At the second Committee meeting in May, 1931 Mr. Lazarus[2] estimated world consumption for 1930 at 153,000 tons and for 1931 at 137,000 tons; he estimated world production in 1931 at 151,000 tons. On these figures a

[1] The figure was wrong. It was around 196,000 tons (including 7,000 tons from China).
[2] L. N. Lazarus of the metal broking firm of Lewis Lazarus and Sons was present as an adviser.
[3] The actual figures of consumption for 1931 proved to be around 141,000, but the various speakers were not always referring to the same thing.
[4] Letter from the Netherlands East Indies Delegation to the chairman of the Committee, 10 September, 1931.

restriction of exports from member countries by 35 per cent of the standard tonnages would be necessary to bring about equilibrium; but to bring about a reduction of stocks would need a restriction to 42 per cent. Houwert of Billiton Co. agreed with Lazarus and thought a cut of 20,000 tons in production would be adequate. Pearce of Williams, Harvey and Co., representing the Patiño interests, convinced Lazarus that world consumption in 1931 would be no higher than 130,000 tons.[3] The Committee agreed to a quota of 77·7 per cent, a figure which would reduce world production to an annual rate of 145,000 tons. The compromise figure was, in fact, an excellent guess on the production side (production over 1931 proved to be 144,000 tons), but nothing seemed able to stop the flow of tin into stocks, and by the middle of 1931 these stood at well over 50,000 tons or four months' supply.

Malaya and the first quotas

In its first few months of operation the restriction scheme creaked badly so far as Malaya was concerned. The standard tonnage of Malaya was 69,400 tons of tin. It was the obligation of the government of Malaya to allocate this tonnage locally so as to provide each producer in Malaya with his assessment. This required local legislation and the working out of details for some 1,200 mines, some of whom had recently very strongly opposed the principle of restriction. In the three months March-May, Malaya exported about 1,000 tons above her quota, and then (even although the quota was then smaller) another 1,100 tons in June, 1,000 tons in July and 1,700 tons in August. In September the Dutch very sharply drew the attention of the Committee to this "over-production in Malaya which has impeded and handicapped a rise in prices and consequently has inflicted loss on the other participants in the agreement".[4] The Malayan defence was that, while the international Malayan assessment for Malaya was based on the 1929 production, it had not been possible to set the local assessments within Malaya on the same basis. The local assessment of each mine was based on the mine's potential output; that had to be assessed by local committees from whose decision there was a right of appeal to a central committee. The certificates of production, first issued in June, covered the six months March-August. These assessment certificates were based on an estimated potential production for the six months of 31,500 tons. The Malayan quota for the period of six months was 24,000 tons of tin; the quota reduction was 7,500 tons or 23·8 per cent of the standard tonnage. To ensure conformity with the international quota the Malayan government lifted its cut to 25 per cent. Unfortunately, the revised and corrected estimate in July gave a figure of 38,000 tons as the potential production of Malaya for the half year. The application of the Committee's rate to this new assessment gave the Malayan miners the right to produce 28,500 tons of tin in the half year against the Committee's quota of only 24,000 tons.

To meet the grievances of the other members of the agreement the government of the Federated Malay States took two immediate steps. It arranged, at its own cost, to hold up the August excess of concentrates for one month with the local smelting companies so that Malayan exports for the whole half year should not exceed the international quota; and it imposed for the quarter September-November a reduction of no less than 60 per cent in the potential output so that the export permissible from Malaya would be only 8,240 tons against a quota imposed by the Committee of 10,942 tons. The exports actually still in excess would be wiped out by February, 1932. In reporting these drastic measures the Malayans stated[5] that the 60 per cent internal cut would mean the repatriation of perhaps 30,000 coolies to China and the taking of special measures by the Malayan government for internal security. The statement added:

> "A large number of the Malayan mines are the property of comparatively poor Chinese; and, since the fall in tin prices, they have been worked by the coolies themselves on co-operative lines. Present economic conditions cannot be survived by such mines but it would be a serious political blunder to enforce a cut (in excess of 60 per cent) which would inevitably close them all together at one moment."

[5] Statement by the Malayan Delegation to the sixth I.T.C. meeting.
[6] 6th and 8th meetings of the I.T.C., 1931.

The British Colonial Office had been disturbed by the strength of the Dutch complaint. It had taken the Malayan government sharply to task and "had impressed on the High Commissioner (for Malaya) that it is essential that Malaya should take adequate steps to re-assure her partners in the international scheme, and public opinion generally, regarding the performance of her international obligations". In particular, the Colonial Office urged Malaya to buy a quantity of tin equivalent to the 4,000 tons exported by Malaya in excess as at September, to hold this quantity off the market and dispose of this holding in instalments equal to the reduction in the excess effected by the 60 per cent cut. The Malayan government regarded this proposal very coldly as likely to lead to acrimonious discussions in the Federal Council on the principle of restriction itself.[6] But the problem was solved. Exports in September fell to as low as 2,558 tons against Malay's quota of 3,781 tons. This wiped out some 1,200 tons of Malaya's accumulated excess. The balance of the excess, about 3,500 tons, was bought by the Malayan government for later release on the Colonial Office's proposed lines. By the end of 1931, Malaya had liquidated another 2,000 tons of the excess; and in the first quarter of 1932, with the liquidation of another 1,200 tons, the sheet became almost clean again.

Quotas in 1931-32

The first quota of the Committee, operative over March-May, 1931, failed to be effective, mainly owing to the failure to check the Malayan output. The price continued to fall to the dismal point of £104 a ton for May. For the six months beginning on 1 June, 1931 the permissible export tonnage was further reduced by one-sixth to 65·4 per cent. But the continuing excess of exports from Malaya very largely destroyed the effect of this reduction. In October, 1931 Pearce (Williams, Harvey and Bolivia) pressed for a further reduction; John Howeson (Tin Producers' Association) was even more critical. Visible supplies had actually risen by about 5,000 tons and he believed that at the end of 18 months they would be back to where they had started.

The Malayans, committed to an internal cut of 60 per cent to wipe out their excess exports, felt that an immediate further reduction in the international quota would be politically impossible and " would be tantamount to asking the miners to commit suicide".[7] A compromise, produced by Malaya and Nigeria, was agreed. The quota was reduced by about another one-seventh to 56·2 per cent from 1 January, 1932 with the object of cutting annual production by 15,000 tons a year. This new quota was to be held unchanged for eight months, but, if at the end of that period, there was not a unanimous recommendation to continue at this rate, the quota would revert to its figure of November, 1931 (65·4 per cent).

The arrangement was almost at once challenged. In January, 1932 at the 10th I.T.C. meeting, Antenor Patiño (Bolivia) pointed out that, after allowing for normal stocks and the tonnages in the hands of the international tin pool, there was still about 10,000 tons of tin hanging over the market. If this surplus stock could be rapidly thrown off he anticipated that "there would be a rapid and sustained rise in price". The thinking of the Committee was now being strongly influenced by the existence of the international tin pool which held about 21,000 tons of the world visible supply of 51,000 tons and which therefore had a very heavy stake in the recovery of the price. A sub-committee reported gloomily in February, 1932. Between 1 March and 31 December, 1931 stocks had risen by nearly 7,500 tons. Consumption was still declining, although it admitted that on the new quota consumption from 1 January, 1932 was running at 2,000 tons a year above the permitted production. At the current rate of restriction it would take the Committee another year of control merely to get back to the position from where it had started in March, 1931.

The sub-committee proposed two steps in tighter control. The first was for a relatively small reduction of some 8,300 tons a year in the quota (that is, about 5 per cent in the standard ton-

[7] 7th meeting of the I.T.C., 1931.
[8] The reduction would not apply to Siam which had been given a minimum flat rate of 10,000 tons a year.
[9] Making a total of 7,700 tons of tin.
[10] 12th meeting of the I.T.C., March, 1932.

nages); the second was for the four signatory countries to with-hold one month's quota of tin (say, 8,000 tons) from the market.[8] The first step would reduce stocks; the second step would wipe out the debit balance that had accumulated and would therefore balance production and consumption. The cut would operate from 1 June, 1932, notwithstanding the decision so recently made by the Committee to leave the quota unchanged for eight months. The Committee accepted the proposed further cut in the quota plus, hopefully, whatever further tonnage the Siamese might (and would not) voluntarily abandon. As for the holding-back proposal the Committee put alternatives to governments. In the first alternative each government would hold back one month's quantity of tin at the proposed reduced rate;[9] this "frozen" month would be released at the rate of 5 per cent of the original quantity each month. In the second alternative a gov-ernmental pool was to be formed. Into this pool each government would put a month's production at the reduced rate. This gov-ernmental pool would be entirely independent of the existing international private pool; but it would adopt the scale of release, the price scale and the general conditions of that private pool.

Malaya, Bolivia and the Netherlands East Indies accepted the proposed cut in the quota. Malaya agreed to neither the month's hold-back nor the governmental pool; Bolivia refused to enter the tin pool but emphasised the holding-back of one month's production and Sr. Vargas (Bolivia) said that Patiño was pre-pared personally to finance the holding back of his own produc-tion and that of the other Bolivian miners; the Netherlands Indies East preferred the pool to the holding back.[10] A sub-committee, which included John Howeson, produced a formula which was sent to governments. Each government would hold back one month's production (at the reduced rate) off the market, by purchase or otherwise, for a period of four months. Thereafter this production would be released at 5 per cent per month. The formula, unusually for the Committee, went into figures of prices. If the monthly average price of tin reached £176 a ton the rate of release might be lifted to 10 per cent; if it reached £187 to 15 per cent; and if it reached £198 to 20 per cent. This formula was not accepted.

The Byrne memorandum, 1932

The price of tin had been given a temporary fillip in the last quarter of 1931, due almost certainly to the formation of the international tin pool, but the stimulus faded in the first quarter of 1932. By April, 1932 stocks showed no decline from their very high level and the tin price had sagged to £109 a ton. The Committee was faced with an even grimmer picture of falling consumption and it was anticipated at the April I.T.C. meeting that the existing quota rate was between 8,400 and 10,800 tons above the annual consumption. The meeting recommended a cut of 20,000 tons (annual rate) in production as from 1 June. The same meeting also considered an ambitious proposal from Mr. Byrne. This Byrne memorandum, already considered informally by the Colonial Office and the Trustee of the British section of the international tin pool, put forward to the I.T.C. three steps.[11] First, there should be a complete suspension of production for two months (June and July, 1932). This would reduce visible stocks (exclusive of the tin pool's holdings) to a normal figure of around 24,000 tons. Secondly, production should be resumed in August, 1932, but only on a quota of 40 per cent. This would mean a reduction on the current rate of 25,000 tons of tin a year and production would then run at 2,000 tons a month below current consumption. Third, the tin pool should raise its minimum selling price to £200 a ton. Production would continue on a 40 per cent quota until the tin price had averaged £200 for a month. The quota would then be raised to a total of 50 per cent.

If the price were maintained at £200 the tin pool would sell 7,000 tons of tin. If the price fell below £200 the quota would be dropped to 40 per cent and the pool would sell. If the monthly price were £210 or higher the quota would be 50 per cent and the pool would sell another 7,000 tons of tin at not less than £210. If, after the liquidation of the pool's second 7,000 tons the price still remained at £210 or higher, the quota would stay at 50 per cent

[11] 13th meeting of the I.T.C., 22 April, 1932.
[12] For the operation of the Byrne plan within Malaya, see Yip Yat Hoong, *Development*, p. 205.

and the pool would sell its last 7,000 tons at £210; if the price fell below £210 the quota would revert to 40 per cent.

This bare-faced subordination of the control scheme to the price interests of the tin pool had some attractions. The complete suspension of production for two months would in itself do as much to remedy the over-supply position as limited production would do in 9 or 12 months; and some delegates appreciated the emphasis on price as the real factor in determining the level of permitted production. But, nevertheless, a complete stoppage for two months would mean serious dislocation of the industry and might force governments to grant financial assistance to small miners (especially in Malaya). This need for financial assistance would arise, of course, at a time when governmental income from royalties on tin production would cease.

The proposal turned the control scheme from being a scheme to regulate production into being a scheme to stop production, which was a very different kettle of fish. Some delegates disliked the I.T.C. quotas being tied to specific tin prices and some of them stressed the inadvisability of taking any action "which might lead to dissatisfaction on the part of consuming countries".

The Byrne plan, nevertheless, was generally accepted with some modifications and the quota from 1 July was fixed at 33·3 per cent. There were some reservations by Malaya. There were to be no further cuts in the quota for twelve months; the minimum release price from the tin pool was to be not less than £165 a ton; no sales should be made by the pool except when the quota was 40 per cent or more. These reservations were accepted by the pool and the Committee.[12]

The crisis overcome, 1933

The level of production at 33·3 per cent was to be applied for the remainder of 1932 and for the whole of 1933. The price crisis and stock crisis had been overcome. The upward movement of the price in July, 1932 began a rise that did not stop until 1934. The decline in visible stocks was to go hand in hand with the rise in price.

The Byrne plan was giving immediate results. The exports for July and August, 1932 averaged only 4,000 tons a month (and were kept off the market); the exports for the second quarter of 1932 were only 7,400 tons a month against nearly 11,000 tons a month for the year 1931. But the Committee faced two other longer-term problems – an excess of exports by Bolivia and the question of outside production.

In Bolivia the government was involving itself in a dispute with Paraguay which was to develop into the dreary bloodletting of the Chaco War; it was perhaps not anxious to police too closely a system of export restrictions which was reducing so seriously the necessary flow of foreign exchange. By the end of 1932, Bolivia had accumulated an excess export of 1,400 tons. The Dutch, ever watchful, drew attention to this excess. The Bolivian excess export was to remain constant through 1933 but was to disappear in 1934-36 as more Bolivian tin miners were pushed out to die in the Chaco.

The problem of outside production from non-participating countries was more serious. During 1931-32 tin production from countries outside the five participating members rose from 16,500 to 20,000 tons at a time when production from the participants had fallen from 133,000 tons to 80,000 tons. These figures convinced the Committee of the dangers from outsiders and laid the basis for the unhappily generous way in which it turned to buy these outsiders into the second agreement.

The Committee had also other things on its mind. Mr. Byrne had turned up with a new scheme – the "Byrne tin finance plan".[13] He pointed out that the original Byrne plan, due to end in mid-1933, had been modified, had raised the price of tin but had not produced results commensurate with the sacrifices of the producers. The total visible supplies of tin had not decreased sufficiently and "in consequence, the possible liquidation of the (international tin) pool's stocks has become more remote". He proposed that the signatory countries should impose a special tax of £50 a ton on all tin exports. At the same time, each individual miner should hypothecate 25 per cent of his output with the

[13]In a letter of 13 February, 1933, to the I.T.C.

I.T.C., such hypothecated tin not being subject to the export tax. The proceeds of the special tax should be handed over to the I.T.C. and the Committee would advance to each miner £150 a ton on the hypothecated tin. The Committee would thus take over 1,400 tons of tin a month (that is, one-fourth of the current permitted rate of output). At the same time the international tin pool would increase its holdings to 30,000 tons of tin. The pool would.agree not to sell tin at less than £200 a ton and the I.T.C. would agree the same for its hypothecated tin. As and when the Committee sold its tin it would pay the miner the full price realised; and the special export tax would be withdrawn when the Committee started selling. Arrangements would be made with the international pool regarding its selling. The proposal had the merit, in the words of its author, that "the Consumer, who has had a long innings of very low prices will, in effect, be finding the necessary finance to make the Plan operative and also to pay the producer a higher price for his Tin".

The Committee admitted the attractions of the new Byrne plan. It provided the producers with a ready-made buffer stock; it emphasised the desirability of removing still more overhanging stock from the market into the international pool; and it tied very closely control and tin pool. On the other hand, the Committee thought that the proposed export tax was most unlikely to be accepted by governments. If prices rose under the scheme suddenly this would stimulate even more outside production. The Committee (perhaps conscious of the element of open exploitation of the consumer in the plan) could not recommend it to member governments.

The demand for higher quotas

Through 1933 the rate of world consumption was rising very steadily and by the third quarter of the year was half as high again as it had been in 1932. In the same period actual world production had knowingly been screwed down until it was little over half the consumption level. The fact that the quota remained unchanged for the whole of 1933 showed how important the Committee felt it to be that the shortage should be met from the accumulated stocks of the international tin pool rather than from the traditional source of mine production[14]. The pool began to release its stocks about the middle of 1933 in fairly generous tonnages and continued to release them until the late spring of 1934. It was not until October, 1933 when the tin price had been above £200 for six months and was now as high as £224, that the Netherlands East Indies requested an increase in the quota to at least 40 per cent. In December, V. A. Lowinger (Malaya) asked for a 45 per cent quota from January, 1934. He stressed the importance of the non-participating countries. He said:

·"Stimulated by the price and the knowledge that control was to continue (under a second agreement) the 'outside' countries are developing in a way they could not have thought of doing if the main producing countries were unfettered. Something is needed to curb their activities. The Committee have seen, from the demands made in the course of recent negotiations, how extravagant their ideas have become; and (he) was convinced that it is in the Committee's interest to increase the outputs from the restricting countries, and to lower the price rather than to encourage the 'outside' countries to develop at a time when the participating countries are prevented from doing so".[15]

A quota of 45 per cent was supported by Sir Frank Baddeley (Nigeria) as being still well below likely consumption. The Netherlands East Indies disliked the possibility of a rise in the quota being promptly followed by a reduction; they thought that some consideration was due to the international pool; and the pool must be allowed a reasonable opportunity to sell the stocks which it held. Against the opposition of the N.E.I., Bolivia and John Howeson of the Tin Producers' Association, the Commit-.

[14]The Statistical Office in The Hague gave the following figures to the Committee:

	Monthly average figures World production	long tons World consumption
1932	7,730	8,332
1933, first qr.	7,078	8,992
second qr.	7,051	10,591
third qr.	6,819	12,851

[15]At the 31st meeting of the I.T.C. in December, 1933.
[16]See figure 12 on stocks and prices, 1923 to 1941.

tee had to be content with easing the quota for the starting of the second agreement in January, 1934 only to 40 per cent.

The principle of universality of membership, to which many members of the Committee attached very great importance, meant in practice initially the bribing into membership of the agreement of countries which were low-cost (like Siam) or new, developing and low-cost (like the Belgian Congo) and able to stand on their own or anxious to work without restriction. For the first agreement the most important was Siam. Siamese production had risen in the five years before 1931 at a faster rate than in any other producer – even Malaya and Bolivia; it was the only major producer which had in 1930 not merely not reduced output but actually increased it; and it held 6 per cent of the world output. Its bargaining position it knew to be very strong. When it came into the first agreement in September, 1931 it was brought in on the basis of a flat rate of 10,000 tons a year – a guarantee half as high again as its actual production had averaged in the last five years. When in July, 1932 Siam was requested to accept as an act of grace a reduction of its flat rate quota at a time when the quota for the other members of the agreement was down well below half, she not unnaturally refused. It was perhaps more irritating to the other members that, in the catastrophic year 1932 when price forced even some Siamese mines to stop, Siam produced over 9,000 tons of tin or 9 per cent of the world total.

The private international tin pool of 1931-34

The ominous inverse relationship between the weight of stocks overhanging the market and the resultant market price for tin was only too clear to the world tin industry.[16] The build up of world stocks through 1929 and 1930 had been phenomenal. By the end of that latter year world visible stocks had almost doubled and the tin price had more than halved. The stock level continued to rise through the first half of 1931. By the middle of the year visible stocks (not including the Malayan carry over) had reached a record 51,600 tons or the equivalent of five months of world consumption. The price continued to fall and in the middle of the year had reached the lowest monthly price in thirty years. It was thus inevitable that an attempt would be made to repeat the success of the British-Dutch Bandoeng pool of 1921-24.

The international tin pool was formed in the latter part of 1931. The British and Dutch private groups which formed it acted throughout with the cognisance and approval of the countries signatory to the international control agreement. The chairman of the I.T.C. was chairman of the pool board, and the life of the pool was tied to the life of the control agreement. Its transactions and price policy were secret. The pool began accumulating tin from the autumn of 1931 and by the end of the year was believed to have a holding of 19,000 to 21,000 tons, a tonnage larger than that held at any time by the Bandoeng pool. The value was then between £2·6 and £2·9 millions.[17]

Through the whole of 1932, the stock held by the pool represented nearly one-half of the world's visible supply of tin metal, and the fact that it remained frozen completely for that period undoubtedly prevented the fall in the tin price during the spring of that year becoming completely ruinous for the industry.

The price policy of the pool is not known, but the indications are that substantial changes were made in its conception of a reasonable price. A reference at a meeting of the Tin Committee in late 1931 spoke of a buying price limit for the pool of £140 a ton; the first "release point" was £150 at which point it was obligatory on the participants to release 5 per cent of their hold-

[17]Details of the contributors and contributions to the pool are not public. Metal stocks in the Netherlands remained unchanged at 5,574 tons for twelve months after April, 1932 against nil in 1930 and preceding years; this figure possibly represented the Dutch contribution to the pool. In the United Kingdom metal stocks averaged 12,000 tons more in 1932 than in 1930. Siam, in joining the international control arrangement in 1931, told the Committee that it would not contribute to the pool. Simon Patiño told President Salamanca of Bolivia in 1932: "I have been able . . . to interest a powerful London metal firm and to form a syndicate . . . (which) has available a capital of £1,500,000 and its first objective will be to increase the price of tin to about £160 a ton. Once this objective is attained . . . we shall consider what further measures will be needed to reach £180 a ton." See C. F. Geddes: *Patiño, The Tin King*, p. 233.

[18]He was speaking before the recovery in consumption became much more marked in the second and third quarters of 1933. In fact, over the full year 1933, when restrictions were kept unchanged at 33·3 per cent, total world consumption proved to be 44,000 tons above production.

ings in the pool; but these figures – if correct – did not stand. Byrne's memorandum of April, 1932 (the "tin finance" plan, which had been previously considered – but not necessàrily approved – at an informal meeting with the Colonial Office and the tin pool), advocated a complicated proposal for moving the Committee's export quotas up or down in relation to a tin price of £200 a ton, to which figure the pool would raise its release selling price.

By April, 1933 the tin market price had risen to £186 a ton. This was the highest monthly average since the winter of 1929. The Committee could well feel that the system of export restriction had been now successful in pulling the price out of the dreadful slough of despond. It was also well aware that it owed a debt of gratitude to the tin pool for maintaining such substantial tonnages of stocks off the market. In February, 1933 the chairman of the I.T.C. knew that the maintenance after 1 July of the export quota at the current very low level of 33·3 per cent would give an annual mine production about 13,500 tons below consumption.[18] The Committee had to consider whether the world requirements should be met by increasing the export quota or by accepting sales from the tin pool at a price point which was almost certainly now held to be above £200 a ton. The Committee decided in May, 1933 that the July quota should be maintained at 33·3 per cent. This policy of starving the market of additional supplies from the mines – a policy repeated when the October quota was fixed also at the same level – handed over the reviving market to the tin pool. The tin price rose for May, 1933 to £220, the highest price for four years, and the pool began to sell, apparently about June-July. From that point up to the end of 1933 the pool sold 13,000 tons. The balance of 8,000 tons was sold in 1934, probably from January to April, so as to ensure that the pool would be fully sold out before the first international tin agreement ended.

In October, 1933 the I.T.C. agreed to raise the quota operative from January, 1934 from 33·3 to 40 per cent. At its meeting in December, 1933 no serious support was given to lifting this 40 per cent quota; and the Netherlands East Indies' representatives stressed the importance of giving consideration to the pool, which must be allowed every support to sell its stocks. The figures in front of the meeting showed that for the third quarter of 1933 mine production was little over one-half of world consumption.

The results of selling by the tin pool were generally beneficial. The maintenance of severe restriction on exports side by side with rising consumption would in itself have produced in the second half of 1933 a very sharp price rise. This was prevented by the releases from the pool.

Between the months of July, 1933 and December, 1933, when export restrictions on tin were very tight and the pool was selling, the monthly average tin price in London ranged between £217 and £228. This range of £11 compared with ranges of £71, £24, £25, £27, £11, £23 and £40 in the seven preceding half-yearly periods. In the period January-April, 1934 the monthly average tin price rose only £9. The pool in its disposals certainly did not exploit the consumer and, in fact, gave a degree of price stability which the market had not seen for a number of years.

All parties had some reason to be satisfied with the pool. Its disposals had almost certainly prevented a runaway price, but had not prevented the price rising to a level reasonably satisfactory to the producers. It had been undoubtedly profitable to its participants.[19]

The second agreement:
Buying up new members, 1934-36

There were problems but no serious difficulties in negotiating the second control agreement which was to follow from the beginning of 1934. The discussions with Siam did not succeed in reducing in any important degree the privileged terms on which she would consider continuing in membership. The Belgian Congo entered on terms which allowed not for restriction

[19]The 21,000 tons of tin held in December, 1931 would have been valued possibly at about £2·5 to £2·9 millions at London prices; sales at an average price of £225 a ton over July, 1933 to April, 1934 would have realised about £4·7 millions. Interest charges on carrying the stock for 18 months or more may have been about £0·3 million.

[20]At the 26th meeting, May, 1933.

but for annual expansion. The standing Malayan complaint on the assessment of standard tonnages was met by the use of a more appropriate figure for the calculation of the tin content of concentrates which gave that country another 2,000 tons of tin a year on its standard tonnage. The N.E.I., which had screwed a concession in standard tonnages at the expense of Bolivia and Nigeria as an essential condition for signing the first agreement (on the reasonable enough ground that it had deliberately held back its own production boom in 1929), now maintained that she would sign the second agreement only if the division on the 1929 basis were maintained. This problem and the irritant of a Bolivian accumulated excess export of around 1,400 tons were removed by a supplemental agreement which provided for 1934 alone an additional fixed quota of 6,626 tons for the four original participants, of which the Bolivian excess exports would be deducted from the Bolivian share. A first step was made at laying down the right of withdrawal and this was tied to the growth of outside production on which the Committee was so sensitive.

The Committee still hankered after universality but resented the price which it felt obliged to pay. In the discussions on the second agreement it was pointed out in May, 1933 that, had Siam joined on the same terms as others, its current output would have been limited to no more than 3,330 tons a year and the Committee felt that "the disparity (which was not contemplated when the scheme was framed or when Siam adhered to the scheme on a flat rate basis) is now so large that the signatory governments consider it imperative that some more satisfactory and equitable arrangements should be come to".[20] It suggested to Siam for the second agreement a flat rate of 7,000 tons a year. Siam was sitting pretty and knew it and the country had a long and successful history of defeating others by inertia in international negotiations. A draft second agreement in October, 1933 suggested a flat rate of 10,000 tons, subject to movement if the export quota rose above 100 per cent. The Siamese hedged and would neither sign the draft nor say on what terms they would sign; and some Committee members were prepared to go ahead without them. The Committee released its feelings again. It resolved:

"The Committee desires to emphasise, with all possible insistence, that unless the new agreement is signed by Siam as a participating country the present control will lapse in the very near future. None of the signatory governments is prepared to sign the new agreement unless Siam signs also; all are ready to sign if Siam adheres. Without that signature the control scheme will very shortly end under existing arrangements; and the Committee are convinced that the consequences would be disastrous for Siam, which is and has been profiting greatly by the very heavy sacrifices made by the participating governments for nearly three years. The delegations of Bolivia, Nigeria, the N.E.I. and Malaya feel that the situation warrants the strongest and most urgent diplomatic representations; and they agree to recommend such action as a matter of grave and urgent importance".[21]

The strong resolution did not affect the Siamese. When they signed the second agreement in October, they signed on the basis of a flat rate of 9,800 tons subject to upward movement if the quota went above 65 per cent.

The Committee was not discouraged. It sent out letters in the middle of 1933 to the Congo, Burma, Australia, French Indo-China, Cornwall and Mexico asking on what terms these would join the second agreement. Naturally, the replies asked for flat rate tonnages as good as or even more generous than the terms which Siam had been given. The Cornish companies were offered a standard tonnage of 3,300 tons with a flat rate of 2,200 tons, rising with increases in the quota above 66·7 per cent. The Cornish Tin Producers' Association haggled for more but in March, 1934 accepted for the three major mines[22] on a standard tonnage of 2,615 tons and a flat rate of 1,700 tons for three years, subject to moving when the quota was above 65 per cent. French Indo-China came in speedily. The French pointed out that, in entering the second agreement, France was departing from her

[21]At a special meeting of the Committee, 10 October, 1933.
[22]South Crofty, East Pool and Geevor. Wheel Reath had dropped out.
[23]M. Picard at the 29th I.T.C. meeting, September, 1933.
[24]31st I.T.C. meeting, December, 1933.
[25]Swaziland was offered a standard tonnage of 270 tons and a flat rate of 180 tons. Mexico did not propose to adhere; the exhaustion of the San Antonio mine had made her tin position unimportant. South Africa was offered a standard tonnage of 1,300 tons and a flat rate of 800 tons.

traditional and consistent policy; to limit the production of a raw material in a colony while the mother country imported large quantities from outside was something new in the economic history of France.[23] The four companies in Indo-China agreed not to exceed 1,700 tons in 1934, 2,500 tons in 1935 and 3,000 tons in 1936.

The Belgian Congo companies wanted better terms even than Siam. The Congo "was a dangerous potential producer; a general development programme undertaken by the government had opened up the country and had made the production of tin from areas which were not formerly accessible a practicable proposition now. The Belgian Government appeared to attach much more importance to the unemployment question than to any other single element in the situation; apparently they were prepared to contemplate the production of tin, even at a considerable loss, provided this gave a substantial increase in the number of men employed".[24] The Congo claimed from the Committee for 1934 and 1935 rising tonnages (the Congo 5,760 tons by 1935 and Ruanda-Urundi 1,000 tons). The Belgians even anticipated that they could reach 12,000 tons by 1936. The Committee reluctantly accepted in April-May, 1934, the adhesion of the Congo and Ruanda-Urundi on a flat tonnage of 4,500 tons for 1934, 6,000 tons in 1935 and 7,000 tons for 1936 with a rise when the export quota was above 65 per cent. Portugal joined early in 1934 on a standard of 800 tons and a flat rate of 650 tons a year.[25]

This marked the limits reached by the control system. The largest producers still outside the Committee were China, Burma and Australia. In September, 1933 the government of Yunnan was approached for membership on the basis of a standard production tonnage of 11,000 tons a year and a flat rate of 8,000 tons. In July, 1934 Yunnan was understood to want a flat rate of 10,000 tons a year; the Committee accepted this last figure but negotiations then faded into oblivion. In the case of Burma the Committee offered a standard tonnage of 3,850 tons and an initial flat rate of 3,080 tons. The Tavoy Chamber of Mines was willing, but not the India Office. The government of India was anticipating rising production in Burma; wolfram prices – so closely tied to tin – were moving up; and the Burmese Assembly felt strongly on the restriction of output. The Committee was forced very reluctantly in January, 1934 to accept a higher basis – an export of 3,850 tons; but in August, 1934 the India Office told the Committee that, although new tin prospecting licences would be refused for Mergui and Tavoy, Burma would not come into the control scheme. Australia was not interested.

The agreement through 1934-36

The second agreement had a relatively easy life. In its three years world consumption of tin swung out of its slump back to the level of the middle 1920s. Almost all countries of any importance in tin production were within the international control scheme and the production of tin was allowed to move upwards at a rate which was rapid and, indeed, in the later years was in excess of consumption. World stocks fluctuated, but not abnormally, and were at the highest no more than 22,000 tons and at the lowest as high as 11,000 tons; neither figure represented a disturbing factor in the world situation. Over the three years total world production at 439,000 tons and aggregate world consumption at 425,000 tons were near enough in balance. Price fluctuations were small. The movement from maximum to minimum daily prices within each of the three years was relatively small and did not show signs of instability until 1936; the range in annual averages from year to year was narrower than in any corresponding post-war period except perhaps the three year period 1925-27. The stocks of the international tin pool ceased to be a factor by the middle of 1934; the Committee's own buffer stock was present over part of 1935, but not as a force of importance. The back of the economic crisis had been broken in 1932-33. The main problem of 1934-36 was to see how far and at what speed the tin control scheme could adjust itself to an upward market.

The second agreement started January, 1934 with a quota rate of 40 per cent. It was difficult, however, to maintain much longer the subordination of the producers' direct interests to the interests of the pool. A proposal to the April, 1934 meeting of the Committee for a 45 per cent quota, supported by Bolivia and

Nigeria, was unacceptable to the Netherlands East Indies who would not agree to a rise in the quota until the Committee's own buffer stock scheme, then under consideration, had been definitely agreed. But in May the Dutch changed their attitude and a quota of 50 per cent was agreed for the six months from 1 April, 1934.

In August, when the quota from 1 October, 1934 came up for decision, circumstances and the Dutch position had changed again. The international pool had sold its stocks so that the world visible stock at 15-16,000 tons was absolutely normal. The Committee now had its own buffer stock, as well as control of exports. In such stabilising circumstances the stocks henceforth required as normal by the world would be only about 12-14,000 tons. To enforce a reduction of world stocks to this level the N.E.I. requested that the export quota from 1 October should be reduced to 40 per cent. Malaya thought that the statistical picture did not support a quota reduction. Lowinger thought (and Bolivia agreed) that the quota should be fixed independently of the buffer stock needs. Malaya, Bolivia and Nigeria offered to accept a 45 per cent quota, but the Dutch insisted on their 40 per cent. This figure was fixed as the quota for the last quarter of 1934. It was to be continued for the first quarter of 1935.

The quota system, however, was not starving the market of tin. The price was the clear indicator of this. It was being maintained through 1934 with a stability greater than at any time for the last quarter of a century and at an annual price level which was much the same as the average level for the fifteen years 1915-29.

Rising quotas in 1935

The work of the Committee had been successful enough over the whole year 1934 in balancing world production at 120,000 tons against world consumption at 125,000 tons. It had also produced a not unreasonable tin price of £230. Many members of the Committee had an uneasy feeling that the swing up in consumption would have to be met by easing the quota system, although curiously this upswing in consumption went hand in hand with a price that was "dead". In February, 1935 the Committee considered contradictory versions of the likely statistical position,[26] but could not settle the conflict between those who believed in the use of the buffer stock or in the use of the quota. The Dutch accepted an increase in the quota to 45 per cent for the quarter April-June, 1935 only most reluctantly and under pressure.

This relatively low quota may have sustained the price, but certainly did not cause it to rocket; it had an effect in bringing in June an unusual visitor to the Committee. This was Sir William Firth, representing the Welsh Plate and Sheet Manufacturers Association (the U.K. tinplate interests).[27] He explained the tinplaters' views on the price. That price had increased by £115 since 1931. The British tinplate industry took 9-10,000 tons of tin a year, and the rise in the tin price since 1931 had added an additional 38s. 6d. a ton to the cost of producing tinplate. The high price of tin had undoubtedly led to a contraction in the demand or, at the best, had prevented the normal expansion of that demand. £230 a ton was much too high a price for tin; the policy of the Committee in maintaining the price at that figure was unwise in its own interests; his Association would use all its influence to resist the stabilisation of the price of tin on that high level.

He suggested, as a reasonable profit for producers, a price figure in the region of £175 a ton of tin.[28] The Chairman of the Committee (Sir John Campbell) pointed out: "The tin control scheme did not set out to fix or regulate the price of tin. Its primary object was to adjust production to consumption". In this spirit, the Committee noted the standing statistical lag of production behind consumption, noted the Dutch policy to meet part of the shortage from the buffer stock, but lifted the quota to

[26]The Statistical Office in The Hague foresaw a consumption in the first half of 1935 of 62,000 tons; Lowinger (Malaya) foresaw 60,000 tons; a sub-committee foresaw 54,000 tons as a pessimistic estimate. The actual result for the twelve months of 1935 was 146,000 tons.

[27]John Hughes (U.S.A., with a nomination concurred in by the United States Steel Corporation, the American Iron & Steel Institute and the National Lead Company) seems to have been the first consumer representative to attend an I.T.C. meeting in November, 1934.

[28]42nd meeting of the Committee, June, 1935.

[29]It was known that Bolivia was already finding it difficult to live up to a 65 per cent quota.

50 per cent for the third quarter of 1935 – an increase that would fill about half the calculated shortage.

The Committee, absorbed in its buffer stock arguments and operations, was now being caught by surprise. By August, 1935 the statistical evidence of the recovery of the industry was beyond dispute. The Statistical Office believed that during the first quarter of 1935 world production had been running 2,600 tons a month below consumption and in the second quarter 2,800 tons below. World stocks had fallen as low as 12,500 tons and the tin price would have risen far above its August monthly average of £233 but for the activity of the new buffer stock. The quota of 50 per cent, already agreed for the third quarter, was retrospectively raised to 65 per cent. This new figure was repeated for the fourth quarter of the year on the assumption that the result would be an increase of stocks to a fairly normal level of around 19,000. This linking of the quota to a concept of the right stock level was not working. Consumption was still being under-estimated – as for many previous quarters – by the Committee.

The new quota level seemed to be merely balancing production and consumption. The buffer stock was being liquidated through September and it would have been a political embarrassment to the Committee if its buffer stock experiment were to fizzle out in a wave of speculation based on inadequate quotas of production. In early October, 1935 the Committee again had to face the question of retrospective action. It agreed a quota of 70 per cent operative from 1 July. In late October a quota of 80 per cent was adopted from 1 October.[29] These re-considerations of the quota figures were no tribute to the soundness of the statistical assessments or to the degree of stability which the Committee was giving to the producing industry. The process of re-adjustment was to continue. In mid-November the Statistical Office estimated an 80 per cent quota as adequate to give an increase in world stocks; a fortnight later it felt that the quota needed to be increased, even although the Netherlands Indies had provided about 1,500 tons of tin in October-November as a special export over and above the quota. In early December the Committee raised the quota to 90 per cent for the first quarter of 1936. In February, 1936 in the light of some increase in stocks and of a fall in the price over the last three months, the quota for the second quarter of 1936 was reduced to 85 per cent.

The Bolivian under-exports, 1935-36

The quota figure was now, however, in some respects unreal. In general, under the control schemes and until the middle of 1935, the participants in the scheme exported in total all the tin granted to them by the quotas. But, in the third quarter of that year, Bolivia began lapsing on her quota. In the first quarter of 1936 Bolivian exports totalled only half her quota and by the end of that quarter she had accumulated a total deficit of over 8,000 tons. The Bolivian industry had been badly affected by the loss of skilled mine workers in the Chaco war. By June, 1936 the Bolivian under-export had risen further to 10,000 tons. The control agreement had made no provision for the problem of under-export (and, indeed, even the post-war tin agreements shied away from a serious solution to the problem). If the quotas were correctly assessed, under-export on any scale would create a shortage and therefore a price rise; and it was perhaps asking too much from the mining industry to penalise a participating country which was producing this happy result or to transfer to other members the right to export, at the expense of the price, the tonnages so unexported. The point was currently sharpened since the tin price, in spite of the Bolivian under-export, was continuing to fall.[30]

In July, 1936 Bolivia (Sr. Martinez Vargas) suggested that 5,000 tons (that is, half the total accumulated shortfall of Bolivia) be transferred to other countries. This ingenious proposal would mean that the world on a 75 per cent quota would, with the surrendered tonnage, produce as much tin as on the 90 per cent quota for the third quarter of 1936 (which the Nigerians and

[30] At the June, 1935 meeting of the Committee, Sir William Firth said that the tinplate companies were investigating the more economic use of tin, including the electrolytic deposition of tin on tinplate.
[31] At the 52nd meeting of the Committee, 6 July, 1936.
[32] 4,467 tons was allocated to the other three signatory countries and 576 tons to the other four adhering countries (including Siam).
[33] Gutt at the 56th Committee meeting, 11 December, 1936.

Malayans had pressed for and the Dutch had resisted in June). In a comment on the method of explaining this to the market, M. Gutt (Belgium) pointed out laconically: "The Committee must remember that the tin market was not perhaps a highly intelligent organisation; it was, however, highly sensitive".[31] The Committee in July, 1936 accepted a compromise. Bolivia renounced the right to export all her accumulated arrears now standing at 10,288 tons (half at once and half for the third quarter of 1936).[32] The quota for the third quarter of 1936 was fixed at 90 per cent for the members other than Bolivia, which was limited to a 75 per cent quota. This did not solve the Bolivian problem. By the end of August the country was still running into fresh arrears (over 2,700 tons) after her clean sheet. In spite of this shortfall, the quota for the fourth quarter of the year was left unchanged (at the October, 1936 meeting). Not unnaturally in face of a rising consumption, this attempt to stabilise the export quota promptly pushed the tin price up, especially when the new control agreement, including Siam, was declared almost ripe for signature. In early November the market price had jumped to £235. The Committee foresaw a runaway price and promptly reacted by increasing the quota, without formality of a meeting, retrospectively to 105 per cent from 1 October. It had, temporarily at least, adopted a much more expansionist policy. It did not disagree with M. Gutt when he made a point which was to become even more important in the post-1956 agreements – that the Committee had no effective machinery which would enable it to increase stocks rapidly and so reduce a price which might be regarded as excessive.[33] It took his advice to aim at too high rather than too low a quota, and fixed the quota for the first quarter of 1937 at a level of 100 per cent on the new and higher standard tonnages of about 200,000 tons due to operate under the third agreement.

The international buffer stock, 1934-35

The tin pool had started to unload its stocks in the middle of 1933 and might have been entirely out of the market as a supporting influence on prices when the second control agreement was operating. That possibility concentrated the minds of the Committee. In December, 1933 John Howeson prepared for the Committee a memorandum on the need for a buffer stock.[34] Under the new agreement, without a tin pool but with stocks reduced to normal, the market might be open to violent attack and to cornering. The efficacy of the quota system required two inseparable pre-requisites: "(a) the maintenance of marginal stocks and (b) their being held by interests which are either co-operating with or controlled by the Committee or the signatory governments". If there were no stock surplus, the market would be at the mercy of speculators who could, in spite of an equilibrium of production and consumption, force up the price. In that event the Committee's only remedy would be an increase in the quota. But such a reply would be to destroy the fair and reasonable relation between production and consumption and thus one of the basic objectives of the control scheme. The necessary time-lag in expanding production would tend to provoke the oscillation in price which it was the second objective of the scheme to prevent. If surplus stocks were permitted to accumulate control of the situation would pass to the holders of such stocks. "Normal" world stocks should be determined at an agreed figure (say, 15-16,000 tons). The quotas under the control scheme should be regulated so as to bring stocks down to normal within, at most, six months. There should be built up a buffer stock. A tonnage (not less than 8,281 tons or 5 per cent of the total standard tonnages) should be taken as the basis for that buffer stock. This should be raised by a special production quota. The stock should be at the disposal of the I.T.C. to be used as an adjunct to the tin control scheme. The sole purpose of this buffer stock would be to prevent rapid and severe oscillations in price. It should be made self-regulative by the fixation of a basic price (invariable or subject to quarterly review).

Sales could begin at a figure above and purchases at a figure below this basic price, possibly with quota increases or decreases becoming operative at further similar price intervals, upwards or

[34]In a printed memorandum dated 7 December, 1933, and circulated to the 31st meeting in December, 1933.
[35]I.T.C. communiqué after the 33rd meeting, April, 1934.

downwards. Each producer would be offered his share of the quota for the buffer stock. The local government should, on the delivery of such buffer stock quota tin, pay the producer a certain amount per ton. The accumulated balance at the end of the operation, whether in cash or in tin, should be distributed pro rata amongst the original contributors.

In April, 1934 the Committee approved a fairly detailed draft scheme, accepting some of Howeson's principles. It stated publicly:[35]

"The Committee are agreed that a buffer stock of tin is essential to the interests of consumers and therefore in the long-term interests of producers. Such a stock should prove a powerful influence in discouraging speculators and in preventing undesirable price fluctuations which are not based upon any major change in the statistical position of the industry. Recent experience also suggests that the existence of adequate buffer stocks will ensure reasonable stability of prices".

Thailand raised objections. She was not prepared to say whether or not she would join the buffer stock, but the scheme was agreed in June, 1934.

The buffer stock was operated by a committee of four members (the Chairman of the I.T.C. was to be chairman of the buffer stock) representing the four signatory governments. The buffer stock committee was to act in accordance with the general instructions of the I.T.C. These instructions included the following:

"1. The primary objective of the Buffer Stock Committee is to assist the International Tin Committee in its endeavours to prevent rapid and severe oscillations of price, and all operations should be directed towards that end.

"2. The general policy should be to give as much information as possible to the International Tin Committee and, through it, to the producers who have provided the tin in the buffer stock, consistently with the primary consideration that no information should be given which would interfere with the satisfactory working of the buffer stock scheme. The International Tin Committee recognise that secrecy as regards the sale or purchase of tin will normally be essential to success.

"3. (a) The buffer stock shall consist of 8,282 tons of metal divided among the four signatory governments in proportion to their standard tonnages . . .

(b) If participating or adhering Governments or Associations desire to contribute to the Buffer Stock the terms and conditions . . . and the tonnage . . . shall be determined by the International Tin Committee. Such tonnages shall be in addition to the total of 8,282 tons . . .

"4. As often as may be necessary, and normally once each week, the four members of the Committee shall . . . decide upon their course of action . . . All such decisions must be unanimous . . .

"5. The proper functioning of the Buffer Stock Scheme requires that the Committee shall have at their disposal at all essential times a reasonable sum in cash, and the operations should be conducted as far as possible in such a manner as to secure this . . ."

The buffer stock committee had some initial trouble. To safeguard the individual position of its members it was proposed to form a limited company (Ibstock Ltd.) to deal with all the commercial transactions entered into by the buffer stock committee and to act as a corporate body capable of suing and being sued. But this suggestion raised the prospect of British income tax being applied to the profits of the buffer stock and the committee accepted the offer of Mr. McKenna (of the Midland Bank Ltd.) to act as trustee for the scheme, with the Midland Bank being the Committee's executive agent to carry out the Committee's decisions as to the operation of the stock.

The buffer stock was raised by a special quota on the standard tonnages. By the end of February, 1935 it had 7,467 tons of tin. Almost immediately it was suggested to use it on the market. Malaya and Nigeria proposed an increase in the export quota to 55 per cent for the second quarter of 1935; the Dutch desired to maintain a lower quota but to meet any unexpected rise in con-

---

[36]Report appended to the minutes of the 42nd meeting of the I.T.C., June, 1935. The blank prices are in the original.

[37]Contributions were Nigeria 544, Malaya 3,597, N.E.I. 1,816 and Bolivia (which was not fully paid up until July) 2,325 tons.

sumption by sales from the buffer stock. The quota was, in the event, lifted to 45 per cent but the buffer stock committee thought it desirable to reach an agreed policy on sales and prices. It agreed:[36]

(i) tin should be sold at not less than £   per ton until one-half of the stock had been sold;

(ii) tin should normally be sold, up to the amount permitted, to prevent the price rising above £   ;

(iii) tin should be bought, if cash were available, to prevent the price falling below £   ; and

(iv) sales or purchases should be made on the market and not by private arrangement with the consumer or with any particular dealer.

By April-May, 1935 the London price had moved up to £236 a ton and visible stocks were beginning to decline. The price level for buffer stock selling was reached. In May the chairman of the buffer stock committee was authorised to sell spot tin and to buy forward tin with the proceeds in order to ease the backwardation on the market. By the beginning of July the buffer stock had sold 3,492 tons of cash tin and had bought 1,240 tons of forward tin. Sales in the U.S.A. of the Malayan tin contribution were disappointing and most was transferred to London. By the beginning of August the stock held 3,440 tons. Heavy sales of 2,100 tons were made in August and the balance was sold in September.

The total contributed to the buffer stock was 8,282 tons.[37] The total sold (including 1,242 tons of tin that had been bought) was 9,487 tons. Most of the stock was contributed in the first two months of 1935 when the London tin price averaged £231; early sales to May were at £230-232 for cash tin; the later sales over June-September were when the London cash price averaged £228; and the average price realised on all sales was £226. The profits, if any, cannot have been important.

It was true that during the period of the building up and sale of the buffer stock, say, the first three quarters of 1935, the tin price was remarkably steady (ranging monthly between £216 and £234). But in the previous 12 months without a buffer stock the price range had been even narrower (£227 to £237). The total tonnage in the buffer stock was less than three weeks' of world consumption. It is probable that the buffer stock did little more than help a market position in the middle of 1935 which was being solved by the easing of the control quota in the autumn of that year.

By no means all producers, however, had been converted to the buffer stock. In June, 1935 V. Lowinger (of the Malayan delegation on the I.T.C.) attacked the inclusion of a buffer stock in the control agreement. He said: "The buffer pool agreement runs until the end of this year (1935), unless it is decided to prolong its life, but I think I am on safe ground in saying that the difficulties of operation are such that there will be no general demand for its prolongation. Its machinery is too complicated and the difficulty of arriving at rapid and necessarily unanimous decisions so great that it has proved in practice an undesirable complication of an otherwise elastic scheme ... I believe that the buffer pool will cease to exist at the end of this year and that it is unlikely that any other official pool will be proposed ... I believe that by keeping reasonable reserve stocks on the mines and by the operation of the quota the production could be adjusted to consumption with sufficient rapidity to prevent the violent oscillations of price that have in the past characterised the tin market."[38]

The chairman of the Tin Producers' Association (Sir George Maxwell) had resigned in 1934 against the acceptance with the second agreement of a buffer stock that would serve only to perpetuate production restriction. The Perak Chinese Mining Association (one-quarter of the Malayan output) had unanimously, and the F.M.S. Chamber of Mines by a majority, rejected the proposed second agreement in 1934 on a variety of grounds (export quotas would have to be restricted to enable the buffer stock to make a profit, high prices would stimulate outside producers, the scheme would alienate consumers). Above all there rankled the profits which the private pool had made side by side with the quota system in 1931-34. The Malayan government was perhaps wise in not going forward with a proposed referendum on the second agreement.[39]

[38]Quoted by the Malayan Chamber of Mines three years after.
[39]See Yip Yat Hoong, *Development*, pp. 223-224.
[40]Lowinger (Malaya) at the 48th I.T.C. meeting, February, 1936.

The third control agreement, 1937-41

The third control agreement ran through 1937 to 1941. It maintained the objectives of relating production to consumption and of preventing rapid and severe oscillations in prices. But it was clear that the attempt of the Dutch under the second agreement to tie the executive and quota-fixing activities of the Committee to a particular level of stocks had led to considerable arguments, had been unsuccessful in result and had been resented, and the third agreement confined itself to the concept merely of "maintaining reasonable stocks". The changes in its machinery were not important, but the unanimity rule for decisions of the Committee was ended and replaced by a voting system (a total of 20 votes) divided only very roughly on the basis of the standard tonnages.

The first problem was an old problem – that of universal comprehensive membership and the price to be paid for it. None of the original members required to be convinced of the desirability of a new agreement, but some of them jibbed at continuing to pay the high privileged price on which both Siam and the Belgian Congo had been bought into the second agreement. Between March, 1931 and the end of 1935 the quotas of the four original signatories had averaged only 50 per cent of their 1929 tonnages, but in the same period the output of Siam remained at 100 per cent of its 1929 tonnage, and the output of other countries had risen from 11,000 tons in 1931 to 24,000 tons in 1935. Siam was now claiming, on the new basis, a standard tonnage as high as 25,000 tons. The Congo had produced in 1931 1,000 tons and in 1935 over 6,000 tons; it now wanted a standard tonnage of 7,000 tons.[40] French Indo-China, even in the favourable circumstances of 1935, had been unable to produce more than half the new standard tonnage it now claimed. Malaya was prepared to offer generous standard tonnages but insisted that these countries should be subject to the same quota percentages as the original signatories.

A delegation of the Committee visited Bangkok. The negotiations with Siam were between a weaker and a stronger party; Siam could afford to stay outside the agreement. The delegation

offered Siam a standard tonnage of 18,000 tons subject to the
restrictions applicable to all participants and without any
guaranteed minimum, or a standard tonnage of 16,700 tons with
a minimum of 10,000 tons; Siam wanted 20,000 tons standard
tonnage with a guaranteed minimum of 15,000 tons. Each party
rejected the offer of the other. After the delegation left Bangkok,
the Siamese offered to take 20,000 tons standard tonnage with a
guaranteed minimum of 14,000 tons; the delegation counter-
offered 18,000 tons standard tonnage and a minimum of 10,500
tons. At the meeting of the Committee the Siamese (through
Phya Rajawangsam) justified their claim. The flat rate quota
which Siam had accepted in 1931 had been accepted purely as a
temporary figure to avoid delay in becoming a member of the
control scheme. At that time an accurate figure of Siam's true
capacity could not be furnished, partly because many dredging
companies had just entered into production, partly because other
companies had temporarily suspended production. The claim
now made was merely an equitable adjustment to give Siam what
she should have received in 1931. In fact, he maintained, Siam
had carried an undue share of the burden of control. He elabo-
rated on the build-up of the new claim. In 1931, prior to restriction, Siam's production had been at an annual rate of 14,221
tons. The output of 15 dredging companies in 1931 had been
3,576 tons. Many of these companies had shown higher outputs
in the past and were therefore assessed on past performance.
Some had not yet reached full capacity or were not working
full-time in 1931; these had to be assessed in relation to capacity.
The 15 dredging companies had therefore now been assessed at a
capacity of 6,272 tons. Seven other dredging companies, not
operating at all in 1931, still had to be granted assessments.

[41]Tin concentrates in Siam were assessed for tin content on the traditional but
incorrect basis of 72 per cent tin. Siam desired to continue this traditional
basis. The true tin content was about 74 per cent. A standard tonnage of, say,
18,500 tons of tin a year on the Siamese calculation would have meant about
19,000 tons a year on the true assay basis.

[42]In the first three years of the third agreement the maximum annual output of
the Congo and Ruanda-Urundi was 9,669 tons in 1938; but output jumped
under maximum production to 12,253 tons in 1940. With Siam production,
with or without effective quotas, at no time reached anything like the capacity
figure quoted by that country; maximum production was 17,116 tons in 1940.

There was to be added a small tonnage produced under washing (dulang) licences; more important were the potential producers, with a capacity of 5,000 tons a year, who were entitled to assessments in view of the fact that they were taken unawares when restriction was introduced. On these arguments the true capacity of Siam in 1931 was 25,500 tons.

This curious sample of Siamese reasoning was rejected by the Committee. Siam then offered, as a last and final compromise, a standard tonnage of 19,000 tons with a minimum guarantee of 12,500 tons – an offer again rejected by the Committee. M. Picard (France) suggested a compromise standard tonnage of 18,500 tons for Siam. The Netherlands East Indies was stubborn. The offer to Siam was too high; the N.E.I. was prepared to repeat the Committee's offer (18,000 tons as standard tonnage with 10,500 tons as guaranteed minimum), but now with conditions attached. The true assay basis should be used for Siam;[41] Siam should become a partner in the international research scheme; no special privileges should be attached to Siam; and the control scheme should be officially accepted by the Siamese parliament.

The beginning of the new control agreement was too near to tolerate more haggling. Siam was accepted on the basis of a standard tonnage of 18,500 tons on a 72 per cent assay basis and a guaranteed minimum of 11,100 tons. The difference between these nominal figures and the figures on a true assay basis (that is, a standard tonnage of 19,142 tons with a minimum of 11,485 tons) would be made up in a separate agreement by a surrender of tonnage by Bolivia and the Congo. The Siamese had won not the argument but certainly the prize.

In the case of the Belgian Congo, there was a very much shorter discussion. The Congo was allotted a standard tonnage of 13,200 tons. It was clear that this tonnage was not likely to be reached in the early years of the second agreement and the Congo was allowed a special carry-over of unexported tonnage to cover this position.[42]

French Indo-China entered the new agreement with a standard tonnage of 3,000 tons and a guaranteed minimum of 1,800 tons. But Portugal was unwilling to continue to adhere unless it were guaranteed an effective quota minimum of at least 1,000 tons a year (a figure twice its recent rate of export); the Cornish companies could not agree on the continuation of the figures of the second agreement; a Committee approach to China and Burma had no results. With a coverage in 1937 of 86 per cent of the world production the control agreement was still comprehensive enough.

Fixing the quotas, 1937-41

The third agreement started in a mood of strong optimism and was to remain in that mood for a whole year. World tin consumption was bounding ahead to reach in 1937 the highest figure yet recorded or indeed to be recorded before the 1970s. The yearly price for 1937 was not extortionate but was the highest for ten years. World visible stocks were constant but not high. The Committee was not initially concerned with the principles and need for a buffer stock – a form of interference with the market to which many of the Malayan producers were hostile. The scent of war and of the expansion of the tinplate industries everywhere to create stocks of canned food was in the air.

For its first quota (the first quarter of 1937), the Committee accepted a rate of 100 per cent or nearly 200,000 tons. This was generous, since the majority of the Committee believed that world consumption for the year would run about 180,000 tons and was well aware that "outside" producers would add perhaps another 25,000 tons a year to the insiders' tonnage.

In fact, the new standard tonnages had been pitched too high. By the end of the first quarter of 1937, Bolivia was again unable to live up to its quota and was producing only half of what it was officially called upon to produce. In lesser degree every other member, even including Malaya, was in arrears with exports and for the quarter as a whole a permissible export at the annual rate of nearly 200,000 tons became an actual rate of about 156,000 tons. The market did its calculations, took account of the tonnage likely to be available from "outsiders", noted that visible stocks were below the level equivalent to 17 per cent of current world consumption, noted that stocks which were stationary in relation to a rising world consumption were therefore moving down in real value, and began to lift the price of tin. In February, 1937 Trench, editor of *The American Metal Market*, cabled asking the chairman of the Committee to permit unrestricted export of tin for at least two months. The chairman agreed that, unless the Committee did something urgently, the tin price might rise out of control. This might have undesirable political results at a time when the League of Nations had commodity control schemes under review. He asked, by letter, in February, authority to fix the quota for the second quarter of 1937 at 100 per cent; the Netherlands Indies objected (apparently to the principle of his acting on such authority); but the Committee meeting (5 March, 1937) accepted his proposal. The prophecy at that meeting of the consumers' representative (Sir William Firth) that " if the existing quota of 100 per cent were only maintained, the price would go up; however, in present conditions, prices would go higher in any event" was realised. On 17 February the London price was £231; on 5 March £256; on 9 March £277; and on 15 March £311 – all this scarcely a tribute to a scheme to prevent rapid and severe oscillations of price. On 9 March the chairman of the Committee, again by letter, asked for authority to raise the quota to 110 per cent and by 12 March was authorised to act accordingly. On 18 March the Committee was asked by the chairman, in the light of the political aspect of the examination of the control agreement, to approve a period of unrestricted export. But neither the Netherlands Indies nor Malaya thought that there would be much difference in results between increasing the quota to, say, 110-130 per cent and a state of unrestricted export, and Captain Lyttelton, Mr. Groothoff, Mr. van den Broek and Sr. Antenor Patiño raised the question of a buffer stock into which any over-production on an increased quota could be diverted. The Malayan government desired an immediate increase of the quota, perhaps retrospective to the beginning of 1937, but the quota of 110 per cent for the second half of 1937 was left unchanged.

The state of under-export continued to be lamentable. In June, 1937 Lowinger (Malaya) asked for a quota of no less than 125 per cent, but the quota of 110 per cent for the third quarter of the year was merely repeated.

The control scheme in 1937 had so far failed by a very considerable figure to meet the world's consumption requirements. For a period of six months (April-September) the monthly average tin price had been above £250 a ton, a price which the consumers' representative on the Committee regarded as much too high. This price was now stimulating production and misleading the Committee. The short-lived consumption boom of 1937 was turning down. The decision of the Committee in September, 1937 to continue the quota for the last quarter of the year at 110 per cent was a mistake in optimism which had to be reconsidered at a special meeting requested by the Bolivians in late October. The price, said Antenor Patiño, was falling rapidly (by which he meant that the October average price would be below £250 a ton); he proposed a substantial reduction in the quota, retroactively from 1 October with measures (the creation of a syndicate or a reservoir of tin or a buffer stock) to be taken to give a stabilising element to the price. Pearce (Bolivia) thought in terms of a quota reduction to at least 85 per cent. Sir William Firth was stirred to speak with some bitterness.[43] He "felt some difficulty in the matter because, so far as he could ascertain, no one paid any particular attention to what he said". Now that world prices had fallen, the tin price had also fallen. The Committee now wanted to climb back, if possible, to a price of around £250-260 a ton. A price of £260 a ton was unquestionably too high; a price of £225 would undoubtedly give producers a very handsome margin of profit. On former occasions when he had discussed the price question he had been told quite politely that that was rather outside his ambit and that the Committee was interested not in price but in controlling on the basis of equilibrium between production and consumption. He noted, however, that when the price of tin fell a special meeting was called as a matter of urgency. The Committee listened as politely as before

[43]62nd I.T.C. meeting, 25 October, 1937.
[44]Eighty per cent of the actual exports of 1937 or 80 per cent of the standard tonnages, whichever was the less. As members during 1937 had substantially under-exported against their quotas, with the exception of Malaya and the N.E.I., this would mean in practice a real export well below the nominal 80 per cent quota.
[45]There was in addition a quota in each of these periods for the contribution to the buffer stock.

and indicated publicly 95 per cent as the likely quota for the first quarter of 1938. But by the end of 1937 the price slide showed no signs of stopping. The December monthly price was down to £191—the lowest figure for a year. The boom was clearly over. In December Malaya proposed for the first quarter of 1938 a nominal 80 per cent quota[44] but a compromise was arrived at on a 70 per cent basis.

The control machinery was now finding it difficult to meet such rapid quota movements. The year 1937, when the Committee had pushed quotas to high nominal levels, had seen many members, perhaps fortunately for the overall stability of the tin position, unable to fulfil their quotas. In the year 1938 quotas were moved downwards so rapidly that by the end of the year they were only half of what they had been at the beginning and less than one-third of what they had been over the previous year. The problem now was to ensure that the tap of production could be turned off as quickly for 1938 as it had been turned on for 1937.

Turning off the production tap, 1938

The turn-round in consumption towards the end of 1937 had been dramatic. World production in the fourth quarter of 1937 had jumped to no less than 20,000 tons a month against a world consumption of only 14,500 tons, and total stocks in January, 1938, including smelters' stocks, were 43,000 tons as compared with 34,000 tons a year earlier. In February, 1938 M. Gutt (Belgium) proposed a drastic cut to a 60 per cent quota for the second quarter of the year; Groothoff (Netherlands Indies) supported him, subject to the Committee arriving at a favourable decision on the formation of a buffer stock; Antenor Patiño (Bolivia) wanted first 45 per cent and then 50 per cent. The result was a compromise 55 per cent.

This drastic cutting of supplies was justified by the price movement. The nine months' slide in prices by May, 1938 had reduced the price by over £100 a ton to £155 and threatened a return to the great depression of 1930-32. But, with a further reduction of the quota to 35 per cent for the third quarter of 1938 and with a continuation of that very low rate[45] until the end of

March, 1939 the tin price was turned upwards. There began in the middle of May, 1938 a slow, persistent but not abnormal rise in price. This rise by midway through 1939 was to bring the price back to a moderate enough level of £227; the price was then passed over for stabilisation to the hands of the buffer stock.

This relatively long-term fixing of the quota at 35 per cent took into account that, in fact, the reduced quotas did not necessarily fulfil their objective of reducing the tonnages actually exported. By February, 1939 over-export was fairly general; the total over-export had reached 7,500 tons, a figure which weakened the effect of nominal restriction on the price and which also affected the ability of the buffer stock to stabilise the market. The most persistent offender on over-export (as it had been on under-export in the previous year) was Bolivia. The Bolivian government was in a more than usually miserable financial position; it could not afford to reduce too drastically the tin production from which it drew royalties and, even if it had wanted to observe its obligations under the control agreement, it found it difficult to drive this point home to the smaller producers.

By June, 1939, with the tin price apparently quite tightly under the control of the Committee or at least of its buffer stock, the Committee (by the vote of the chairman) lifted the quota to 45 per cent for the third quarter of the year. The chairman's decision took into account the political need not to tighten restriction at a time when much outside political opinion thought the price had already been squeezed too high. But with the outbreak of war in September, the Committee showed how quickly it could act. On 1 September the quota of 45 per cent for the third quarter was changed to 60 per cent and the quota for the fourth quarter fixed at the same 60 per cent; on 12 September the third quarter quota was lifted again to 80 per cent and on 18 September to 100 per cent. Further action on 11 October raised it to 120 per cent with retrospective effect. The fourth quarter quota was raised to 70 per cent and again on 1 November to 100 per cent; the quota for the first quarter of 1940 was fixed provisionally at not less than 60 per cent. But in December, 1939 the Committee agreed to revert to 100 per cent for the last quarter of 1939, and to increase the figure for the first quarter of 1940 to 120 per cent.

The third control agreement ended effectively, although not nominally, in 1939. Its end was with some justification for its work. It had restricted output in both 1938 and 1939 but in neither year had the world market been starved of tin. In fact, in each year the world production, largely controlled by the Tin Committee, proved to be about one-tenth above consumption. The average price for the year 1937 at £226 a ton in London was admittedly double the starvation price to which tin had dropped in the great slump in 1931; but even the industrial developed societies were paying little more for their tin in 1939 than they had paid on the average over the preceding 20 years.

The international buffer stock, 1938-39

Through 1934 and 1935 the control scheme seemed to be justifying itself. In the two years the price had a degree of stability not seen for 20 years. Perhaps more important, that steadiness was at a price always well above £200 a ton. Visible stocks were steadily moving downwards. In such circumstances, a revival of the buffer stock idea could scarcely be expected to appeal to producers. But, in the spring of 1936, the price began slipping downwards until by the middle of the year it was well below £200. In these new circumstances the concept of a buffer stock (which in the minds of most producers meant a stock buying to push up the price) was revived.

In December, 1936, at the I.T.C. meeting, J. van den Broek (Netherlands) raised the question of a buffer stock. Experience, he maintained, had shown that the quota machinery was not sufficient to secure a diminution of the oscillation in prices. A solution might be found in a buffer stock or in the holding of stocks at various centres. At the January, 1937 meeting a special sub-committee was appointed. The scare on producers about prices had now passed and had been transferred to consumers. The price, in spite of almost unrestricted production, had jumped to £289 for April; this was the highest price since early 1927. For the moment both producers and consumers were alarmed. On the I.T.C. even A. Patiño considered that the recent rapid price rise had created a most dangerous position; if the current high quota of 110 per cent were continued it was absolutely essential that any excess production should be diverted to the building-up of a buffer stock. Dr. Groothoff (N.E.I.), Captain Lyttelton (Nigeria) and Gutt (Belgium), but not Calder (Nigeria), thought a long-term fixing of the quota and the creation of a buffer stock went hand in hand.[46] Sir William Firth wanted the Committee to have an adequate stock to control prices. A price range of £200 to £225 would be more satisfactory from the point of view of the consumer and from the long-term interest of the producer than were current prices (then £241 a ton). The normal consumer found it difficult to believe that the prevention of market oscillations was, in fact, one of the main objects of the tin control scheme; he, perhaps not unnaturally, thought the control scheme was primarily designed in the profit-interest of the producer.

The sub-committee on a reserve or buffer stock reported to the Committee in June, 1937. It approved the principle. There should be a buffer stock of 10,000 tons, to be collected through a production quota additional to the quota already in operation in accordance with the control scheme. The stock was to be at the disposal of the I.T.C. and should be held at the consuming centres. A manager of the stock should be appointed by the I.T.C. to operate on the general lines laid down by the Committee. The reasonable range of price in which the stock would buy or sell tin should be £30 a ton (that is, £15 on either side of a pivotal unstated price).[47]

Encouraged perhaps by the more favourable tone of references to buffer stocks at the meeting of the League of Nations Committee for the Study of Problems of Raw Materials and stimulated

[46]At the 59th I.T.C. meeting, March, 1937.

[47]At the 60th I.T.C. meeting, 11 June, 1937. Wilcoxson (Malaya) objected to the principle of a buffer stock as well as to the practice of the stocks being held at consuming centres (perhaps naturally as he was a smelter). Sr. Vargas (Bolivia) thought that buffer stock operations should be run by a sub-committee of the I.T.C. representing contributing governments and that the stock should be formed from current quotas. The Congo, N.E.I., Bolivia and France supported the buffer stock; the Malayan delegate thought the Malayan government would not accept.

[48]At 63rd I.T.C. meeting, 10 December, 1937.

certainly by a sharp drop in the tin price during the last quarter of 1937, the sub-committee (re-named the Price Stabilisation Sub-committee) reported again in December, 1937.[48] There should be a buffer stock. The object should be to reduce to narrower limits the large price ranges which had occurred in recent years. The price basis should be from £200 to £230. The stock would buy tin at a price below £200 and would sell tin above a price of £230; between those prices it would not operate. The price basis would be subject to adjustment up or down if there were a serious change in the U.K. Board of Trade price index. The buffer stock should be raised by a 7½ per cent quota additional to the existing quotas. The buffer stock should last for the life of the control agreement. No country could terminate the buffer stock scheme prematurely. The operating body should be appointed by the I.T.C. The initial amount of the buffer stock should be published but once it had started its operations were to be kept secret. The I.T.C. should not be informed of day-to-day operations.

Malaya and the buffer stock

The main obstacle to the acceptance of the buffer stock scheme was the strong resistance of Malaya. Many miners there had always been suspicious of tying the control scheme to a buffer stock. When in February, 1938 the U.K. Colonial Office asked for the comments on the latest buffer stock proposals, the Malayan Chamber of Mines raked up its old grievance in London – that the standard tonnage allocated to Malaya by the control scheme was far below her proved productive capacity; that as quotas were raised some signatories failed to produce their higher tonnages and that signatories which could have increased their output were not then allowed to make up the deficit of the others. In consequence, there followed a shortage of stocks and a time-lag in production at a time when consumption was rapidly increasing. In the eyes of the Chamber, the buffer stock fundamentally changed the objective of the control scheme "from that of adjusting production with consumption to that of price fixing. Price fixing postulates a static industry – the tin industry is not static". The pool was to last for the period of the control agreement; but if it failed it might create disastrous conditions within the industry. Malaya could in no circumstances agree to the cancellation of her right of withdrawal in the control agreement, a right which was the only safeguard if "outside" production continued to increase or if the buffer stock should fail. The Chamber doubted whether the buffer stock could find the right management or whether the secrecy of its operations could be guaranteed.[49]

But the initial Chamber reaction did not represent the views of all Malayan producers. A minority of the Chamber (claiming, however, to represent a majority of the tonnage on the Council of the Chamber) cabled its case separately to the Malayan government. They asked for a sympathetic consideration of the buffer stock idea and for a referendum of the whole of the Malayan industry. The claim for a revision of Malaya's standard tonnage was entirely distinct from the question of the buffer stock. The instability of tin prices during the latter part of the life of the pool (that is, the second half of 1935), which the majority in the Chamber regarded as a proof of the failure of the tin pool, was in fact in a period when the pool was being liquidated or had ended all market operations. The majority in the Chamber repeated their objections.[50] Only a reduction in supplies to a level below deliveries, it maintained, would raise the price of tin. The pool formed in August, 1931 had contained 21,000 tons out of a total world visible supply of 63,000 tons. There was no price correction until April, 1932 when supplies were reduced as a result of a lower export quota; the price had risen immediately, notwithstanding the big excess stocks outside the pool, and had continued to rise while stocks were being reduced towards a more normal level. The I.T.C. could have smashed the pool interests by merely increasing the quota of exports; and that was one of the chief reasons why the pool interests forced the pool on the control scheme against the wishes of the Malayan producers. If a

[49]Opinions conveyed to the government of Malaya in March, 1938.
[50]Cable of 12 April, 1938, from the Malayan Chamber of Mines to the High Commissioner for the Malay States.
[51]At the 64th meeting of the I.T.C., 18 February, 1938.

new buffer stock were formed, it was probable that unofficial pools would also be formed taking advantage of the quota machinery. If the pool interests claimed that the pool had controlled the price position between 1932 and 1935 there was nothing now to prevent those interests forming a new pool.

In the referendum in Malaya in May, 1938 the industry (with the Chinese-owned mines heavily in favour) voted for participation in the buffer stock.

The Netherlands Indies had become restless at the delay and in February, 1938[51] Groothoff told the I.T.C. that it was extremely difficult in present circumstances to fix the quota without reference to some stock stabilising scheme. The two were inevitably closely linked. If a buffer pool could not be constituted the N.E.I. warned all delegations that they must resume complete liberty of action as regards their general attitude towards tin control.

The Nigerian Chamber of Mines in March, 1938 approved the buffer stock scheme, as Bolivia and the Congo had done, but wanted the operating body to have some degree of latitude in varying the proposed buying and selling limits in order to discourage speculators. As late as June, 1938 the position of Thailand remained indeterminate.

The consumer representative on the I.T.C. raised a last objection. Mr. Todd (U.S.A.) approved a reasonable stability of price but objected to the pegging of the price, especially at so high a level as £200-£230 proposed for the buffer stock. He was supported unexpectedly by Groothoff who was instructed by his government to state that the proposed price range was too high and should be dropped to £180-£210; he could not sign the agreement on that too high price. Sr. A. Patiño retorted that he could not sign the agreement on any figure other than the £200-£230. These last minute comments did not prevent the agreement coming into effect.

Buffer stocks and the U.S.A.

The I.T.C. was aware of a larger, more important and more consistent objector. The United States government had bestirred itself on the question of the proposed price range in the buffer stock. It wrote to the British government[52] regarding its own interest in the cost of production, stocks and price of tin. The world supplies of tin above ground were still very small. The U.S.A. had a natural interest in the maintenance of adequate stocks for its defence and its industrial needs. If the purpose of the buffer stock were to make accessible larger total stocks of tin it would be welcomed. But the emergence of the scheme at this time, coupled with a further lowering of production quotas, suggested that its inevitable effect, if not its essential purpose, was to immobilise a large part of available stocks rather than to increase them. Further, if the scheme raised prices to £200 the liquidation of private stocks and of the excess stocks held under the scheme implied that quotas would remain low for a considerable period of time. The £200-£230 price range seemed ill-suited to the requirements of consumers over a long-range period; even the lower of these prices was substantially higher than the prevailing price (of £163). The letter trailed off weakly. The U.S. government believed that the proposed range was unwarranted by the present level of costs, but its information was not sufficiently complete to permit it to form a reliable judgement as to whether the price range might be considered a fair objective of production regulation, that is, one that would satisfy the general criterion of prices formulated at the World Monetary and Economic Conference of 1933. The U.S. government hoped —vainly, as time showed—for a full and comprehensive study of the costs of production of tin with a view to securing reliable indications of a reasonable price objective.

The stock in action, 1939

The buffer stock agreement, as an inter-governmental adjunct to the export control scheme, was opened for signature in June, 1938.[53] The signatories comprised the full membership of the control agreement (Belgian Congo, Bolivia, Nigeria, French

[52]Letter of 17 June, 1938.
[53]See a summary of the buffer stock agreement in I.T.S.G.: *Statistical Year Book, 1949*, p. 231.

Indo-China, Malaya, Thailand and the N.E.I.). No country not in the I.T.C. came into the buffer stock scheme and the buffer stock contained no consumer members.

The stock was initially fixed at 10,000 tons, lifted later by the I.T.C. to 15,000 tons in November, 1938. To provide the stock the I.T.C. fixed a special quota of export, additional to the normal export quota; and each producer signatory was entitled to contribute to the buffer stock *pro rata* to its standard tonnage. This additional quota was fixed at 10 per cent (the Dutch disagreeing both with the lift to 15,000 tons and with the 10 per cent). Contributions (slowly in the case of Siam) were completed by the summer of 1939 – a total of 15,611 tons from the seven signatory countries. During the building-up of the stock the tin price had kept within the price range of £200-£230 a ton applicable to buffer stock activities. Up to June, 1939 the buffer stock executive limited itself to purely tactical activity. The original stock had been contributed in tin metal (half in warrants for standard tin on an officially approved L.M.E. warehouse, about half in Straits tin represented by hypothecation orders for tin in warehouses in Singapore-Penang). The immediate objective of the buffer stock managers was to mobilise its resources in tin in the form most readily negotiable and most immediately effective on the market price (that is, as Standard Tin warrants) and to convert some of its metal holdings into cash by selling tin in the East against simultaneous purchases of Standard Tin in London or by large exchanges direct with London dealers. A second objective was to enable the L.M.E. to work more smoothly. As the buffer stock began to acquire greater control over the London market, a backwardation (the amount by which the cash price fell below the forward price) appeared which increased the cost of hedging, especially to the smelters, on the market. To ease this problem the buffer stock executive lent tin to the market at a charge of £2.50 a ton. The third objective – to turn some of the buffer stock metal into cash – began to be realised in the summer. In June, 1939 the market price rose to the buffer stock's selling point of £230 a ton. The buffer stock then became a free seller of cash tin. With the outbreak of war in early September, sales practically exhausted the buffer stock within a fortnight. The stock was replenished temporarily by the re-purchase of Standard Tin which the executive had previously sold forward. This acquisition of metal enabled the executive to control the market steadily at a price of £230 a ton through October and November, 1939 (at the temporary ceiling of the United Kingdom government).

By February, 1940 the buffer stock reported itself as being almost entirely in cash. In April, 1942 the I.T.C. agreed to liquidate the stock and in June-July, 1942 the balance of £2·3 millions in the buffer stock (which was additional to the £1·0 million previously distributed) was divided amongst the contributors.[54]

The total contributions to the buffer stock of 15,611 tons represented only about 10 per cent of world tin consumption for that year; but the buffer stock as an addition to the amount of tin available lasted for only about six months. It had a very important and stabilising result on prices. In London in the first half of 1939, in spite of the low quotas under the control scheme but mainly because of the knowledge of the creation of the buffer stock, the monthly average price of tin had moved only from £214 to £227 or by 6 per cent. In the second half of 1939, the war pushed up the New York price as high as 75.00 cents a pound, a figure which had been exceeded previously only in the war boom of 1917-18, but in London the buffer stock kept the price almost straight between July and December at £230, a figure very little above the annual average price appertaining over the previous five years.

The total purchases and sales (including exchanges) of the buffer stock amounted to 66,428 tons. The net tin price which it realised was £230.85 a ton. The London market price (monthly average) at the time the tin metal had been contributed had ranged between £194 and £218 a ton. The profit of the buffer stock had therefore been substantial but the amount saved to consumers by its activities in checking a war-inspired rocketing of the price had certainly been far in excess of the profit it had made.

[54]The amount due to Siam, then at hostilities against the United Kingdom, was handed over temporarily to the Custodian of Enemy Property.

The contracts with the U.S.A., 1940-41

The last great service of the International Tin Committee was, ironically enough, to provide the United States, which had so persistently damned the Committee for making tin scarce and dear, with massive quantities of cheap tin.

In May, 1940 the U.S. government expressed to the British Colonial Office its desire to obtain as much tin as quickly as possible. The request was passed on to the Committee. The tin would be bought on U.S. government account and would be held for seven years (subject to being drawn upon in emergency). If the stock were liquidated at the end of the seven-year period that would be done in an orderly manner and in consultation with the governments and producers concerned. Any tin beyond commercial demand would be bought by the government to keep prices at the present level (around 50 cents a pound). Lowinger and van den Broek negotiated and signed the agreement with the U.S.A. in June, 1940. This was the biggest tin deal yet negotiated in history. The U.S.A. through Metals Reserve Co. bought until June, 1941 all the tin available under the quota at 50 cents a pound.

By April, 1941 Metals Reserve was reported to have bought 60,000 tons and the government stock was 37,000 tons. In order to ensure that enough tin would be available the Committee agreed to raise the quota under the control agreement up to 130 per cent for the year from July, 1940.

In February, 1941 Jesse Jones of the U.S.A. asked for the continuation of the 130 per cent rate and in turn guaranteed purchases at the current rate and price for another year. In May, 1941 a contract on the same terms (including terms as to liquidation) was signed.

In September van den Broek and Lowinger were discussing an even wider agreement with the U.S.A. under which Metals Reserve would become the sole buyer and distributor of tin in the U.S.A. from November, 1941. It would buy all tin offered until its own stock was 100,000 tons; thereafter it would buy for export and consumption requirements. Before its stock target was reached it would consult with the Committee so that an appropriate control quota could be fixed. Differences between production and demand could be met by temporary releases from or additions to the stock and by the operation of a buffer stock pending the corrective influence of the Committee's quota. The price would remain at 50 cents. The agreement was signed by the Committee and the Americans during the month of October, 1941.

The over-running of Malaya and the Dutch-owned Netherlands East Indies early in 1942 made it physically impossible to enforce the production on which the American contract was based. The foundations of the American strategic stockpile had been, however, well and cheaply laid.[55]

[55]The question of the ultimate disposal of these American stocks always justifiably worried the Committee. In the June, 1940 contract, the U.S.A. agreed to hold the tin for not less than three years from January, 1941. The stocks could then be liquidated on written notice, but not more than 5 per cent (with a maximum of 5,000 tons) in any three months' period. Tin, however, might be released on the request of the U.S. government for its own use or for private consumers in the event of a national emergency. The agreement, or rather the extension of the agreement in May, 1941 was on the same terms.

The agreement of October, 1941 showed that the I.T.C. had done more thinking. Metals Reserve would buy all tin offered until its total stocks (including those bought under the June, 1940 and May, 1941 agreements) reached 100,000 tons. Thereafter, the Company would buy tin sufficient to meet requirements for consumption and export in such a way as to maintain its stock at 100,000 tons as closely as possible. The Company might at its own option fix its own stocks at more than 100,000 tons.

On the termination of the agreement the Company would continue to hold its stock of 100,000 tons for a period of three years from January, 1941 or for one year after the termination of Anglo-German hostilities (whichever was the earlier). The liquidation of the stock was not to exceed 5 per cent in any three months' period. In the event of a price rise, the rate of liquidation might be increased. The I.T.C. would fix the quota at a level necessary to enable Metals Reserve to reach a stock of 100,000 tons. During the period of the liquidation of the stock the I.T.C. would adjust export quotas so as to help to make orderly the process of liquidation. Metals Reserve and the Committee would consult regarding future demand so that the quota could be fixed accordingly. The Committee would consult with the U.S. government how best to effect suitable and adequate American consumer representation in the new control agreement then under discussion.

[56]Drawn up by Lowinger, Mills of the buffer stock executive and F. Burgess. The only major difference was that the appointment of the buffer stock executive should lie with a sub-committee of the I.T.C. consisting of the chairman, the vice-chairman and the Bolivian delegate.

The proposed buffer stock, 1941-42

The exhaustion of the buffer stock tonnages, the unrestricted output advocated by the International Tin Committee and the over-running of south-east Asia by the Japanese made the buffer stock scheme pointless. In December, 1940 the I.T.C. determined, however, to keep the buffer stock organisation in being and to consider a draft agreement for a continued scheme. It even discussed in December, 1941 such a draft scheme,[56] but the international control agreement, with which the buffer stock was so closely linked, had not been fully renewed and in April, 1942 the I.T.C. accepted the liquidation of the buffer stock.

The results of control, 1931-39

The control and buffer stock arrangements of 1931-39 had three main objectives – to regulate the relationship between production and consumption, to prevent severe price oscillations and to maintain reasonable stocks. They operated over a period which was initially an economic cyclone and later an economic boom; they covered a great divergence of low-cost and high-cost producers; they had relatively little disciplinary machinery with which to enforce the compliance of governments with their recommendations; and their members did not always agree on the choice of weapons. The most that could reasonably be expected from the International Tin Committee was that it would follow some of its objectives from time to time. It had variable success.

The control arrangements first enforced, almost by rule of thumb, a shortage of world production over the four years 1932-35. The sum total of that shortage was about 62,000 tons or one-seventh of world production. Most of that shortage of production was enforced, however, in one year—1933. But in the next four years, 1936-39, the control arrangements permitted a surplus production which reached an accumulation of 70,000 tons or one-tenth of world production. This surplus production was spread over those years much more equally than had been the shortage and in none of the four years was it less than 13,000 tons If the Tin Committee starved the market in the first four years, 1932-35, it certainly overfed the market in the second four years, 1936-39; and in that latter period showed itself able to meet, with ample capacity in hand, the highest figure of annual consumption yet reported—in 1937.

The industry showed a general better long-term relationship between production and consumption than had been seen in the pre-control 1920s or even than was to be shown under the International Tin Council between 1956 and 1971. In the nine control years 1931-39 average annual production and consumption roughly coincided at around 144,000 tons; in the ten years 1920-29 production averaged at 145,000 tons a little above consumption at 140,000 tons; and in the 16 years 1956-71 average production at only 155,000 tons a year lagged well behind an average consumption rate of 165,000 tons. The control of 1931-39 could claim that its worst over-production year (25,000 tons in 1936) was better than the pre-control record (36,000 tons in 1921) and little different from the worst that it was to be under the International Tin Council (20,000 tons in 1957). On the

TABLE 10
Control and no control, 1920-71
in long tons

Period	Production			Consumption		
	Lowest tons	Highest tons	% range	Lowest tons	Highest tons	% range
1920-29 (a)	116,000	196,000	69	80,000	189,000	136
1931-39 (a)	88,000	206,000	134	105,000	189,000	80
1956-72 (c)	116,000	192,000	62	136,000	184,000	35

Period	Annual prices			Daily prices		
	Lowest £	Highest £	% range	Lowest £	Highest £	% range
1920-29 (a)	159	296	87	139	419	201
1931-39 (a)	118	242	105	100	311	211
1956-72 (c)	735	1,554	111	723	1,715	137

(a) Free of export control and buffer stock. (b) Export control always and buffer stock often in operation. (c) International Tin Council always in existence, export control and buffer stock sometimes in operation.

~ther hand, the worst consumption excess under control (45,000 tons in 1933) was heavier than before the 1930s (8,000 tons in 1923) or under the Council (27,000 tons in 1960). In the 1920s the oscillation between highest and lowest annual productions was between a minimum of 116,000 tons and a maximum of 196,000 tons (or 68 per cent); in the control period from 1931 to 1939 between 88,000 and 206,000 tons – a range of 134 per cent. The range in consumption in the 1920s was between 80,000 tons and 184,000 tons or 130 per cent and in the 1930s between 105,000 tons and 189,000 tons or only 78 per cent.

In terms of price, it is probable from its history that the International Tin Committee, although full of short-term concepts particularly in relation to the buffer stock, had no agreed views as to a satisfactory long-term price on which to base a long-term policy for tin. It thought in terms of a year or even in terms of quarters, as have in practice almost all others who have had to have a hand in tin pricing policy. That the average price of tin under control (£196 over 1931-39) was one-sixth lower than it had been in the 1920s (£234 a ton) showed that tin kept under control was still cheap tin, but was perhaps irrelevant to the thinking of the control machinery. The range in price from year to consecutive year was more important. That range in the 1920s had been between 0·7 per cent and 79 per cent; under control in 1931-39 it was by no means as extreme (from 2·1 per cent to 43·6 per cent). On a comparison between the lowest and highest daily prices in each year, the range in the 1920s was between 22·8 per cent and 115·1 per cent and in 1931-39 between 10 per cent and 72·1 per cent, but the greatest fluctuation in the 1920s (115·1 per cent in 1920) was much wider than the greatest fluctuation in the 1930s. In that sphere the control years had some success.

As regards stocks the degree of success of control was measured not by the fluctuating volume of stocks listed in the statistics at any given moment but by the degree to which these stocks were under control or were running wild. The work of the international tin pool in 1931-34 and of the buffer stock in 1939 was undoubtedly successful in influencing prices, particularly in the latter instance in checking the sharp upward movement of prices.

The control arrangements had, however, overall only one general check on how far they served the interests of the tin producers who had formed them, who accepted the sacrifices and who expected a return. Their benefit lay in selling less tin at lower prices and more tin at higher prices. This the scheme enabled them to do. In the four early years of control 1932-35 the income of world miners from the production of the restricted amount of tin they sold was perhaps about £25 millions a year; in the four less restricted years 1936-39 the value of their higher production was perhaps of the order of £35 millions. Most producers were bound to regard this result as a fairly adequate justification for the control schemes.

VIII

The End of the International Tin Committee, 1940-46

THE Committee had little to do in the six years after 1939. Its loyalties were initially disrupted. Britain and France with Malaya and Nigeria were at war against Germany from September, 1939 but Belgium and the Netherlands were not until 1940. The Netherlands East Indies had perhaps no real sensation of war until they were attacked by the Japanese at the end of 1941. Siam was a neutral until it unwillingly joined Japan in 1942. Bolivia had no sentiments but depended on its tin concentrates being smelted in the United Kingdom and later in the United States. The Japanese attack fused the Committee but lost it its major producing areas.

The first work of the Committee was to maintain its system of quotas not really as a means of controlling production but in order to preserve the principle of allocation. It had to take growing notice of pressure from the U.S.A. In February, 1940 the chairman of the Committee, who had already raised the quota for the first quarter of 1940 to 120 per cent, told the Committee[1] that he had been informed of the views of the U.S. government. That body considered the Committee's estimate of consumption too low; it would regard a quota for the second quarter of 1940 which did no more than meet consumption as exceedingly dangerous, since there could now be no help in tin from a buffer stock without tin; and it expressed its hope for a quota at or near 100 per cent for the second quarter. The members of the Committee could scarcely be expected to accept cheerfully this advice from a U.S. adminstration which had recently been so strongly opposed to the Committee's activities. One of the less committed members (Houwert of the Netherlands East Indies) even wanted the quota of 120 per cent for the first quarter to be cut retroactively; he advocated a 70 per cent quota for the second quarter. Sr.

[1] 73rd meeting of the Committee.

Antenor Patiño (Bolivia) was prepared to meet the U.S. view-point by going as high as 80 per cent; Lyttelton (who had become head of the Non-Ferrous Metals Control in the United King-dom) urged that, although the anticipated demand could be met by a quota of 75 per cent, it would be wise to be liberal with a quota of 80 or even 85 per cent. The committee agreed on a figure of 80 per cent – an extremely tactless decision which, in the eyes of the United Kingdom and the U.S.A., exposed it to the charge of deliberately reducing the supply in wartime of an essential raw material.

By May, 1940 the Committee was aware of the desire of the U.S.A. to acquire as quickly as possible a substantial réserve stock of tin for the armament programme. In the light of this knowledge and to further this objective, the Committee first (in May) raised the quota for the third quarter of the year to 100 per cent and later (in July) modified this to a quota of 130 per cent to run for one year from 1 July, 1940. The same 130 per cent was decided also (March, 1941) for the second half of 1941. The fixation of the quota was becoming a formality, with perhaps dangerous undertones, which was nothing but a nuisance to the United Kingdom and the U.S.A. Quotas were solemnly fixed for the first half of 1942[2] but in September, 1942 the Committee accepted the prevailing circumstances and announced:

> "The policy of the Committee is to encourage during the war period the maximum possible production of tin in all ter-ritories not under enemy control. The quota has been con-tinued (until further notice) at 105 per cent of the standard tonnages.
> "The Committee have, however, decided that, if any territory should be able to exceed this amount, the quota will be increased, retrospectively and generally, in order to cover the actual production of that territory . . .".

[2] At 105 per cent, which, on the new standard tonnages operative under the fourth agreement, was much the same as the 130 per cent under the third agreement.

[3] The percentage quota of the Committee was based on standard tonnages reflecting actual production in 1929; the internal quota was related to tonnage capacity. The natural difference persuaded the Malayans that they were being victimised.

The second work of the Committee was to wind up formally the buffer stock.

This was completed in 1942.

The fourth control agreement, 1942-46

Consideration of the fourth tin agreement began as early as December, 1940. It was now generally accepted that the practice of the last ten years had proved the value of the international control of exports to all producers and there was no argument on principle.

There was, however, substantial argument over a long period on the details of the new organisation – the fourth control agreement – which was due to operate from the beginning of 1942.

The first argument was, as usual, on the assessment of the standard tonnages to be allocated to each participant. The original division on the basis roughly of actual production had been disturbed by the need to bribe "outsiders" to enter, by the persistent failures of Bolivia to observe her standard tonnages (either in shortfalls or in surpluses), by the insistence of Malaya that the standard tonnages understated both her production and her willingness to produce, and by the determination of the Netherlands East Indies that no concession should weaken her own comparative position.

In March, 1938 Malaya, irked by the difference between the rate of quota decided by the Committee and the rate applied internally to the Malayan miners[3], had proposed brightly that, to ease the labour situation, the Malayan mines might be allowed to produce an additional 2,900 tons for the second quarter of 1938, such tonnage being held in bond in the Malayan smelters and being deductible from following quarters. The Dutch objected and the Malayans reverted to the revision of the standard tonnages. The productive capacity of the Malayan industry was over 99,600 tons of tin compared with a standard tonnage of 71,940 tons in the international agreement and with an actual permitted export in 1937 of 77,542 tons.

Malaya and the Netherlands East Indies had been the only countries during 1937 fulfilling their quotas. A reduction in the quota was, therefore, a real handicap for Malaya, but for other countries was really a relief from pressure.

The Netherlands Indies did not object to a revision "provided that such (revision) takes place in favour of the N.E.I. and Malaya and that the other partners are prepared to make a sacrifice".[4] The N.E.I. had never accepted the 1929 production as a fair and satisfactory basis for the standard tonnages; it had, in fact, proposed the average of the eight years to 1929. The 1929 basis was advantageous to Bolivia and Malaya but not to the N.E.I. and in recognition of this an extra standard tonnage of 2,736 tons had been allocated to the N.E.I. during the course of the first agreement.

The Bolivians stated that the nature of their mines meant that reductions in the quota produced greater labour difficulties there than in Malaya. The recent over-production of tin was due entirely to the high quotas fixed by the Committee and Bolivia had ceded part of her standard tonnage temporarily to cope with a situation she had not produced. The Belgian Congo, safe in her guaranteed tonnage, was entirely opposed to a revision, but proposed a compromise to meet Malaya and the Netherlands East Indies without hurting in any way herself – that those countries which in 1937 had produced 107·5 per cent of their standard tonnages should be given that rate of increase in their tonnages. This was approved.

Battle was renewed on the standard tonnages in late 1940. Since all countries were running at record or almost record current rates of production agreement now was not difficult. Malaya (W. J. Wilcoxson and V. Lowinger) supported the actual tonnage for the year to June, 1941 as the basis; so did Nigeria and the Congo. Bolivia preferred the existing unchanged standard tonnages but would take as a basis the six or nine months to the middle of 1941. The Netherlands East Indies (van den Broek) was more cantankerous. It challenged the position of Bolivia. It was clear, from long experience, that Bolivia could not export at a rate higher than 80 per cent of her standard tonnage. The current rate of exports from Bolivia, including her exports on

---

[4] At the 65th meeting of the I.T.C., 2 June, 1938.

contract to the U.S. Metals Reserve Co., was abnormal. Provision should be made in the next agreement for the right of withdrawal to a participating government if another member (he meant Bolivia) seriously over-exported against its quota.

In May, 1941 the Committee reached provisional settlement on the standard tonnages, using the actual exports for the year to June, 1941 as the basis. A modification lifted the Siamese standard tonnages by about one-tenth above the actual 1940-41 output and, in consequence, all other tonnages were raised correspondingly to give the six countries a total standard tonnage of no less than 251,600 tons. This figure was clearly well beyond anything which the participants could reasonably think of producing at any time in the future. On one thing – more special treatment for Siam – the rest of the Committee was now firm. It refused to allow that country an annually rising standard tonnage and it refused a guaranteed minimum. In December, 1941 the Netherlands East Indies, Malaya, Nigeria and the Congo agreed that they would have a fourth agreement, if necessary without Siam; they were even prepared to go ahead without Bolivia (which ultimately declared her willingness to sign subject to the standard tonnages and voting rights being reconsidered when the Netherlands East Indies and Malaya were freed from Japanese control).

But the problems of the fourth agreement were now academic. The new agreement was a cleaner and more flexible document than had been its predecessors. Its representation of producers was, by necessity, narrower but its concession to consumers – three persons, as against two in the third agreement, to tender advice regarding world stocks and consumption (but nothing on prices) – were ludicrous in the light of the changes taking place in world political thinking about the form and use of commodity control agreements.

By March, 1946 the Committee had to make up its mind as to the creation of a fifth agreement operative from 1 January, 1947. The Netherlands delegation, much less in touch with American thinking and American pressure, thought that any new ideas could be fitted into an amended control agreement. The position of the consumer representatives might be strengthened. In the immediate post-war years tin would be in short supply, so that the restrictive export measures to which so many consuming countries objected would be unnecessary. It would be wise to include a buffer stock in the next agreement. It would be desirable to include a Study Group – the currently favoured panacea for commodity problems – within the Committee.[5].

Other delegates (Sr. Penaranda for Bolivia, Th. Heyse for Belgium and J. van den Broek) were unwilling for the Committee to commit suicide, but hope really ended when the United Kingdom (Sir Gerard Clauson) stated categorically that there was no prospect of the United Kingdom government signing a new agreement. The Committee, under pressure from almost all its member governments, met in March, 1946 to maintain that a tin surplus would arise in the future; to recommend that international co-operation in tin should be continued; and to ask members of the Committee to consult with the more important non-member consuming and producing countries to frame acceptable international institutions. The Committee met for the 88th and last time in December, 1946.

The international tin conference of 1946

With the resistance of the International Tin Committee broken an appeal could be made for the establishment on a wider basis of a new international organisation for tin. That appeal required the full support of the United Kingdom, still politically in control of over half the world's production, and of the U.S.A., economically in control of the financial resources on which the rehabilitation of much of the world depended. In October, 1946 the British government convened an international tin conference in London. Invitations were sent to the countries represented on the International Tin Committee (Belgium, Bolivia, the United Kingdom, France and the Netherlands) and to other major tin importing or producing countries (Siam, China, the U.S.S.R. and the U.S.A.). All the countries, except the U.S.S.R., accepted.

---

[5] P. H. Westerman (Netherlands) at the 86th meeting of the Committee, 1946.

The conference had two main concerns. The first was to ensure that the International Tin Committee would be dead beyond chance of revival; the second was to recognise that, for the time being at least, there would be a new organisation fully in line with the principles of the new Havana charter for an international trade organisation and as toothless as possible. On the first question, the United Kingdom – only too well aware of the attitude of the U.S.A. – repeated that it was inconsistent to maintain the Committee and at the same time to support the new principles. Even if the export control agreement were to be continued the future of tin was so uncertain that only the skeleton of an agreement could be devised. The U.S.A. (Mr. D. D. Kennedy) stressed that the U.S. was committed to a policy of the expansion of world trade and employment under conditions which made possible the fullest use of private channels of trade and which permitted world trade to take place on a multilateral basis and in a non-discriminatory way. This involved the reduction or elimination of restraints to the flow of goods in world commerce and, in particular, of international arrangements, whether private or governmental, which restricted markets or fixed prices.

Only the Netherlands fought for a continued life, even in modified form, of the International Tin Committee. The Netherlands accepted in general the basic principles of the international trade organisation, but disagreed with a machinery so complicated that it would make urgent action on commodities impossible. The Committee should be remodelled and renamed the World Tin Council. The Council should aim at stability in both supplies and prices; restrictive measures should be applied when other efforts, such as to increase consumption, had failed; voting power should be accorded to producers and consumers with due regard to their interests in tin; and the Council should co-operate with and eventually take over the allocating activities of the Combined Tin Committee in Washington.

On the question of a successor organisation to the International Tin Committee, the U.S.A. recognised that burdensome surpluses of tin might arise and would be prepared to co-operate in an international solution if the basic principles for the new international trade organisation were observed. It advocated the setting up of a study group whose function would be to discuss problems and special difficulties on the production, smelting and use of tin; to initiate such studies on tin as would appear to be wise and necessary; to consider possible solutions to tin problems as these arose; to make appropriate recommendations to member-governments and to keep other interested governments informed. The United Kingdom agreed on the setting up of the study group, on the grounds that when the tin surplus came there would be an existing body to discuss solutions, including the regulation of output and buffer stocks. Bolivia (Sr. Vargas) was hesitant but felt that, within the spirit of the world trade organisation, preferential consideration should be given to the restoration of complete freedom for the marketing of tin (a back-handed slap at the U.S.A. and United Kingdom who were controlling, and were to control for the next few years, the world price of tin). Siam was prepared to co-operate if the price of tin were well and truly controlled.

There was now no serious objection to the establishment of the tin study group or to its terms of reference.

The third key question for the conference was: When would the tin surplus which, it was almost automatically assumed, was inherent in tin, be realised? The Statistical Committee of the conference foresaw a sharply rising curve of world production which would show a small surplus even as early as 1947 and a very heavy surplus by 1949. The south-east Asian producers inflated their estimates of production grossly. They overlooked the conflict of their requirements for steel and tools with the same requirements of western Europe and Japan and they were eager to strengthen their individual bargaining positions in any future scheme regulating supply. The consumers, and especially the U.S.A., equally inflated their estimates of consumption.

IX

Tin Control During the War, 1939-45

BEFORE the outbreak of the Second World War Germany had been consuming tin at the rate of about 13,000 tons a year. A very small proportion of this came from domestic mine production in the very old mining districts of Altenberg-Zinnwald and Schneeberg-Annaberg in Saxony, a higher proportion from the treatment of low-grade tin concentrates from Bolivia, some refined metal from the treatment of tin-containing scrap and the remainder as imports of tin metal from Indonesia, Holland and Malaya.

The war immediately cut Germany off from her Malayan supplies and the invasion of the Netherlands in 1940 from the Dutch and Indonesian supplies and she was left, until her invasion of the U.S.S.R. in 1941, with a small flow of imports from that country. For the rest of the war, Germany had to depend for new tin supplies on stocks looted in France and the Netherlands, on high-priced purchases from Spain and Portugal and on occasional shipments through the Allied blockade from Japanese-held areas in south-east Asia. The availabilities of tin were reduced to one-half or one-third of the pre-war tonnage; the shortage – which never seems to have been as acute as the British authorities led themselves to believe – was met mainly by a shift-over from tinplate to blackplate for canning.

Japan found herself in a very different position. By 1942 she had conquered Malaya, the Netherlands East Indies, Thailand and Burma, countries with a total production of no less than 150,000 tons of tin a year; her own industrial requirements were between 8,000 and 10,000 tons a year. Without any great pressure for production on the conquered areas and subject mainly to her shipping capacity she had no serious war time worries on tin, and, in fact, was left at the end of the war with accumulated stocks of 10,000 tons of tin which were then seized and shipped by the U.S.A.

So long as Malaya was free, there was no serious problem, other than the problem of shipping, for the United Kingdom or the U.S.A. The former had, in any case, its supplies of concentrates for smelting from Nigeria and Bolivia, the latter had succeeded in buying very substantial tonnages of tin for its stockpile through the contract it had negotiated with the International Tin Committee.

The atmosphere changed in 1942, when south-east Asia was lost and the United Kingdom was expected to meet the heavy tin requirements of her new ally, the U.S.S.R.[1] There was put into full operation by the U.S.A. and the United Kingdom a series of measures on prices, control of usage, sharing of supplies and an international allocation of metal which was to continue until well after the war which called them into existence.

The initial price steps of the United Kingdom government had been mild enough. The government laid down between September and December, 1939 a temporary ceiling price of £230 a ton (which was the same as the ceiling price on which the buffer stock of the control scheme had worked) and temporarily subjected tin to export license. During 1940, with ample supplies of tin available, the London Metal Exchange was left free enough and the tin price ranged between £232 and £290; and during 1941 buying and selling operations by the government kept the price fairly stable around £256-260 a ton. In December, 1941, on the beginning of the Japanese war, the London Metal Exchange was closed, private dealing was banned and all tin metal and tin concentrates became subject to requisition by the British government.

The government, through the Ministry of Supply and the Non-ferrous Metals Control, became the sole buyer and seller of tin and determined from time to time (and in fact until 1949) the selling price of metal. The Ministry also entered into contracts for the purchase of all the tin concentrates produced in Nigeria and the main Cornish mines and agreed the prices to be incorporated in the long-term private contract for the supply of the Patiño concentrates from Bolivia to the British smelter.

[1] Over 1942-44 the United Kingdom shipped to the U.S.S.R. a total of 22,400 tons of tin.

In the United States, the government had agreed in 1940 to buy up to 50,000 tons of tin from the producers in the international tin agreement at a minimum price of 50 cents a pound. At the same time, the offer by the Metals Reserve Co. (a government body) to take up all tin on offer established in effect a fixed market price. In November, 1940 Metals Reserve agreed to buy all the available Bolivian output of concentrates (other than the Patiño material on contract to the United Kingdom) for five years; in 1941 the contract with the International Tin Committee was renewed for another 50,000 tons and another year; and from August, 1941 an official ceiling price of 52 cents a pound was imposed on the New York market.

The problem of price had two aspects – the prices fixed for sales of metal to domestic consumers and the price paid for metal or concentrates imported. Neither in the U.S.A. nor in the United Kingdom was there any necessary connection between the two. Both governments desired a low and stable price for domestic sales, since in practice a great part of the metal so sold went into government contracts. The U.K. selling price for metal remained unchanged or almost unchanged at £276.50 or £301.50 for 4½ years (February, 1942 to September, 1946); the official U.S. maximum price stayed at 52 cents over five years (August, 1941 to November, 1946). When circumstances compelled them to increase their domestic selling price – the weakening in the post-war period of the need for stability or the rising cost in their import purchases – the increases were often very substantial. The U.K. price was moved up by about £55 a ton in March, 1947, by about £77 a ton in December, 1947 and by another £50 a ton in June, 1948, and the U.S. price moved similarly.

From the end of 1941, the U.K. government bought Nigerian concentrates on a contract with each producer under which the producer received his costs of production and his pre-war profit per ton produced; in consequence, each producer received a different price. From January, 1946, however, purchase was on a flat price per ton of tin in concentrates f.o.b. Nigeria. When Malaya was re-opened the government initially bought through a Tin Ore Buying Agency all the concentrates available from the producers at a fixed metal price and had the concentrates smelted on toll at Penang and Singapore. As from mid-1946 the Malayan smelters were allowed to revert to their traditional practice of purchasing the concentrates and the U.K. government bought the metal from the smelters.

The U.S.A. current supplies during 1942-45 came mainly from Bolivia and the Congo. The price paid to the Bolivian producers was stable at 49.50 cents per pound f.o.b. South American port through 1942-43, but was increased in 1944-46, and again very sharply to as much as 99 cents in the middle of 1948. The price paid by the United Kingdom for Patiño concentrates was generally the same as the American price for the non-Patiño production.

Within the United States, the Texas City smelter (which covered all the internal production of primary tin metal) was operated by the Billiton Co. of the Netherlands for the account of the governmental Reconstruction Finance Corporation and the latter body was also responsible for selling metal against allocations made by the Department of Commerce.

Elsewhere, a high degree of governmental control continued until and sometimes after 1949. In France from 1945 all tin metal was imported by a Groupement operating under the control of, but not buying on account of, the government; import of tin into and sales of tin within India were all on government account. In Australia the commonwealth government fixed the prices paid by the smelter to the tin miner and the price of the metal bought by the consumer from the smelter.

The Combined Tin Committee

During 1942-45 the general policy of the allied powers – Canada, the United Kingdom and the United States – on the allocation of raw materials for war purposes had been determined by the Combined War Materials Board. Within that Board, the case of tin was comparatively simple and was concerned primarily with the building up within the United States of maximum stocks for the purposes of that government only. The end of the war marked the end of the need for the Board and its allocating authority, but created more problems for tin. The mining industries in south-east Asia would require massive rehabilitation (although more as a consequence of neglect than of destruction); the tin-using industries of western Europe required new stocks and a fresh flow of raw material, and tin would therefore remain in world short supply. It would be necessary to ration tin supplies on a basis and amongst claimants wider than the cosy, almost family treatment which had been sufficient for the purposes of the combined Raw Materials Board. Before its own death the Board set up in November, 1949 in Washington the Combined Tin Committee. This new body was:

> "While the emergency period of shortage continues . . . to keep the tin metal position under review and to allocate supplies of tin metal to member and non-member countries in such manner as may be agreed among the countries represented on the Committee."

Membership of the Committee was fairly wide. It covered initially five governments with a substantial interest in the production or consumption of tin metal, the producing countries being the United Kingdom (on behalf of the U.K. and Malaya), the Netherlands (on behalf of the Netherlands East Indies) and Belgium (on behalf of the Belgian Congo) and the consuming countries being the U.S.A. (also a large smelter, but not an exporter, of tin metal) and France. In 1946, membership was extended to include China (an exporter of tin metal) and two less important consumers, Canada and India, the latter with an embarrassingly high conception of its claims as a consumer. An invitation given to the U.S.S.R. was not accepted.

Any government outside the membership could – and many did – claim an allocation. Allocations were made by the Committee on the basis of a six months period. The tonnages approved were the result of questionnaires on consuming requirements, estimated consumption and current and future stocks and on the availability for export from producing countries. The allocation was not a guarantee of supply but merely the right to buy. The importing country could buy its allocation from whatever source it desired. The exporting countries restricted their exports in order to prevent over-purchase against the allocations and to prevent duplication of supply to an importing country which might not itself control imports.

The Committee confined its work to the allocation of tin metal available for export. It therefore was excluded from dealing with the largest stock of tin metal in the world (the U.S. strategic stockpile) and it did not deal with the supplies of tin concentrates from Bolivia, Malaya and Nigeria which were under contract to the United Kingdom or U.S. governments.

The Committee allocated 48,800 tons of tin in 1946, 56,800 tons in 1947, 83,200 tons in 1948 and 61,500 tons in 1949. In its most active year, 1948, it was making allocations to a total of 39 countries, including some of the socialist East European countries (but not the U.S.S.R.).

By the end of 1947 there was actually a rough world balance between mine production and the limited commercial consumption. At its first two meetings in 1948 a new international organisation – the newly established International Tin Study Group, which had no connection with the Combined Tin Committee— was talking of the possibility of a surplus of world production of tin within a couple of years. In 1949 the Study Group was anticipating a surplus that year. The restrictions on the use of tin in the United Kingdom were lifted and even in the United States they were not to last beyond December, 1949. In the eyes of the United States the Combined Tin Committee seemed to have fulfilled one of its primary functions. The allocations of metal to the United States had covered no less than 127,000 tons of the total 250,000 tons allocated by the Committee and had therefore ensured for the United States that its requirements for building up a strategic stockpile should not be subordinated in any way to the needs of a war-damaged European economy. In those circumstances, and with some relief at no longer having to haggle over the inflated claims of consuming countries, the Committee recommended its own dissolution in December 1949. The London metal market resumed trading activities for tin in November, 1949.[2]

---

[2] For the allocations and a brief statement on the Combined Tin Committee, see I.T.S.G.: *Statistical Year Book, 1949.*

X

The International Tin Study Group, 1948-56

WORLD opinion on the question of international agreements in the sphere of raw materials was changing. The beginning of that change had been clear even before the outbreak of the Second World War and the general, if conditional, acceptance of a new outlook found expression in the very wide discussions, even before the end of that war, on a new international trade organisation.

As early as 1937, the League of Nations, through its Committee for the Study of the Problems of Raw Materials, had so far advanced in its economic thinking as to give some approval to intergovernmental schemes for regulating the supply of raw materials, although it still remained critical of early types of schemes. It even discussed the desirability of buffer stocks. In 1945, the victorious countries were creating the United Nations Organisation as a means of both satisfying the political problems between countries and of ensuring the minimum change in the new post-war establishment which their victories had secured. They, and in particular the U.S.A., turned also to ensure the acceptance of a new body of conduct in the regulation of the world's economic relations.

Proposals for the application of this new economic philosophy were made by the U.S.A. to the United Nations at the end of 1945. The proposals were supported by the United Kingdom, still a factor of the first importance in world trade. The proposals were intended to make effective an international scheme on trade and employment which would provide for expansion in world production, employment, exchange and consumption. They covered an equitable basis for dealing with the problems of measures affecting international trade; a curbing of restrictive trade practices resulting from private international business arrangements; and a mechanism governing the institution and operation of intergovernmental commodity arrangements. The principles were naturally slanted in favour of a United States economy now feeling itself able to dominate the world (outside the various communist countries), but the attitude towards intergovernmental commodity arrangements showed that governmental, if not private, thinking within the U.S.A. had taken quite long strides in less than twenty years.

The Havana Charter for an International Trade Organisation, drawn up at the international Havana conference after the greatest of labour in 1947-48 was never ratified by any of the major countries, rich or poor, which had been present. Nevertheless, the views which it expressed on raw materials were to remain gospel truth to governments for many years. All the recognised intergovernmental study groups or agreements or proposed agreements on commodities over the last 30 years – in tin, wheat, sugar, coffee, cocoa, rice, fats – have been obliged to acknowledge their debt to the Havana Charter subject only to the modifications, interpretations and extensions which have been made by the United Nations Economic and Social Council and by the United Nations Conference on Trade and Development.

As regards raw materials, the Charter accepted the principle that, in certain cases and under certain conditions, commodity agreements were appropriate to deal with long-term surplus supply. These agreements were to be intergovernmental in character. Before they could come into existence they would require a set of findings through an intergovernmental commodity conference or through the International Trade Organisation. The set of findings required a positive answer to a series of questions:

(a) that a burdensome surplus of the primary commodity has developed or is expected to develop;

(b) that, in absence of specific governmental action, this would cause hardship to producers (including small producers);

[1] Belgium, Bolivia, British Colonies, Canada, China, Cuba, Czechoslovakia, Finland, France, India, Italy, Mexico, the Netherlands, Portugal, Thailand, Sweden, Switzerland, United Kingdom and the U.S.A.

[2] The headquarters of the Study Group were fixed for The Hague, partly on the grounds that the pre-1939 Statistical Office of the International Tin Research and Development Council was in the offices of the Billiton Company there. The choice had no other good reason. The Hague was not within a producing country; Dutch tin consumption was relatively low; and the Netherlands had no metal market.

(c) that – since price reduction did not readily increase consumption or decrease production – normal market forces would not operate in time to prevent such hardship;

(d) that widespread unemployment or under-employment had developed which could not be corrected by normal market forces in time to prevent widespread and undue hardship to workers because price reduction led to a reduction in employment and because the producing area did not afford alternative employment opportunities for the worker involved.

A commodity agreement would also be appropriate to maintain and develop the natural resources of the world and to protect them from unnecessary exhaustion.

Many of the tin producers looked with some suspicion on the Havana conclusions. They had still in existence an International Tin Committee which was intergovernmental and which had dealt successfully with the pre-war problems of overhanging stocks, of surplus production and of a buffer stock. They thought that this committee could be adapted easily enough to the Havana principles on commodity agreements. They felt that the industry had already passed beyond the stage of an international study group with merely discussive powers; and to establish such a study group now would merely postpone, and was intended to postpone, the time of an effective international agreement. But any public expression of these dissenting views died with the International Tin Committee.

The setting-up of a study group

The London tin conference of 1946 was followed by an invitation to a wider range of governments to co-operate in the formal establishment of a study group. The mechanism of the Havana Charter was strictly observed. Invitations were sent widely to governments and not merely to those who were substantially interested in tin as producers or consumers. At the first meeting of the International Tin Study Group held in Brussels in April, 1947 delegates or observers attended from 20 countries[1]. The purposes of the meeting were largely formal – to set out what the Havana Charter and the London conference had already agreed[2]

upon and to arrange a headquarters and very simple organisation – a Management Committee of seven of its original 14
member governments, a secretariat with offices in The Hague
and full Group meetings at intervals. The terms of reference
contained one key point:

"It shall be the responsibility of the Group to consider possible
solutions to the problems which are unlikely to be solved by
the ordinary development of world trade in tin".

The group was to hold eight full conferences, to await the
work and reports from four working parties, to produce four draft
international tin agreements, to see two United Nations Conferences on tin and to wait no less than eight years for an international tin agreement to become accepted[3].

The history of international study groups in the commodities
tempts one almost to believe that the primary purpose of some
member governments has been to prevent rather than to stimulate further action towards effective international agreements.
Admittedly, the study group in tin became the international tin
agreement, but took long years to make the transition. The
International Cotton Advisory Committee, a body equivalent to
a study group, arose from a meeting of the major exporting
countries in 1939, met annually after 1947, studied an international cotton agreement over 1951-54, abandoned the idea in
1954 and is still an advisory committee. The Wool Study Group
was the outcome of a conference in 1947, set up a sub-committee
on an agreement in 1948, agreed in 1950 that a system of reserve
prices would be appropriate but still ineffective and is now of
minor importance. The Rubber Study Group was formed in
1944, set up its sub-committee on an international agreement as
early as 1952, failed to agree and is still a study group. The Cocoa
Study Group, established in 1956 under the auspices of the Food
and Agriculture Organisation, has had a series of conferences on
an international agreement but had just reached one by 1972.
The Lead and Zinc Study Group, established in 1959, was

---

[3] Throughout the whole period the Chairman was G. Péter (France) and the
Secretary-General William Fox.

[4] I.T.S.G.: *A statement on the position and prospects of the tin industry.* The
Hague, 1950.

unchanged in 1972. The International Sugar Agreement wandered from time to time between agreement and a study group as the effective clauses of its agreement were cancelled or restored. Even the International Wheat Agreement has had its teeth removed.

The draft international agreements of the Study Group

In its earlier years the Tin Study Group worked hard and fast; in its later years it was essentially a body waiting for an international agreement to mature. It considered in October, 1948 its first draft agreement (the Belgian draft) and recommended the report of a working party which stated that an agreement would be appropriate and practicable. It produced from a drafting committee a second draft in Washington in December, 1948 (the Washington draft). Another Working Party in the Hague produced in November, 1949 a detailed justification for an agreement[4] and a third draft agreement. This was amended and modified in Paris in March, 1950 to become the Paris draft. At the same time the Study Group called for a formal United Nations Conference on a tin agreement. This met and failed in October-November, 1950. With rather diminished ardour, another working party met in Brussels in June, 1953 and another drafting committee drew up another draft agreement (the London draft) in August, 1953. This was submitted to the resumed United Nations Tin Conference in November-December, 1953. That conference approved the first international tin agreement. The parties waited two years for ratification, and in July, 1956 the agreement became effective. It seemed a very long wait, but in fact proved to be less than the wait which most international study groups in other commodities had to endure. The wait had at least one merit; the agreement which was to emerge was built soundly enough to remain almost unchanged in machinery and principles for at least twenty years.

The Study Group operated from 1948 to 1956. Its major objective was to convert itself from the talking shop of a discussive and statistical group into an international commodity body wider in conception than and at least as effective as the International Tin

Committee had been. Most of its members were agreed on the conversion; but, after an initial promising period, the attitude of its most important member, the U.S.A., hardened. The mind of the U.S. State Department was shifting to sympathy on an international agreement during the three years before the middle of 1950 but this was not true of other agencies of the U.S. government and was certainly not true (then or at any later time) of American business opinion. The prospect of an international agreement being reached in the first two years' life of the Study Group was erased by the Korean War. That war pushed the U.S.A. into a frantic stockpiling policy for raw materials, and the interests of the U.S.A. in tin became a purely strategic interest. The price of tin was pushed up spectacularly from the middle of 1950; and the charge of "price gouging" was a poisonous element in the U.S. political relationship with tin producers for many years. The U.S. policy of strategic accumulation guaranteed a market for surplus tin production and stimulated production to a level shown clearly to have been too high when the[5]

TABLE 11: Study Group Estimates
World Mine Production

long tons

Estimates for	London 1946	Brussels 1947	Washington 1948	The Hague 1948	London 1949	Paris 1950	Actual (a)
1946	94,400	—	—	—	—	—	89,500
1947	141,600	117,000	115,200	—	—	—	112,500
1948	197,000	163,000	150,000	150,000	—	—	151,500
1949	218,500	201,000	170,000	170,000	170,000	162,000	162,000
1950	—	—	200,000	190,000	190,000	172,000	166,000
1951	—	—	—	200,000	205,000	191,000	167,500
1952	—	—	—	210,000	215,000	199,000	171,000
1953	—	—	—	210,000	220,000	206,000	177,000

The estimated over-all position 1949-53

long tons

	1949	1950	1951	1952	1953
World mine production	170,000	190,000	205,000	215,000	220,000
World metal consumption	138,000	158,000	162,000	166,000	170,000
Excess of production	32,000	32,000	43,000	49,000	50,000
Stocks at beginning	134,000	166,000	198,000	241,000	290,000
Stocks at end	166,000	198,000	241,000	290,000	340,000

(a) Excluding non-commercial accumulation.

Source: I.T.S.G.: Fourth Meeting, London, June, 1949 and Fifth Meeting, Paris, March, 1950.
Figures under the last column (actuals) are figures arrived at in 1955-56.

[5] See Table 11.

strategic purchases stopped. Building-up the stockpile was a justification for the U.S.A. opposing and not entering an international agreement; the maintenance of that stock from outside international interference was a reason for her not joining the first agreement; the disposal of a substantial part of that stock was a reason for not entering the second or third agreements.

The Belgian draft tin agreement, 1948

Many members of the Study Group hankered for a machinery tied to the concept of an inbuilt surplus in tin, but the London conference of 1946 has relegated the possible over-production of tin to "some future date". The first Study Group meeting in Brussels in 1947 concluded, more cautiously, that well into 1948 there would continue to be a shortage of supply in view of the difficulties for the south-east Asian producers on mining supplies, on coal and on rice for the labour force. But the arithmetic showed that the anticipated over-production was nearing. The 1947 meeting in Brussels and the 1948 meeting in Washington grossly inflated estimates of world tin production for the fairly near future – at the first, 201,000 tons by 1949, and at the second 200,000 tons by 1950. The Netherlands East Indies thought it would reach an output of 50,000 tons by 1951 and Malaya at least 55,000 tons by 1949. The consumers were equally optimistic as on their requirements. The American agencies calculated that currently and for the near future world industrial requirements would be 200,000 tons with no likelihood of a burdensome surplus in the period up to 1950. In general, at all its meetings during 1947-50 the Study Group seemed rather obsessed by a production target of around 200,000 tons a year which was always due to come about two years ahead of the time the assessment was being made. This target was not to be reached, at least within the next quarter of a century; and the same was true of the consumption target. But it is to be remembered that there had been reached only a few years earlier an actual production figure (nearly 230,000 tons in 1940-41) well in excess of that now anticipated and a consumption figure (189,000 tons in 1937) not far below the figure now anticipated.[5]

In October, 1948 the Belgians put a fairly comprehensive draft agreement before the Study Group. This was in no sense a mere copy of the pre-war control and buffer stock system. The draft accepted the doctrine that the tin available should be the subject of allocation; that the producing countries should accept basic obligations to produce; that the consumers should accept basic obligations to buy; that a set of basic prices should be laid down and that a buffer stock should be established. This involved a line of thinking in tin new in very many respects. It would establish, for the first time, international organisation in the trade, as distinct from the production of tin; it would impose a discipline on both consumers and producers far beyond anything conceived of by the International Tin Committee. In its implicit guarantee to meet the consumption requirements of the United States the scheme had an obvious attraction for that country. More generally, it tried to meet both the shortage of tin, which it admitted to be a current problem, and the surplus of tin, which it believed to be a near future problem.

An International Tin Council, open to all members of the International Trade Organisation and to all consumers and producers substantially interested in tin, was to be established. Each group should have an equality in aggregate voting power (consumers based on tonnages bought or obliged to buy, producers on deliveries and obligations to deliver). A producers' committee would give each year estimates of production over the next five years and a consumers' committee similar estimates of requirements.

The producers would give formal undertakings to place on the market in the next two years a fixed quantity of tin (not above the maximum price), the consumer would give formal undertakings to buy in the next two years a fixed quantity of tin (not below the minimum price). Basic maximum prices (fixed at 115 cents a pound) and minimum prices (moving from an initial 100 cents down in equal steps to 90 cents) were fixed for five years; these might be subject to adjustments.

---

[6] Composed of the U.S.A., the United Kingdom, the British Colonial and Dependent Territories (that is, Malaya and Nigeria), the Netherlands, Belgium and Bolivia.

The proposals then put forward machinery for dealing with a shortage. In that event the Council could establish an allocation system for the distribution to consumers of the tonnages offered by the producers. In particular, the Council could approve arrangements between the producers' committee and the U.S.A. which would guarantee a minimum amount of tin to the U.S.A. for a particular period. In a surplus period the consumers' committee would call on all members to abolish limitations on the usage of tin; would work out a plan to increase consumption; would make suggestions for disposals of tonnages on special arrangements at agreed prices. At the same time the producers' committee could suggest a system for the regulation of production, the fixation of prices and the setting up of a buffer stock under the Council.

The owners of non-commercial stocks could purchase outside the agreement provided they had first fulfilled their obligations under the agreement.

The Belgian proposals were very new and far-reaching. They were not, so far as they were new, much to the liking of many producers. A system of allocation would give consuming countries in effect a lien on tin that they might never take up (Westerman for the Netherlands East Indies). The proposal for higher production might, it was suggested, drive the producers (as it had done between 1941-45) into long-term unsound mining practices.

The principles of the Belgian draft agreement and the more general principles approved by the Study Group were turned over to a drafting committee[6] which met in Washington in December. This committee considered further versions of an agreement.

There were two drafts – the first (by Clarence Nichols of the U.S. State Department) which reflected fairly closely the principles of the Study Group, and the second (by Sir Gerard Clauson of the U.K. Colonial Officer) which took into more account the Belgian draft and which drew for some of its language fairly heavily on the obligations and commitments provisions of the International Wheat Agreement, then a popular horse in the intergovernmental stables.

The Washington draft, December, 1948

The main discussion was on the methods by which the votes of the categories of members could be divided; on the machinery for the allocation of tin metal in times of shortage; and, quite inevitably, on prices.

Both drafts agreed on an equality of total votes for each category. The Nichols' draft wanted consumers' votes divided on the basis of net imports and the producers' on the basis of net exports, with votes subject to change during an agreement so as to avoid the freezing of status as under the International Tin Committee. The Clauson draft wanted to base voting power on purchase targets (for consumers) and on sales targets (for producers); it remembered the old pre-war squabbles amongst producers about the division of votes or tonnages between the individual producing countries and said nothing on that question.

On the allocation of tin metal, the Nichols draft proposed that the new Tin Council should continue temporarily the allocation work of the Combined Tin Committee (which was still active) but should maintain allocation thereafter only if it found the need proved. Each country would receive a basic allocation and could receive additional allocations. This would mean that countries which put forward high purchase commitments would have prior rights to some of the tin available for allocation. It was, of course, public knowledge that the U.S.A. wanted, for stockpile reasons, more than her industrial proportion of the tin available. The Clauson draft did not diverge much on this point.

On price, Nichols proposed that, when international allocation of tin metal was operating, the price of tin should be fixed. At other times there would be one fixed price at which the producers would offer tin in accordance with their export commitments (in effect, a maximum price) and another fixed price (a minimum price) at which importers would offer to take up metal in accordance with their purchase commitments. The Clauson draft proposed a fixed price for the first two years of the agreement and maximum and minimum prices in the next three years.

[7] France, the Netherlands, Thailand, Canada, Australia, China, Czechoslovakia and India.

Both the producing and consuming members of the proposed Council would accept obligations. The producers would remove all obstacles to the production of tin; they would endeavour to produce to the maximum; and they would state their production targets for each year. The consumers would state their total purchase targets for each year; they would state the tonnages they would buy each year from the producing countries at the minimum price (the purchase commitment); and they would remove restrictions on their use of tin.

Tin metal would be allocated by the Council for the first year of the agreement and all metal bought and sold by the contracting governments would be bought and sold accordingly. Tin not taken up could be bought by a country which had already fulfilled its obligations.

Prices would be fixed for the period of allocation; for each later year there would be a minimum and maximum price. Producing countries would not sell above the maximum price and consuming countries would not buy below the minimum price.

The members of the Study Group were asked whether they would participate in a United Nations conference to consider a tin agreement on the lines of the Washington draft. Many of the less important members of the Group were in favour[7]. Bolivia had qualifications – that the minimum price should be a fixed and not descending one and should be at a reasonable level in the light of rising material and equipment costs, that an arrangement should be reached within three years to control production and – most interesting of all as a piece of forward thinking – that the stockpiling countries (that is, the U.S.A.) should give guarantees to the producing countries as to the stability of prices when the stockpiles were being liquidated. The U.S.A. had decided to take a more direct path to solve its tin difficulties; it was opening or about to open bilateral talks with the main producing countries for the long-term purchase of concentrates or metal. Many of the producing governments thought, with very considerable truth, that such U.S.A. purchases would postpone for some time ahead the emergence of a surplus in the industry. The action of the U.S.A. – even if the negotiations with Belgium and the Netherlands were prolonged or with Malaya never came to fruition –

removed from tin the sense of urgency which had been the driving force in all the negotiations and draft agreements of 1948. In 1949 the U.S.A. told the Study Group that it would attend a U.N. tin conference but would prefer a further meeting of the Study Group; in April, 1949 the United Kingdom welcomed the U.S. suggestion.

The London draft, 1949

The quiescence of the Study Group did not last long. Most of its members desired an international agreement. The United Kingdom, fully aware of its political and economic dependence on the United States, had accepted a Study Group meeting as a postponement but was aware of its previous commitments on the principle of an agreement, was bracing itself towards the ultimate independence of Malaya and was beginning to appreciate the innate importance of the concept of under-development in the old colonial areas. More immediately, although in the opinion of the Study Group the U.S. was likely to soak up all surplus tin production for the next four or five years, the British felt that the assured market for tin would last for a much shorter period. There lay a conflict of views. The producers aimed at an agreement to deal with a future surplus; the Americans would look only at an expansionist agreement whose primary objective was to fill the U.S. stockpile; the producers wanted to know what would happen when that stockpile was filled.

The statistical assessment of the Study Group in June, 1949 emphasised the future fears of the producers. Over the five years 1949-53 the total surplus of commercial production over commercial consumption (before taking account of the U.S. stockpile purchases) was estimated at a total of no less than 206,000 tons. What such a surplus might mean in terms of social unrest in Malaya was clear and ominous.

Another working party on an agreement – the most hard-working yet – met in The Hague in October, 1949. A preliminary memorandum from the British Colonial and Dependent Territories stated the Nigerian and Malayan cases for an agreement; the Belgians did the same for the Belgian Congo; both emphasised the social and economic results of surplus production and the effects on unemployment in those under-developed areas.

The meeting drew up another draft agreement, at least on matters still not satisfactory to the U.S.A., and pushed it forward to the Paris meeting of the Study Group in March, 1950. There it became the Paris draft of 1950.

The Paris draft, 1950

It was clear that the U.S.A. could itself look after the immediate needs of the stockpile, but, unless the U.S.A. demand was to be rapid, substantial and long-term, a world surplus would arise. In fact, in spite of the reopening of the free tin markets, commercial consumption in 1949 proved to be lower than in almost all of the twenty years before 1939. General opinion in the Study Group swung back to the older concept of export control and buffer stock.

The Paris draft did not retain the obligations as to production targets, purchase targets or purchase commitments; it abandoned the range of maximum and minimum prices; it maintained an International Tin Council with equal voting power for producers and consumers; it provided for a system of export control, applicable on a two-thirds majority vote, in which the producers' percentages would be subject to annual change. A buffer stock was to be created on producers' money, operating fully within the range between floor and ceiling prices; that buffer stock would operate only after the control of exports had come into effect. The liquidation of governmental non-commercial stocks should, as in the Washington draft, be carried out on due notice and consultation to avoid serious disturbance to the world tin market.

The U.S.A. was the only serious opposition to an agreement. C. Nichols lifted the discussion to a higher level – the question of economic adjustments for the industry and the transfer of mining activity to more economic sources of supply, brought the terms of the preamble closer to the wording of the Havana Charter and avoided the emotional phrase "burdensome surplus." He asked and obtained agreement that the voting rights of consuming countries be based on imports (which would make the U.S. voting power dominant); he could not agree to a system of export quotas before the ending of the U.S. stockpiling purchases; he limited action by the buffer stock to the floor and ceiling prices' (and not between); and he opposed the convening of the U.N. tin conference. The Americans lost, and by a majority the request for a conference was carried forward[8].

The Geneva conference, 1950

The United Nations conference on tin[9] met at Geneva in October-November, 1950. For a conference dedicated to discussing the means of dealing with a surplus no more unhappy moment could have been selected. The opening of the free tin metal market in November, 1949 had been preceded by a mistimed lift in the fixed official price and the market produced a price well below anticipation; it was also, in spite of the belief in a potential surplus, almost a straight line price at this lower level through the first half of 1950. Hardly had the Paris discussions closed before the outbreak of the Korean war began to push the price up. In July the price jumped by £70; in August by another £113; in September by another £98. When the Geneva conference opened it had reached what was regarded as a fantastic level. During the conference it moved up to over £1,000 a ton. Every market evidence indicated not a surplus but an intense, if temporary, shortage. The U.S. government, in particular, was clamouring for supplies of tin at almost any price.

At the conference, which took the Paris draft as its basis, the

---

[8] The voting was nine in favour (British Colonial and Dependent Territories, Belgium, Bolivia, Indonesia (late the Netherlands East Indies), Australia, Canada, Netherlands, India and the United Kingdom). Three abstained (China, Thailand and France) and one was against (the U.S.A.).

[9] The chairman of the conference was G. Péter; the chairman of the main operative committee was G. C. Monture of Canada. The case of the British Colonial and Dependent Territories was stated by Sir Hilton Poynton, of the U.S.A. by Clarence Nichols, of the Belgian Congo by M. van den Abeele and H. Depage, of Indonesia by Dr. Subardjo (also vice-chairman of the conference) and of France by Pierre Legoux.

[10] And therefore to be the average of two averages (that of the previous thirty days and that of the previous eighteen months).

producers re-stated their now familiar case. The stockpile
purchases were unpredictable and temporary; their scale and
duration were outside the control of the producing countries
which would suffer most when they were cut down or stopped
(Sir Hilton Poynton). In mining, especially in underground mining, price fluctuations might lead to the permanent loss of tin-
containing ground, and an agreement was necessary to conserve
the world's mineral resources of tin (J. A. Dunn of Australia).
The inevitable end to stockpile buying created a danger all the
more serious to producers since that buying in itself would have
raised world production to a maximum (M. van den Abeele for
the Belgian Congo).

Argument centred first on the operation of any control quota
system, and on the relation of this to the buffer stock. The U.S.A.
stood firm on this point. Control of exports was to be the exceptional case, imposed on grounds of proved necessity and for short
periods by a majority vote. No tin should be placed in the buffer
stock until there was tin available in excess of demand. The
buffer stock should hold an adequate tonnage before export
control operated (C. Nichols).

The second argument was on the size of the proposed buffer
stock. The desirable size was at least 30,000 tons, to be in stock
before export control was applied (Nichols). The producers did
not believe that any specific tonnage should be required in the
buffer stock before export control could be applied; it was
anomalous to expect producers to put 30,000 tons of tin into a
buffer stock that might not be required (Sir Hilton Poynton) or to
expect them to finance the whole of the buffer stock themselves
(H. Depage).

The third argument was on the price for buffer stock activities.
The U.S.A. thought of a range of 15 to 20 per cent between the
floor and ceiling prices (C. Nichols). The Belgian Congo (H.
Depage) thought that a basic price should be determined each
year to take account of short-term and long-term fluctuations[10]
with the floor price 10 per cent below and the ceiling price 10 per
cent above the basis. This pricing formula would give a basis
price of £803 a ton or nearly 100.50 cents a pound. This was
acceptable to Indonesia. Bolivia (R. Querejazu-Calvo) thought

price determination in a field of changing prices should be left to the Council.

The haggling on the price continued. The U.S.A. found that, on a formula in relation to the prices of other raw materials, the basic tin price might be as low as £550-575 a ton. Sir Hilton Poynton put forward 95 to 100 cents a pound as floor price (£760 to £800 a ton), on the grounds that the floor price was the lowest price which would produce the marginal quantity required from the highest cost efficient producer, which would provide for a sufficient reserve of productive capacity to meet the growing tin requirements of the world and which would permit the replacement of worn-out plant and the building-up of mining reserves by prospecting.

The producers failed to reach agreement among themselves on the allocation of their votes on the Council (Indonesia standing firm on its proposal to use either the maximum production figures of 1940-41 or quite unrealistic forward production estimates); the consumers failed to reach agreement on their votes.

The smell of failure was in the air. The two major delegations found it necessary to justify themselves. The British Colonial delegation (Sir Hilton Poynton) pointed to the concessions it had made – that the producers' buffer stock be 15,000 or 20,000 tons and that 15,000 tons be held in the buffer stock even at the ceiling price; but he could not accept the right of any member (that is, the U.S.A.) to absorb the Council's buffer stock into its strategic stockpile at any time it liked, so destroying at a stroke the delicate and elaborate mechanism for stabilising the tin industry. The U.S.A. (C. Nichols) pointed out that quotas should come into effect only when the price fell below the floor price; that, when a quota was imposed, steps should be taken to ensure the full total being exported; that the U.S.A. might think in terms of a buffer stock lower than 30,000 tons; that the gap between the U.S.A. floor price of 65 cents a pound (£680) and the producers of 90 cents (£720) seemed unbridgeable; but that one point was crucial – the U.S.A. believed if there were not sufficient tin in the buffer stock to meet commercial and non-commercial demand the U.S. stockpile should have the prior claim. On those notes, the last of which particularly stuck in the gullet of the producers, the meeting accepted that further review by governments was necessary and went home.

The London draft, 1953

The failure of Geneva had no immediate influence on the tin price. For the next three months it continued its breakneck rise until for February, 1951 it stood at £1,470 or double what it had been before the conference. It was to be broken in two stages – the first mildly in the summer of 1951 and the second drastically in the summer of 1953. The latter stage brought it back to a point even lower than it had been before the 1950 conference. The abnormal price rise of 1950 had been largely the result of panic buying by the Americans; the reduction had resulted very largely from the unilateral action of the U.S.A. in offering tin for sale in order to keep the price lower. The U.S.A. alleged that an international cartel was responsible for the high tin prices. C. V. Aramayo of the Bolivian delegation told the Study Group in September, 1951 that the U.S.A., in creating a "buyers' cartel", was guilty of "economic aggression".

By 1953 the Korean cyclone was spent. Commercial consumption was beginning to recover. Free trade in tin had been restored in the U.S.A. in August, 1952 and the controls over usage removed in February, 1953. In March, 1953 the U.S.A. warned the Tin Study meeting in London that the unexpectedly large supplies of tin bought for stockpiling in 1952 had altered the position; the present objectives of the stockpile might be reached on existing contracts; and the stockpile might be completed by not later than June, 1954 and perhaps even by the end of 1953. The warning to producers was very clear. The market emphasised it by cutting the price by nearly £200 a ton in April, 1953 and by another £160 over the next three months.

The Study Group roused itself again. The producers at the March, 1953 meeting accepted an obligation to raise a buffer stock of 15,000 tons (if more were required the consumers could find it) and they accepted that a substantial quantity of tin should be in the stock before export control could operate. The majority of the consumers (but not including the U.S.A., Federal Germany and Canada) favoured an agreement. On the size of the buffer stock, opinions ranged between 15,000 and 30,000 tons (the U.S.A. standing on the latter). The producers then raised their buffer stock figure to 20,000 tons; it was understood that there would be no restriction of exports until the buffer stock held 10,000 tons of tin.

In March, 1953 the Group appointed another working party. This met in Brussels in June, 1953[11] and, in turn, set up a drafting committee to re-shape the Paris draft in the light of two years' discussion[12]. The result was yet another draft from the hardworking delegation of the British Colonial and Dependent Territories – the London draft of August, 1953.

The most important problem was the statement of the price. The committee compromised by writing alternatives – one a simple statement of floor and ceiling prices, the other a proposal from M. Depage (Belgian Congo) that there should be laid down only a ceiling price with the right of the Council to fix or alter the floor price by a simple majority. No figure was, of course, as yet inserted for any price.

The Geneva Conference, 1953[13]

The London draft, submitted to member governments at once, was presented to the Geneva tin conference which resumed its meetings in November, 1953.

[11]M. Péter was chairman of this as of all working parties of the Study Group.
[12]The members of the Drafting Committee were Belgium, Bolivia, British Colonial and Dependent Territories, Indonesia and the United Kingdom, but not the United States.
[13]Delegations were present from 23 countries covering about 113,000 tons of tin consumption out of a world total of 130,000 and a production of 161,000 tons out of a world total of 170,000 tons. These delegations included Federal Germany and the U.S.A. (both members of the Study Group). The U.S.S.R., Hungary and Yugoslavia, absent in 1950, were present as observers. The chairman of the conference and chairman of the Economic Committee was M. Péter; the vice-chairman of the conference G. C. Monture; and the chairman of the Administrative Committee H. A. Harding (British Colonial and Dependent Territories).
[14]Bolivia put forward January, 1949 to September, 1953 or October, 1950 to September, 1953 or November, 1949 to October, 1952. Indonesia put forward the three years October, 1950 to September 1953. Malaya asked for 1950-52, Nigeria for the period October, 1950 to September, 1953 and Thailand for January, 1950 to September, 1953.

Many of the proposals in the London draft had already become generally acceptable but were still open to extensive comment or were being defended even more extensively as articles of deep-rooted faith.

On the question of votes, the London draft had proposed the use of net imports but for a period as yet unspecified. Each delegate was concerned not, of course, with the principle but only with the result of any formula. A mass of calculations was then produced – consumption in 1952, net imports over 1950-53, net imports (allowing for re-exports) over 1950-52, etc. On actual consumption the U.S.A. would suffer if any figures were used covering 1945-52 since consumption had then been officially restricted; if the base period included 1952 the Netherlands would gain because that year was the year in which its use of tin for high-grade tin alloy exports to the U.S.A had been artificially high; the United Kingdom would suffer if the base period did not include 1946-48; net imports would favour the U.S.A. by bringing in her stockpile buying. On actual consumption over 1950-52 the U.S.A. would hold 472 votes out of the consumers' total 1,000; but on net imports she would hold no less than 609. The smaller countries, who on the net import principle would have only negligible votes (nine of those present would have received ten votes or less each), pressed for and obtained a concession of a separate basic 10 votes each as governments. A compromise on no principle fixed voting on the average of consumption and net imports. In result, the U.S. vote fell again to 546 (although still a majority) and pressure by the U.S.A. brought the basic vote for each member back to five. The U.S. vote was trimmed again, however. The Belgians had always pressed for an overall limitation on the voting power of any member (that is, the U.S.A. in consumption and Malaya in production) so as to prevent one country holding a majority and thereby being in a position to dominate or paralyse the Council, and the United States was held to a maximum of 490 votes.

Amongst the producers argument was less public but equally bitter. Each producer[14] put forward his best period. The producers' committee accepted generally the 1950-52 basis but with some adjustment in favour of Indonesia and Bolivia at the expense of Malaya and the Congo. At a stage so late that her request had to be granted to ensure an agreement Indonesia raised her claim for tonnage. Bolivia was pressed to make the necessary concession but resisted strongly and it was spread over all the producers.

On the types of majorities required for the Council's actions dispute was as long and confused as it had been for the last three years. It was the general intention of the conference that neither the U.S.A. (holding about one-half of the consumers' votes but only one-quarter of the total combined producers' and consumers' votes) nor a combination of the United Kingdom on the consumers' side with Malaya and Nigeria on the producers' side would dominate the Council, although the smaller consumers realised that strength for the U.S. was necessary, since in tin the interests of U.S.A. were in many respects their own interests. In the long run the agreement ended with a very confusing variety of different types of majorities.

The price range on which the buffer stock would operate expressed, in effect, the judgement of the Tin Council as to a reasonable price for tin; it therefore formed the core of the argument. The Belgian Congo promptly proposed a ceiling price of £964 a ton or 120.75 cents a pound (M. Depage) and Bolivia (J. Nunez Rosales) a floor price of £824 a ton or 103 cents. The producers justified their position in a long memorandum which wound up with a request for a ceiling price of £925-950 and a floor of £725-750, that is, a range of £100 around either side of £825-850. Amongst the consuming countries, Belgium, the Netherlands and the United Kingdom, not unnaturally in the light of their mining interests, favoured a floor price of £725-750 but the majority of the consumers objected. A good offices committee[15] recommended a ceiling price of £840 or 105 cents and a floor of £680 or 85 cents. The proposal was objected to by Canada as a consumer (she wanted a floor of £625) and by Bolivia as a producer (she would take a floor price of £640 if the ceiling price were £880) and the situation was saved by the United States which, hitherto very quiet in the whole discussion, suddenly and

[15]Consisting of G. Péter, G. C. Monture (Canada) and A. P. Makatita (Indonesia).

without any apparent logic, decided to accept the Bolivian offer of £640/880.

The size of the buffer stock had been raised in the London draft by the producers to 25,000 tons. Bolivia staged but failed in a last-ditch stand for a maximum of 15,000 tons. As regards the operation of the buffer stock manager the London draft had provided for power to sell between the middle and the ceiling and to buy between the middle and the floor; under U.S. pressure the two sectors became three and the very important principle of a middle sector, in which no action took place and the market was left free, was accepted.

The London draft had carefully dodged any reference to the proposals for action in the event of a tin shortage which France (P. Legoux) had put forward in 1950. Some of the smaller consuming countries (in particular, Turkey, Switzerland and Spain) regarded the "tin shortage" article as an essential part of the agreement. But the French proposals, now repeated, were felt by the producers as being too rigid and imposing too great obligations on them, and the agreement adopted only a series of fairly meaningless recommendations.

One announcement remained. In the prices discussion the U.S.A. said that by March, 1954 the U.S.A. would have 38,000 to 40,000 tons of tin in excess of its present requirements for the strategic stockpile. No decision had as yet been reached concerning the disposal of this surplus; the U.S. government was still studying the question of its disposal with the aim of minimising the effect on the tin market. The wheel had turned full circle; the disposal and not the accumulation of the stockpile would at some time in the future be a primary objective of American tin policy.

XI

The Building-up of the United States Strategic Stockpile

THE United States has had only a negligible domestic mine production of tin. If we can trust the very intensive and detailed geological investigations so far made, including those in the only promising areas of Alaska, the U.S.A. is very unlikely to change this position in the future. She is dependent for her tin supplies entirely upon imported tin, either as tin concentrates for smelting or as metal. Tin is regarded by the United States authorities as an essential material in the prosecution of a war. This complex dependence on outside tin has always deeply irked the Americans and is aggravated by the sharp contrast with almost every other metal required for war purposes. In iron, copper, lead, zinc and nickel the U.S.A. can fall back on important sources of supply from domestic mines or from neighbouring and friendly countries where in many cases, even if there is no U.S. territorial control, there is ample and influential U.S. investment of mining capital. With tin, the nearest sources of supply are Bolivia and Nigeria, some 4,000 to 5,000 miles away. The main world supply in south-east Asia, which is the area normally drawn upon by U.S. industry, is 10,000 miles and up to six weeks away. In none of the major tin-mining countries has there been enough private capital investment by U.S. interests to ensure pressure for a direction of some supplies at all times to the U.S.A.

The first steps in stockpiling

The desire to stockpile tin was fairly long-standing in U.S. official thinking. In 1934, the President's Planning Committee for Mineral Policy recommended the purchase of such a stockpile. In the Navy Appropriation Act of 1938 there was included a fund of $3·5 millions for an accumulation by that department of reserve supplies of strategic and critical raw materials. Additional funds of $500,000 in each of the next two years were similarly provided; and the reserves acquired included some tin. The Strategic Materials Act (Public Law No. 117) of June, 1939 was intended to provide the general machinery for the stockpiling of all the designated raw materials which the Reconstruction Finance Corporation would certify of value in time of war. Action under the latter Act was to provide very useful tonnages from many deposits of other metals, both inside and outside the United States, but had negligible results on the production of tin.

By the end of 1939 the U.S. government had bought about 3,680 tons of tin (additional to the small Navy reserve) or less than one month of the total U.S. consumption. The basis of the tin stockpile was to be laid, however, by the contracts between the International Tin Committee and the Metals Reserve Co. in 1940-41 in which the Americans contemplated a stockpile of 100,000 tons or more – a figure largely attained before the loss to Japan of the south-east Asian tinfields in 1942.

In addition to its purchases from Malaysia and Indonesia (which helped to make the total intake of metal into the U.S.A. from these two countries no less than 230,000 tons in the two years 1940-41) Metals Reserve also bought tin, for current usage or for stockpiling, from the Belgian Congo (almost all of whose output during 1942-45 was shipped to the U.S. as metal or as concentrates). In the case of Bolivia, early agreement was reached with the United Kingdom, which had been smelting the great bulk of the Bolivian tin concentrates, for a division of Bolivian exports, roughly half and half between the United Kingdom and the United States.

By the end of 1945 governmental stocks of tin (in metal or concentrates) were reported to be about 61,000 tons or over one year of restricted war-time consumption. In addition, there was a further "sterilised" stock of metal held by the Navy and the Treasury (7,433 tons by 1941 and 12,140 tons by the end of the war).

The stocks could not be regarded as reasonable enough to keep the U.S. economy going in a period of future emergency of any duration. The U.S. administration had seen one of its major forebodings – the cutting off by enemy action in south-east Asia of its major world source of tin as well as of rubber – become horrible reality in 1942. It had scarcely recovered before it began to anticipate dangerous designs by the U.S.S.R. or by China on the same south-east Asian field; in any case, it feared internal Communist subversion in Thailand and Malaysia. It prepared to shift from the temporary policies of purchases, which had been so successful in meeting the immediate requirements of the recently concluded Second World War, to a longer-term policy of reinforcing the strategic stockpile for the Third World War.

The basic policy in U.S. buying

The bulk of the world output of tin was in the territories of countries (the United Kingdom, the Netherlands and Belgium) which participated in the benefits of the Economic Co-operation Act 1948. The State Department pointed out to the Senate[1]:

"The United States has used substantial quantities of certain of the (deficiency raw) materials in furnishing assistance to Europe during the post-war period and it will use further quantities in furnishing assistance under the recommended program of European economic recovery... It is proper that in partial return for the very considerable assistance provided (to them) by the United States, the participating countries should give reasonable help in replenishing stocks of materials expected to be in long-term short supply in the United States.

"Some among (the participating countries) have resources of this nature either within their own territory or that of their colonies, territories or dependencies... In (some) instances, it appears that, under an aggressive plan of exploration, development and expansion of productive facilities or by other actions, additional supplies could be produced or made available.

"The administring (U.S.) agency should be authorised to

[1] *Outline of European Recovery Program: Draft legislation and background information.* Submitted by the Department of State for the Senate Foreign Relations Committee, 19 December, 1947.

[2] *Foreign Assistance Act of 1948: Report of the Committee on Foreign Affairs on Bill S. 2202.* March, 1948.

[3] *Cmd. 7649. Economic Cooperation Agreement* 6 July, 1948

help to increase production. Procurement by the United States for stockpiling purposes of a fair proportion of available quantities of the materials desired by the United States should be facilitated by the countries concerned . . .

". . . As a further possible step . . . loans made by the U.S. agency . . . might contain a provision specifying that, in the event (that) circumstances make the probability of repayment of the loan in dollars at its maturity date doubtful, the participating country may tender or the U.S. Government may require delivery of materials expected to be in long-term short supply in the U.S. . . . in such amounts and at such times as are mutually agreed as being equitable in full or partial fulfilment of the loan obligation."

The report on the Bill by the Committee on Foreign Affairs to the House of Representatives[2] said with respect to the Bill's stockpiling provisions:

"One of the points to be covered in the bilateral agreements to be undertaken as a condition for participation in the benefits of this recovery program is stated as 'facilitating the transfer to the U.S. by sale, exchange, barter or otherwise for stockpiling purposes' . . . of materials that are required by the U.S. because of deficiencies or potential deficiencies in its own resources . . . Paragraph 9 makes mandatory upon the participating countries the recognition of the principle of equity in respect to the drain upon the natural resources of the U.S. and of recipient countries, through providing a schedule of availabilities for U.S. purchase of strategic materials at world market prices, in order to ensure an equitable share of such materials for U.S. industry . . .

"Insofar as is legislatively possible . . . the Bill may be said to encourage stockpiling through development, purchase or repayment through strategic materials – those deficient in the U.S."

The Economic Co-operation Act of 1948 incorporated provisions on these lines, and the participating countries included the obligations regarding delivery to the U.S. stockpile in the individual agreements into which they entered with the U.S.A. Thus, the U.K./U.S.A. agreement[3] provided that:

"The Government of the United Kingdom will facilitate the transfer to the U.S.A. for stockpiling or other purposes, of materials originating in the United Kingdom which are required by the U.S.A. as a result of deficiencies or potential deficiencies in its own resources upon such reasonable terms of sale, exchange, barter or otherwise and in such quantities and for such period of time as may be agreed to . . . after due regard to the reasonable requirements of the United Kingdom for domestic use and commercial export of such materials. The Government of the U.K. will take such specific steps as may be necessary to carry out the provisions of this paragraph, including the promotion of the increased production of such materials within the U.K. and the removal of any hindrances to the transfer of such materials to the U.S.A.

"The Government of the United Kingdom shall, when so requested by the Government of the United States, negotiate where so applicable (a) a future schedule of minimum availabilities to the U.S.A. for future purchase and delivery of a fair share of materials originating in the U.K. which are required by the U.S.A. as a result of deficiencies or potential deficiencies in its own resources at world market prices so as to protect the access of U.S. industry to an equitable share of such materials either in percentage of production or in absolute quantities from the U.K. . . . and (c) an agreed schedule of increased production of such materials where practicable in the U.K. and for delivery of an agreed percentage of such increased production to be transferred to the U.S.A. on a long-term basis in consideration of assistance furnished by the U.S.A. under this Agreement . . .

"This Agreement shall extend (also) . . . to the territories specified in the schedule (being the British colonial area)".

In the interpretative note attached to the agreement it was stated that:

"It is also understood that arrangements negotiated under

---

⁴ A private statistical calculation, based on published figures of imports, metal production, commercial consumption, etc., indicated a total tin metal stock in the U.S.A. at the end of 1945 of around 120,000 tons (of which private owners held some 16,000 tons) and around 33,000 tons of tin-in-concentrates (practically all government-held).

Article V might appropriately include provisions for consultation in accordance with Article 32 of the Havana Charter for an International Trade Organisation, in the event that stockpiles are liquidated."

The general U.S. policy on the holding and release of stock-piling materials was established in the Strategic and Critical Materials Stock Piling Act of 1946 (Public Law No. 520).

The size, present or proposed, of the strategic stockpile of tin (or of any other strategic material) was to be kept highly secret for the next 15 years. In 1948 the Chairman of the Munitions Board, speaking of stockpiling generally, stated that the amount of materials to be acquired for stockpiling was assessed by selecting the 12 months' greatest usage during the last war (of 1941-45) of a material that was used both by the armed forces and the civilian economy (after deducting the entire amount that would be available in the western hemisphere) and by assuming a five years' war.

This would seem to indicate a target for tin at that time of the order of perhaps 160,000 tons but, even if such a figure were originally so calculated, events were soon to prove it meaningless.[4]

The quarrel over "price gouging", 1951

The extremely tight control which had been maintained over world tin supplies and world tin prices by the United Kingdom and United States governments, operating during the latter part of the war, began to be eased only when by 1949 the rehabilitation of the south-east Asian tin mines seemed to be well on the way.

This long-postponed move back to a free market economy came at a most unfortunate time. The Korean war began an immediate world scramble for raw materials, where the U.S. government buying agencies led the field, sometimes in competition against one another. The tin price rocketed. In the eight months between June, 1950 and February, 1951 the London price rose quite remarkably by no less than £868 a ton; and the New York price followed suit to a peak of about 183 cents a pound.

This price movement intensely exasperated U.S. political public opinion. A Senate sub-committee[5] in March, 1951 denounced the price rises which were resulting from the almost hysterical scramble for tin. If, they said, the U.S. allies in Europe and elsewhere expected the U.S.A. to supply large quantities of food and armaments, etc., the U.S.A. must call on the allies who controlled the tinfields to do their part in helping the U.S.A. to stockpile more tin at a less burdensome cost. The efforts of the U.S.A. to prepare for the common need in face of the unresponsiveness of the producers had furnished the impetus for pushing up the price of tin to its then unreasonable level. The U.S.A. must make it plain to the governments of the producing nations that it was anxious to acquire more tin; that action was required by these governments to stimulate increased production; that government control over supplies and an international tin allocation system must be instituted by the non-communist nations to ensure that the U.S.A. obtained all the tin that could be spared over their actual needs; and that governmental control over tin prices should be instituted so that the U.S.A. was not gouged by the mine-owners and speculators in the tin producing countries for the privilege of defending them. If this kind of co-operation were absent, the United States must reconsider its tin stockpile objective.

The sub-committee made very sweeping proposals. It suggested an international tin conference to improve production, to allocate all exports of tin metal, to restrict in all countries the non-essential uses of tin, to close the London, Singapore and New York tin markets, to have fixed, stable and reduced prices for tin negotiated by governments (at levels based on fair profits for producers rather than on the necessities for U.S.A. defence) and to request each tin-producing government to report on its maximum productive capacity and the reasons for the failure to

---

[5] Preparedness Subcommittee of the Armed Services Committee of the U.S. Senate, *Report*, 1951. Summarised in *The American Metal Market*, 6 March, 1951.

[6] The International Tin Study Group was not invited.

[7] *The American Metal Market*, 16-17 May, 1951. The same paper printed a statement on the Malayan position from F. S. Miller, vice-president of the Pacific Consolidated Tin Corporation, one of the very few U.S. firms owning tin mines in Malaya.

reach the peak 1940-41 levels of production. The U.S.A. should assure the producing governments that any tin obtained from increased capacity would be added to the U.S. stockpile. After completion of that stockpile the U.S.A. would not use the additional amounts of tin made available to it as a means of manipulating the world market price; the U.S.A. would not – except in a war emergency – dispose of its tin in excess of specified amounts or percentages in any one year.

Pending the negotiations of agreements on these lines at reasonable prices, the U.S.A. should withdraw, as it legally could, from further purchases on its Indonesian and Congo tin contracts. When tin prices had reached a satisfactory level the U.S.A. should resume stockpiling at a rate consistent with the world surplus of production over consumption and on a basis preventing any further price gouging.

The U.S. government, through the National Production Authority, should allocate tin internally to the American consumers and should impose more stringent conservation measures. Through the Emergency Procurement Service, it should authorise long-term contracts with the tin producers of sufficient duration and stable prices to encourage such producers to increase capacity so as to fill definite needs at reasonable prices.

This flat-footed and intemperate assertion of the role of the tin-producing countries merely as fulfilling whatever requirements the U.S. set forth was not perhaps the happiest of atmospheres for the proposed international tin conference. This was convened by the U.S. in Washington in March, 1951. It was not fully international; it was confined to the U.S.A. and the major tin producers[6]. It produced no result on international allocation or on shutting the tin markets or on the freezing of prices. But the U.S.A. itself, which had already reverted to tighter limitations on consumption in January, 1951 extended its restrictions in April.

The quarrel with the tin producers was made all the more bitter by the chairman of the Senate Armed Services Committee Lyndon B. Johnson, later U.S. President. In a letter to *The American Metal Market*[7] he repeated the U.S. argument. Must the U.S.A. pay its allies exorbitant tin prices to build up the reserves necessary to resist aggression and to maintain peace for its allies as well as for itself? The tin companies had not acted with any special sense of urgency in production. They had taken advantage of a desperate world situation; they had wanted the perpetuation of their most prosperous period. They must be prepared to negotiate with the U.S.A. for reasonable prices based on reasonable returns and on full information as to costs.

In July, W. S. Symington, the administrator of the Reconstruction Finance Corporation, added more fuel[8]. It was certain that the tin producers and traders on the world tin markets – including the speculators – were exploiting, to the fullest, the profit possibilities of the situation. The R.F.C. knew by experience that a tin cartel existed.

The major producers replied to these attacks without much reservation. The F.M.S. Chamber of Mines[9] categorically denied the charges. The report of the Preparedness Sub-committee contained numerous errors of readily ascertainable fact and there were grave omissions and distortions. Tin production in Malaya had been restricted, under the international tin committee scheme, only during the years 1931-39. Very large expenditures were being made to restore productive capacity as quickly as possible. Notwithstanding internal communist violence the rate of production was being increased; the fact that it had not been increased by the record-breaking prices obtained for metal in recent months indicated that the industry was already working at maximum capacity in the circumstances. The producers had no control over prices and the charges of gouging by the producers were absurd.

The Bolivian producers[10] replied generally that the depression in the tin price (as the results of price cuts by the U.S. administra-

[8] In a statement to the Preparedness Subcommittee of the Armed Services Committee of the U.S. Senate. *The American Metal Market,* 24 July, 1951.
[9] *Notes on Tin,* 1951, p. 89.
[10] In a statement by the Hochschild group, the Aramayo company, the Associacion Nacional de Mineros Medianos and the Representacion de la Minera Chico. *Mining Journal,* 13 April, 1951 and *Metal Bulletin,* 17 April, 1951.
[11] In his letter to *The American Metal Market,* 16 and 17 May, 1951.
[12] The paper was quoted by Lyndon Johnson in his letter to *The American Metal Market.* 16 and 17 May, 1951.
[13] *The American Metal Market,* 26 February, 1954.

tion) appeared to be incompatible with the U.S. Point IV Programme, since the U.S. could not at the same time wish to develop areas through direct dollar aid and also to reduce the price of the principal commodity of those areas. Specifically, they stressed that through late 1949 and early 1950 the tin price for Bolivia of 75 cents a pound had meant practically all the mines in Bolivia working at a loss; Bolivian mining costs were higher than elsewhere; the grade of ground had dropped; labour costs had quadrupled; and the costs of smelting Bolivian tin ore at the U.S. Texas City smelter were $30 a ton higher than in the European smelters.

F. S. Miller stressed[11] the effect of the anti-communist emergency on Malaya's costs and on its ability to expand, develop and equip mining properties. The world supply of tin was adequate to meet all demands for normal industrial consumption. It was only the governmental purchasing for stockpiling in anticipation of future requirements that had made tin appear scarce. It was not the responsibility of tin producers to provide tin for stockpiles unless stockpile objectives were known and unless some reasonable and systematic arrangements were made for their purchases. The Subcommittee's report had confirmed the worst fears of the Malayan tin producers as to the serious threat which the U.S. strategic stockpile constituted for them. The *Straits Times*, the most influential paper in Malaya, declared in a more outright manner that "until there is some safeguard against a collapse in prices when stockpiling ends, Malaya owes America nothing" and appalled Lyndon Johnson[12].

The lines of parallel or divergent argument were to continue for many years to come. In July, 1952 when the dust was settling, the U.S. Senate Preparedness Subcommittee (chairman Lyndon Johnson) still regarded the then price of tin as too high and again charged that the tin producers were an international cartel. In February, 1954 Senator Johnson, in a letter to President Eisenhower[13] referred to the Texas City smelter as a potent weapon in America's battle against the price-gouging international tin combine. There was no assurance that the combine would not return to its price gouging at the first opportunity.

The ugly word "gouging" and the ugly temper of the Senate's subcommittee which used it were long to be remembered on both sides.

The U.S. buying strike, 1951-52

Although the international tin conference convened in Washington in March, 1951 had produced no international price fixing or allocation, the U.S.A. administration had a wide enough sphere within which to operate. It acted in much the same spirit as it had waged war during 1941-45. On 6 March, 1951 all purchases for the strategic stockpile of tin were suspended until the price of tin had again reached reasonable levels. On 12 March, the Reconstruction Finance Corporation was governmentally authorised to be the sole buyer and seller of tin in the U.S.A. The Corporation fixed a lower price, reduced it twice in June, again in July (to 103 cents) and held it there until January, 1952. By that latter date the disciplining had become effective (the Malayan price had promptly followed the New York price downwards) and the Korean panic was easing with the R.F.C. calmer. On 22 January, 1952 the R.F.C. price was raised to 121.50 cents.

The R.F.C. reinforced its attack on the world price by ostentatiously keeping away, as far as it could, from new long-term buying contracts. The 1950 Bolivian contract for tin concentrates expired on 31 December, 1950. It was extended only to 28 February, 1951 and then in effect for three months until 31 May, 1951.

For the next three months no tin was bought from Bolivia, and a further and last contract was only for thirty days ending in October, 1951 and only at a price of 112 cents a pound. The existing contracts with Indonesia (at 9,000 tons of tin a year) and with the Congo (at 1,500 tons a year) ran to the end of 1951 with options to renew, and the contracts with Thai producers for about 3,000 tons of tin in the year 1950-51 were continued on a

---

[14]Tin-in-concentrates stock 21,100 tons at end-1949, but only 10,800 tons at end-1951; metal stocks at Texas City 22,400 tons end-1949 but only 8,600 tons at end-1951.

spot basis. From March, 1951 to January, 1952 the R.F.C. boycotted purchases of foreign metal.

The non-buying policy was intended to be a serious blow to Malaya. U.S. imports of tin metal from Malaya were reduced from 53,000 tons in 1950 to as little as 7,000 tons in 1951 and imports from the United Kingdom from 8,400 tons to under 2,000 tons. The total U.S. imports of tin metal from all sources fell from $153 millions in 1950 to $73 millions in 1951. But these exaggerated the effect of the R.F.C.'s non-buying policy.

The intake of metal from the U.K. and Malaya had been abnormally high in 1950 ánd the average imports over 1950-51 were little below the 1949 figure. The R.F.C. was still tied to contracts for the purchase of tin concentrates and the value of U.S. imports of concentrates actually rose from $47 millions in 1950 to $82 millions in 1951. But Bolivia – the weakest of the producers – was badly affected.

The restrictions on U.S. consumption of tin, tinplate and tin-containing alloys, applied in 1951 with war-time intensity, had war-time results, and U.S. consumption of primary tin was brought back to the 1942-45 level.

Towards the end of 1951, the virtual cessation of purchases of tin metal was beginning to tell upon the disclosed U.S. reserve position. In November, tin was defined by the National Production Authority as a "most critical" item and tinplate was declared in short supply (that is, insufficient for defence and essential civilian needs). At the same time, and in view of the inability of the U.S.A. to obtain tin at fair and reasonable prices, further additions of metal to the strategic stockpile (arising from the smelting of concentrates at Texas City) were discontinued at once; all tin metal purchased (that is, from the R.F.C. smelter) would be made available for industrial consumption, and additional conservation measures would be enforced.

The R.F.C. stocks of tin-in-concentrates were almost halved and stocks of tin metal at Texas City more than halved by the end of 1951. The total stocks of metal with the R.F.C. and with private users (but excluding, of course, the strategic stock) were melting. They dropped from seven months' usage at the end of 1950 to under four months' usage at the end of 1951.[14]

The U.S. pressure oṅ the world market had been effective enough in breaking the unprecedented prices shown at the beginning of the year, yet the average New York price for 1951 still worked out at one-third higher than for 1950. In Bolivia, mine production had been higher in 1951 than in 1950 but the curtailment of U.S. buying meant that stocks of unsold concentrates were being piled up in the Pacific ports. Also, Bolivian labour was very restive on the nationalisation issue. The Belgians in the Congo and the Indonesians were alarmed as to the future of their tin contracts with the U.S.A., which stood to be completed at the end of 1951.

The United Kingdom was anxious that the current unpopularity of tin in the U.S. should not react on Anglo-American political relations. The producers were aware generally that the world surplus of tin in 1952 was likely to be as great as – and perhaps even greater than – the very substantial surplus of 1951.

The resumption of strategic buying, 1952

The deadlock was broken in January, 1952 by an agreement between the British and the U.S. governments. The United Kingdom agreed to deliver to the U.S.A. 20,000 tons of tin at 118 cents a pound f.o.b. Malaya or the United Kingdom. While the agreement was in force the U.S.A. would refrain from bidding directly or indirectly on the London or Singapore markets without prior consultation with the United Kingdom. The U.S. government would remain the sole importer of tin into the U.S.A. during the currency of the agreement and would consult the United Kingdom before relaxing import controls.

This tin deal was part of a wider Anglo-American bargain. The U.S.A.[15] was facing a very difficult position on aluminium supplies where mobilisation requirements for the Korean war were running ahead of domestic production; the United Kingdom was short of steel and dollars. The overall arrangements

[15]U.S. Defence Production Administration, Press release, January, 1952.

[16]The price paid by the U.S.A. for the tin (118 cents f.o.b. or roughly 121.50 cents at New York) was cheap. Most of the tin was actually bought in Malaya by the U.K. government, which lost £143,000 on the tin deal.

[17]La Razon, La Paz, 28 November, 1951 quoted in Notes on Tin, 1952, p. 267.

meant the sale by the United Kingdom of 55 million pounds of aluminium as well as the 20,000 tons of tin and the sale by the U.S.A. of one million tons of steel.[16]

The way was now clear for a resumption or extension of R.F.C. buying from other tin sources. The R.F.C. re-entered the market at the beginning of 1952. In March a long-term four-year contract started with the Belgian Congo producers covering a minimum of 7,000 tons of tin a year or over half the anticipated Congolese annual output. At the same time, a three-year contract with Indonesia was signed taking 18-20,000 tons of tin a year (again also over half the anticipated annual production). The fixed price in both contracts (120.75 cents a pound delivered U.S. ports) was in line with the price of 118 cents f.o.b. arrived at in the Anglo-American deal. On the shorter-term aspect much of the tin metal which had passed into the hands of dealers in 1951 was now glad to find a home in 1952 with the government, and in the first half of 1952 the R.F.C. bought no less than 23,500 tons of such metal.

Bolivia was a more awkward problem. The Bolivian delegation at the General Assembly of the United Nations stated that the price of tin, which constituted the foundation stone of the Bolivian economy, had been submitted over many months to the sole judgement of a buyer (the U.S.A.) against which Bolivia lacked all power to defend itself. This buyer had imposed on Bolivia, through methods which were employed by the strong against the weak, an unfair and artificially low tin price which was incompatible with either standards of fair dealing or co-operation.[17] But the hand of the U.S.A. had been very considerably strengthened by its arrangements with the United Kingdom and the other tin producers. Inside Bolivia, a new government under Paz Estenssoro, committed to the nationalisation of the three main mining groups, took power in April. In early June the R.F.C. bought the bulk of the 15,000 tons of tin which represented Bolivia's unsold accumulation. It was firm in refusing preferential price treatment to Bolivia and the price was on the now standard terms of 118 cents a pound.

By the summer, therefore, the R.F.C. had set the current tin supplies of the U.S.A. back on the recognised rails and had re-established its stockbuilding programme. In August, 1952 the R.F.C., with the purchases from the United Kingdom practically completed, ceased to buy metal, and the private import of tin into the U.S.A. was again made possible.[18] Later in the year restrictions on the use of tin and tinplate were removed.

The R.F.C. could now concentrate on its primary function of obtaining tin for the strategic stockpile. The backbone of the task had been broken. In the year to June, 1953 R.F.C. received from all sources 34,800 tons of tin metal and 20,500 tons of tin-in-concentrates of Bolivian, Indonesian, Congolese and Thai origin. The excess stocks of metal were transferred to the stockpile. When it reported for the year 1952-53 the R.F.C. could claim that the U.S.A. stockpile goals were assured, if as yet publicly unknown.

The contract with Indonesia was extended for another year to February, 1955 and was to be extended again for smaller tonnages and quarterly periods in 1955. Until these contracts were completed Indonesia held up her ratification of the First International Tin Agreement and thus delayed the coming into effect of that Agreement until mid-1956. The Congo contract for smaller tonnages continued until late 1955 and the purchase of spot parcels from Thailand finished in March, 1955.

Purchases of tin concentrates from Bolivia proved more troublesome. These purchases by the U.S.A. were necessary to maintain in operation the government-owned Texas City smelter. There seemed now to be no defence or stockpile argument in favour of retaining the smelter, whose public ownership was an offence to the ideas of private enterprise. On the other hand, the smelter was a long-established employer in a Texas where Lyn-

[18]The R.F.C. still remained available as a seller of metal to U.S. firms until 31 March, 1953 but the R.F.C. price was maintained at 121.50 cents a pound when the New York price ultimately dropped to under 100 cents. Actual R.F.C. sales were therefore small.

[19]The Reconstruction Finance Corporation ended on 1 July, 1954 and was, in effect, replaced by the Federal Facilities Corporation.

[20]My examination of U.S. official and public statistics over the eighteen years 1939-56 inclusive, taking account of published stocks of tin metal and tin-in-concentrates, imports and domestic production of tin metal, exports and consumption of tin metal showed an overall surplus supply of about 350,000 tons. After 1956 about 8,000 tons of barter tin was added to government stock. The total figure officially revealed at the end of 1961 was 349,000 tons.

don Johnson was strong. Through 1955 there went on arguments on the cessation of Texas City operations and the spasmodic purchase of Bolivian concentrates by the Federal Facilities Corporation.[19]

A contract for 10,000 tons of Bolivian tin was completed about the end of 1953, another 12,000 tons was contracted for delivery over June, 1954-April, 1955, and this was followed by another 2,500 tons. It was a still greater offence to much U.S. opinion that a nationally-owned enterprise should be buying concentrates from a revolutionary, nationalising Bolivian government. The operation of the Texas smelter was approved up to the middle of 1955, but it was formally offered for sale in August, 1955.

In the middle of 1954 it was revealed by the stockpile administration that the minimum strategic stockpile target for tin had been reached but that, nevertheless, it had been decided (for unstated reasons) to raise that target by 60,000 tons (part of which had already been received).

But one additional fling of U.S. policy remained to come. However extravagant the U.S. short-term policy had been in the building up of the strategic stockpile, it had none of the long-term implications internally of its price support policy for the agricultural interests. By 1958 the accumulated surplus agricultural holdings which could not be sold were becoming a most acute internal domestic problem. The Department of Agriculture conceived a bright plan to ease at the same time both the strategic and political problems. It would dispose outside the U.S.A. of part of the country's agricultural surpluses; it would obtain in exchange through the machinery of private traders materials – including tin – required for the strategic stockpile, even although that stockpile was brimming over the rim. In 1958-59 Bolivia, Thailand and the Congo accepted the proposed barter deals and another 8,000 tons of tin came into the supplemental stockpile.

The size of the strategic stockpile remained officially secret and was not to be revealed until 1961. It was easy enough to estimate in 1957 on available statistics that the stockpile then held about 350,000 tons of tin.[20]

The completion of the stockpile

In building-up that stockpile, the U.S.A. had created the unprecedented production boom of 1940-41 and had brought into production capacity which was not to be required again at any time for at least the next thirty years. The buying for the stockpile in 1950 had been of a panic character, and the fact that the U.S.A. had not been alone in its panic did not absolve her from the major blame for an unprecedented flare-up in prices. The U.S. cessation of buying and its deliberate cutting of the tin price in 1951 was in the worst tradition of high-handed action and not far removed from economic blackmail. The scarcely-hidden U.S. attitude that the objective of a tin-producing country – not as yet dignified under the title of a developing country – should be to ensure, for the safety of the U.S.A. and at a price lower than the market price, the creation of a strategic stockpile whose ultimate size was not to be made known, jarred badly on outsiders.

But the U.S.A. was not solely to blame when its continued purchases of tin for the stockpile in 1952-54 maintained an over-stimulus to world tin production which helped to create the surplus crisis of 1957-58. The producers were clamorous for more U.S. purchases, although it required little consideration to realise that these purchases, while postponing the day of surplus, were also guaranteeing that, when the world was free of stockpile buying, that surplus would be overwhelming.

Neither party, when stockpiling ceased in 1955 and the U.S. was left sitting on top of around 350,000 tons of tin (the equivalent of over two years' total world production or of seven years of U.S. annual consumption), had yet seriously considered what would happen if the absurdity of these figures were to be fully realised in the U.S.A., and if any part of that gigantic holding were to be released, however gently, on the world market.

XII

The Principles and Machinery of the Tin Agreements since 1956

THE International Tin Council, set up by the first international tin agreement of 1954 and continued by the later second, third and fourth five-year agreements[1], was a body created at a conference convened by the United Nations. Its formal existence depended upon a resolution of that conference and the very language in which its principles were expressed depended on a draft Havana Charter (which had never come into formal force) and on the interpretations and modifications of that draft charter at later meetings of the United Nations and of the United Nations Conference for Trade and Development (UNCTAD). But the Tin Council was not an integral part of the United Nations or of UNCTAD. In some of the agreements membership of the Council did not depend on membership of the United Nations, nor did a termination of the agreement before its formal ending rest upon action by the United Nations. The policy decisions of the Council did not depend on the United Nations, although the Council never, in fact, had the need to take an important decision which could be regarded as completely contrary to U.N. policy. Whatever may have been the debt for its existence which it owed to the United Nations, the Council was not part of the massive United Nations machinery. At no time in the first sixteen years did the United Nations feel that it had any right to interfere in the work of the Council. The Council had an economic function to fulfil and an economic agreement to operate.

By its concentration on this task, the Tin Council built up its reputation and continued its existence without any serious challenge and so helped itself to avoid the interminable flood of politico-economic arguments which have always threatened to

---

[1] The first agreement ran from July, 1956 to June, 1961, the second from July, 1961 to June, 1966, the third from July, 1966 to June, 1971, and the fourth from July, 1971.

overwhelm UNCTAD. In this concentration the Council, of course, lost something. Anxious to maintain its prestige as an effective, continuing body with authority and teeth, it tended to follow rather rigidly the same path of export control and buffer stock, and it never looked seriously at any other possibilities. It had some justification for the rigidity in its thinking; there have been so few winners in the field of inter-governmental commodity control agreements that, given a successful runner in tin, no one has desired to change its riding rules and riding habits.

The preamble of the agreements

The preamble of the first tin agreement of 1954 recognised that widespread unemployment or under-employment in the tin producing and tin consuming industries might arise out of the special difficulties to which the international trade in tin was subject. The special difficulties included a tendency towards persistent disequilibrium between production and consumption and the accumulation of burdensome stocks. The agreement expected a burdensome surplus of tin to develop; and it expected that surplus to be aggravated by a sharp reduction in purchases (by the U.S.A.) of tin for non-commercial stocks. It believed that this situation could not be corrected by normal market forces in time to prevent widespread and undue hardship to workers and the premature abandonment of tin deposits. It recognised the need to prevent the occurrence of shortages of tin and to take steps to ensure an equitable distribution of supplies in the event of a shortage.

This preamble in general repeated the wording of the draft tin agreements of 1949-50 and 1953, which themselves merely copied the draft Havana Charter. The reference to purchases for non-commercial stocks was, however, new; it arose from the U.S. announcement as to the forthcoming cessation of its purchases, an action which was to expose nakedly the massive over-production hitherto concealed from the market.

The preamble suffered only one modification in the second agreement. This related to non-commercial stocks. The problem had changed in time. It was no longer a problem of the building-up or the cessation of building-up of the U.S. strategic stock but of the threat of liquidation of that stock and the consequent disastrous effect on the tin price. The second agreement referred to the effect of uncertainties on the disposal of strategic stocks unless consultation took place.

The wording of the third agreement was affected by the changing vocabulary and concepts of the United Nations in the middle 1960s and, in particular, the concepts of import purchasing power and economic growth among the developing countries. It was also affected by a shift of certain points from the preamble of the agreement, which was, after all, only a recognition of principles and good intentions, to the objectives where the obligatory character was more pronounced. The third agreement accepted that commodity agreements, by helping to secure short-term stabilisation of prices and steady long-term development of primary commodity markets, could significantly assist economic growth, especially in relation to the developing producing countries.

Continued co-operation between producing and consuming countries within the framework of an international commodity agreement would help to resolve problems in tin – a platitude with which no one would dare to disagree. It accepted the "need to protect and foster the health and growth of the tin industry, especially in the developing countries" (a most sweeping phrase which the U.S.A. accepted, perhaps unnecessarily, in its discussions with the Tin Council as one of the determining factors in its own policy of disposals from the strategic stockpile). It noted the need to ensure adequate supplies of tin to safeguard the interest of consumers in the importing countries (a reflection of the shortage in production in the early 1960s which had followed the period of export control over 1958-60) and it reverted again to the new United Nations concept of maintaining and expanding the export purchasing power of producers.

The preamble of the fourth agreement of 1970 repeated the preamble of the third agreement, with one addition. The addition expressed the desirability of expanding tin consumption in both developing and industrialised countries, an admirable but not very meaningful sentiment.

The objectives of the agreements

The rotund, official sentiments set out in the preambles were expressed more crisply in the list of objectives of the Council. That list expanded with time. The first agreement stated four specific objectives, the second five, the third jumped to twelve and the fourth had ten. This was, however, not so much a delegation of wider powers and authority to the Council as a process of editing.[2]

The heart of the objectives of the tin agreement as set out in the first agreement was:

"(a) to prevent or alleviate widespread unemployment or under-employment and other serious difficulties which are likely to result from maladjustments between the supply of and demand for tin;

(b) to prevent excessive fluctuations in the price of tin and to achieve a reasonable degree of stability of price on a basis which will secure long-term stability between supply and demand;

(c) to ensure adequate supplies of tin at reasonable prices at all times."

The first two of these objectives were repeated in the second agreement, but that also modified (but without clarifying) the third objective to make it read: "at prices which are fair to consumers and provide a reasonable return to producers". No attempt was made, or indeed could be made, to define the officially hallowed and meaningless terms "fair" and "reasonable".

The third agreement, obliged to go further so as to cover the general shift of United Nations theory, transformed the second of the objectives into a clumsy double-barrelled formula:

"(c) to make arrangements which will help to maintain and increase the export earnings from tin, especially those of the developing producing countries, thereby helping to provide such countries with resources for accelerated economic growth and social development while taking into account the interests of consumers in importing countries:

---

[2] Much of the editing of the fourth agreement was done by M. C. Bué (France).

(d) to ensure conditions which will help to achieve a dynamic and rising rate of production of tin on the basis of a remunerative return to producers which will help to secure an adequate supply at prices fair to consumers and which will help provide a long-term equilibrium between production and consumption."

With every question so begged and every viewpoint admirably squared the fourth agreement was glad to continue this version unchanged.

The other objectives, brought in mainly under the third agreement but carried forward into the fourth agreement, attempted to balance action in the event of a surplus and a shortage of supplies, to deal with the problem of sales from strategic stockpiles and to bring the Council closer to mining problems. Of these the most important was the problem of strategic stocks. Under the first and second agreements, relatively small tonnages of stocks of tin which had been accumulated by countries members of the Council had come up for disposal. One of those stocks (from the United Kingdom) had been sold on behalf of that government by the buffer stock Manager of the Council in a manner which did not affect the market price adversely; stocks sold by the Canadian and Italian governments had been sold by those governments for internal consumption at tonnage rates known to the Council.

But these three stocks were as nothing compared with the dynamite in the strategic stockpile of tin held by the United States, amounting by the early 1960s to over two years of total world supply. No action on the release of these stocks had been taken by the U.S.A. during the first agreement although that agreement had included provisions (not, of course, binding on the U.S.A. as a non-member) for consultation with the Tin Council on disposals. In essence, the work of the Tin Council under the second agreement was one long and not always united struggle on the amounts, prices and timing of U.S. disposals, and one of the major successes of the third agreement was to see the voluntary cessation of U.S. disposals. By the third and fourth agreements the objective of the Council on stockpiling had become clearer:

"to review disposals of non-commercial stocks of tin by governments and to take steps which would avoid any uncertainties and difficulties which might arise."

The agreements have always maintained as a principle the need to serve the interests of both producers and consumers. It was the Council's duty to concern itself with both surplus and shortage. The initial broad objective of the first agreement to ensure adequate supplies of tin at all times became in the third and fourth agreements more pointed with a double obligation:

"in the event of a shortage of supplies of tin occurring or being expected to occur, to take steps to secure an increase in the production of tin and a fair distribution of tin metal in order to mitigate serious difficulties which consuming countries might encounter; (and) in the event of a surplus of supplies of tin occurring or being expected to occur, to take steps to mitigate serious difficulties which producing countries might encounter."

This admirable balancing of interests in the objectives bore no relation at any time to the action of the Council under any of the agreements. In the fourth agreement itself the relevant article on export control has 21 clauses; its general tenor is strict and compulsory; the parallel article on action in the event of a tin shortage has two clauses and the compulsion there on the Tin Council is merely to enquire, to recommend and to observe. A real balance of interests and obligations between producers and consumers would have required a change in the basic lines of the tin agreements which there was little likelihood would ever take place. In a surplus, the Council was authorised to take steps which, in the opinion of the producing members, could mean nothing but the equitable restriction of exports until the emergency had passed or the excess tin bought by the buffer stock; the first action might be seriously harmful to their economies, the second might be beneficial in the long run to the price but the cost would be carried by producers alone. In a shortage, consumers would expect an increase in production

---

[3] Since the Geneva conference of 1950, Pierre Legoux (France) has fought so brilliantly and tenaciously for this principle of equality for producers and consumers that he may ultimately be successful.

(without any reference to the responsibility for the financing of such extra production) or they would have to accept "a fair distribution of tin metal" which would mean some system of allocation of supplies. The Council had not at any time as yet shown itself willing to embark on restrictions on the movement of supplies of tin metal with the possible disruption of the Malaysian and London metal markets which this might have involved.[3] Fortunately, although shortage did develop in the 1960s, the Council found a solution provided to it not by the exercise of the provisions of its own agreement, but by the U.S.A. in disposals from the stockpile.

The last of the objectives of the fourth agreement:

"to keep under review the need for the development and exploitation of new deposits of tin and for the promotion, through *inter alia* the technical and financial assistance resources of the United Nations and other organisations within the United Nations system, of the most efficient methods of mining, concentration and smelting of tin ores" carried on in some more detail from the first agreement the general interest which the Havana Charter had shown in the most economic use of raw materials. It also reflected the work of the Council in organising conferences on production in 1967 and in 1969 and on consumption in 1972. But until the Tin Council reaches a more influential position where there is a large supply of money available for stimulating exploitation and efficiency, we can regard this part of the Council's objectives as being little more than good advertising.

The membership

The qualifications for membership were stated at their simplest and widest in the first agreement. It was there laid down that "any government, whether represented at the 1953 session of the United Nations Tin Conference or not ... may ... accede to this Agreement...". This opened and was intended to open membership to governments inside and outside the United Nations. The principle was repeated in the second agreement. There was a change, however, in the third. Indonesia, a member of the first two agreements, had quit the United Nations just before the 1965 (New York) Tin Conference which drew up the third agreement and was not present (except actively in the corridors) at the proceedings. Indonesia was, however, probably essential to any international arrangements in tin. To cover her, the third agreement therefore provided that any participant in the second agreement should be entitled to membership in the third.

However politically accommodating the agreement could be on Indonesia (whose absence from the United Nations was only temporary) the argument was different on other countries. The United States, which took part in all the Tin Conferences, was grimly determined that the People's Republic of China (not then a member of the United Nations or of any of its specialised agencies) should be specifically excluded from membership of the Tin Council; Federal Germany (also a participant in the U.N. Tin Conferences, but not a member of the United Nations) was equally determined that membership of the Tin Council should not be open to the German Democratic Republic. On the question of China the south-eastern Asian tin countries might have regarded the possibility of her adhesion to the agreement more seriously if the exports from China to the free market had been on a larger and more damaging scale than they actually proved to be in the years after 1956. In any case, their political relations with China, especially in the cases of Thailand and Indonesia, steadily worsened through the later 1960s. The third and fourth agreements laid down that, in general, membership of the United Nations or its specialised agencies was a necessary qualification for membership of the tin agreement. This ban on the possible (even if improbable) entry of China into the tin arrangements lost its point when China joined the United Nations in 1971.

The ban against the entry of East Germany was accepted by some at the 1965 U.N. tin conference as being the sweetener necessary to bring Federal Germany into the tin agreement in the same way as the ban on China was necessary to ensure the requisite degree of benevolent neutrality from the United States. These hopes were disappointed. Federal Germany did not enter the Council until the fourth agreement in 1971; and it can scarcely be believed that the U.S.A. approach to the Tin Council was seriously affected by its policy on China.

One other problem of membership came up late for the agreement. In 1970, the European Economic Community (then covering six countries) applied for an invitation to the Geneva United Nations Tin Conference meeting to draw up the fourth agreement. The U.S.S.R. and a number of East European countries strongly objected to the acceptance at a conference of sovereign states of a body such as the E.E.C., which was not in itself a sovereign state and which was not in any sense a member of the United Nations or its associated agencies. Some of the producing countries also doubted the validity of the E.E.C. application and felt some alarm at the weight and concentration of the voting on the side of consumers which would follow if the E.E.C. members all voted, as they almost certainly would, in the same direction on the Council. The conference, against the voice of the socialist centrally-planned group and with the producing countries generally abstaining, accepted in the fourth agreement an article (Article 50, significantly the only article in the agreement for which a title was omitted) which avoided the question of membership and said with a most diplomatic lack of clarity: "An inter-governmental organisation having responsibilities in respect of the negotiation of international Agreements may participate in the International Tin Agreement. Such an organisation shall not itself have the right to vote. On matters within its competence the voting rights of its member states may be exercised collectively."

The fourth agreement was subject to signature. The depositary government for the purposes of receiving signatures was the Government of the United Kingdom, then in the process of framing application for membership of the European Economic Community. There may have been a conflict between its impartial juridical position under the agreement and its political desire for membership of the E.E.C. It decided, against the legal advice of the United Nations and without further consultation with the participants in the United Nations Tin Conference, to accept the signature by the E.E.C., but its actions were later endorsed by the Tin Council.

From the beginning, the tin agreements accepted a simple distinction between producing and consuming countries (unlike some other commodity agreements where the distinction was between exporting and importing countries).[4] This ease of classification was helped by the clear distinction in fact between producing countries which consumed little or no tin and consuming countries which produced little or no tin. This distinction was blurred on only two occasions when the country concerned was substantial in both fields. Australia was on balance a consumer of tin until 1964; thereafter, a revival of production pushed it over the balance and in the fourth agreement Australia entered the list as a producing member on the basis not of production but of exports of tin. In the case of the U.S.S.R. – a producer and consumer of unknown tonnages but an importer of known tonnages – the problem of category was solved also in 1971 by accepting her (and, in principle, any other country in future in a similar position) as a consuming country with membership and vote based on her imports of tin. The decision as to category has always lain with the country applying for membership.

The producing members

The producing members of the first three agreements were six (Bolivia, Malaysia, Thailand, Indonesia, Nigeria and the Congo-Zaïre) and of the fourth agreement seven (these six and Australia). The six, as members of all or most of the pre-1939 control schemes, were familiar with the principles and machinery of international agreements; Australia was in a tin scheme controlling exports and paying to the buffer stock for the first time.[5] These producing members normally covered around 90 per cent of the mine production of tin in the world (exclusive of the U.S.S.R. and China). Control of the tonnage of tin entering the world market was therefore always fairly complete, and

---

[4] The terms "importing countries" and "exporting countries" were used in the abortive draft Washington agreement in 1948.

[5] Although in 1968-69 when the Tin Council was applying export control on its producing members, Australia, then a consuming member, accepted a Council resolution and voluntarily limited its tin exports.

the post-1956 agreements were, in general, free from the constant fear underlying the pre-war agreements that the restrictions of output on the participating countries would mean a permanent increase in the loss of output to the "outside" countries. The major threat to the post-war control of markets by the members of the tin agreements came, in fact, from abnormal quarters – sales from the U.S.S.R. and the U.S.A. – which could not be regarded as normal producers.

The major tin producing countries which have stayed outside the four tin agreements are, in rough order of importance, Brazil, Argentina, South Africa, Rwanda, Laos, Portugal and Burma. These countries had in 1956 a total mine production of about 8,000 tons or 5 per cent of world production, and in 1971 a total of about 11,000 tons or 6 per cent of world production. One of them – Rwanda – was included with the Belgian Congo in membership of the first agreement. In 1961, when it had become independent, Rwanda refused to follow the example of the Congo Republic in entering the second agreement. The Rwanda mines had, like all the members of the first agreement, suffered intensely under the export restrictions of 1958-60 and the new government felt politically that it could not continue to accept the colonialism which it associated with the tin agreement. It has shown no interest in the agreements since. Its economic connection with the Congo industry was, however, not affected. Brazil did not develop any substantial tin mining industry until the Rondônia territory was opened up after 1960. The Brazilians, with a new and expanding industry, made it clear at the U.N. tin conference of 1970 that their membership should carry no restrictions on their exports. The negotiations broke down on the timing of entry and the tonnage conditions.

South Africa, with a volume of tin production little higher than its normal metal consumption, has at no time shown any great interest in the tin agreements, and any question of her joining the agreement would certainly have raised bitter political opposition from some members. Rhodesia, a smaller tin producer, had been invited to the 1965 United Nations Tin Conference, perhaps in error, and her presence had aroused strong political representations from the other producers.

Laos, as part of former French Indo-China, had been inside the pre-war control agreements, but that connection, broken in 1941, has not been renewed. Argentina, like South Africa, had a tin production much in line with its consumption and had no political interest (except general Latin American support for Bolivia) in an international agreement. Portugal, after a brief membership of the pre-war control agreements, had in recent years seen its tin production dwindle into unimportance. Burma in the 1930s had refused to join the control agreements because of the disastrous effect of tin restrictions on its mining labour problems and on its associated wolfram production, and showed no interest in the post-war agreements; her tin industry had lapsed into inefficiency and isolation.

The consuming members

The first agreement started in July, 1956 with ten consuming countries in membership. Another four countries ratified the agreement in 1956-57.[6] This initial membership covered under 40 per cent of total world consumption (outside China, the U.S.S.R. and Eastern Europe). This relatively low coverage was a persistent factor in the first three agreements. Uruguay, which had entered as a general South American gesture of goodwill towards Bolivia, failed to continue its membership after 1958.

There were slight differences in membership for the second agreement. Japan (the most important new accession to date) and Mexico joined and brought the consumer coverage in 1961 to 76,000 tons or 50 per cent of the world total as compared with 61,000 tons or 37 per cent of the world total in 1960. No changes took place during the second agreement.

[6] The ten (in date order of ratification) were Canada, Denmark, Australia, United Kingdom, India, Ecuador, Belgium, Spain, France and the Netherlands; the four were Italy, Turkey, Israel and Austria. The Republic of Korea (South Korea) joined in 1958. Israel failed to rejoin the second agreement, apparently in error; joined the third agreement in October, 1967; withdrew in 1970, and did not join the fourth agreement.

[7] The figure of 109,000 tons includes only the import tonnages for the U.S.S.R. The Republic of China (Taiwan), which had joined the fourth agreement was forced to withdraw in March, 1972, after the United Nations had recognised the People's Republic of China as representing China.

The sixteen consuming countries which started the third agreement included one of some political, but not numerical, significance. Czechoslovakia had been a member of the International Tin Study Group but had thereafter shown no interest in the earlier work of the International Tin Council. She was now the first of the European socialist centrally-planned countries to join the Council. She was joined by Poland in May, 1969, by Hungary in November, 1969, and by Yugoslavia in January, 1970; Bulgaria and Rumania were to join the fourth agreement. By the end of the third agreement the Council had 19 consuming members.

The fourth agreement in 1971 added two more substantial new figures. The first was the Federal Republic of Germany and the second the U.S.S.R. The total consumption of the consumer membership was now well over 109,000 tons or about 60 per cent of world consumption.[7]

The motives of the consuming countries in joining the tin agreements were very mixed. Some, such as Denmark (a member from the beginning) and Austria, were pure consumers, with little political weight and no capital connections with the mining industry; but Denmark has one large firm trading in the treatment of tinny materials and in the production of tin-containing alloys for the international market and is therefore extremely sensitive to changes in the price of tin. Italy was also a pure consumer with a tinplate industry growing rapidly during the 1960s. India, again a pure consumer but herself an Asian and developing country, had perhaps some political motive in supporting an international commodity agreement of such direct importance to her other developing neighbours in south-east Asia. The Republic of Korea (South Korea), although a consumer, also had a direct political interest in holding possession of a seat on the Tin Council (as on similar parallel bodies) and thereby underlining her acceptance internationally against North Korea.

Three of the original members of the post-war agreements – Belgium, the Netherlands and the United Kingdom – had been members of the pre-war control agreements. Belgium ratified the first agreement in 1955 also on behalf of the Belgian Congo and

Ruanda-Urundi as the producing member; the United Kingdom ratified in 1954 on behalf of the Federation of Malaya and the Federation of Nigeria, although later these producing countries had separate representation. The three countries were consuming countries with no or (in the case of the United Kingdom) little domestic mine production, but each was heavily involved with the producing countries. The Netherlands through the Billiton company had had ownership of a substantial proportion of the mining industry of Indonesia; it was also smelting large tonnages of Indonesian tin concentrates. The Belgians, even after the independence of the Congo, still owned the major part of the capital of practically all the tin operating companies within the Congo; they owned the tin smelter there and also the smelter at Hoboken (Belgium) which ran on Congo material.

The United Kingdom had very substantial capital interests in tin production in Malaysia, Nigeria and Thailand; and the most important tin smelter in the United Kingdom (controlled, however, by the Patiño interests) treated after 1956, as before 1956, the major part of the Bolivian tin production. It was inevitable that – for reasons which were quite apart from a genuine general desire to aid the developing countries through an international commodity agreement – these three countries should have had an immediate and continuing interest in subscribing to the tin agreements; it was also inevitable that their voting power as consumers in the agreement should not always be thrown in the balance on the consumers' side.

Canada was in perhaps a more curious position. She was a producer for world markets of all the other important non-ferrous metals – copper, zinc, lead, platinum, nickel and uranium – and therefore peculiarly exposed to the problems of price fluctuations which lay at the root of the tin agreements or which arose from disposals by the U.S. stockpile; but in tin she was a consumer almost purely and she had no ties of capital investment or smelting links with the tin producing countries. On the other hand, she was deeply committed to the principle and practice of aid to developing countries. France, amongst the consuming countries, had perhaps the clearest philosophy on commodity agreements and the developing countries. This was perhaps in part the result of personal accidents. The first chairman of the Tin Council over 1956-62 – Georges Péter – had been a member of ICCICA (the Interim Committee for International Commodity Arrangements); the delegate of France to the Council from 1956 onwards – Pierre Ch. Legoux – was for many years by far the most able and eloquent member of the Council and was closely connected with the advisory work of the United Nations on mining policy in developing countries. It was also a matter of general policy. The government of France aimed not only in the Tin Council at being an essential link between the industrialised and the developing countries, an essential link which would bring France status and prestige and might help developing countries to act within the limits of economic reason.

The five socialist countries in Eastern Europe – Czechoslovakia, Poland, Hungary, Bulgaria and Yugoslavia – were all pure consumers, with the exception of the first which had a very small and old-established domestic mining industry. In the 1950s some substantial part of their tin supplies came from the U.S.S.R. or from China, but in the later 1960s the supplies from the U.S.S.R. dried up and imports from China dwindled. It was clearly attractive for the socialist countries to have a voice – if not a very strong voice – in the international body which had such power over the supplies and price of tin. They could not object to the principles of the tin agreement, since they were rather loudly committed at UNCTAD conferences to the principles of commodity agreements and they could scarcely object to the practice of inter-governmental interference in the operations of the free tin market. They were also aware, from the relatively long experience of Czechoslovakia, that politics did not enter too strongly into the Tin Council's discussions. In general, they behaved as consumers on matters of price movements in the Council.

In fact, all the consumers were very conscious that within the woolliness and weakness of the article in the agreement dealing with shortages there lay the germ of a possible guarantee to consumers (and perhaps only to consumer members of the agreement) of the equitable distribution of the available supplies of tin in times of shortage; and, in that event, membership of the agreement might return dividends.

From the beginning in 1956, the Tin Council officials hoped to persuade three big consuming countries – the U.S.S.R., Federal Germany and the U.S.A. – into membership. With the U.S.S.R. the Council's relations were at their worst in 1957-58 when that country was unloading very large tonnages of tin on the world market at a time when the price was already being killed by over-production or when the producing members of the Council were under an intense limitation of their exports. That dispute was settled in 1959 with an agreement between the two parties; but the quarrel had produced an anti-Russian front amongst members of the Council which the Russians clearly did not forget. In the later 1960s, however, the U.S.S.R. found herself without tin supplies from China and began to buy on a fairly substantial scale from the free market economies. She had been loquacious and public in support of UNCTAD principles; she was perhaps embarrassed at not acting on these principles by joining one of the few UNCTAD successes in commodity organisation. She was also aware of the likelihood of the dominant position in the voting power on the Tin Council likely to be held by the countries of the European Economic Community. In 1970 the U.S.S.R. took an active part for the first time in a United Nations Tin Conference and in 1971 entered the fourth tin agreement on a voting formula highly satisfactory to herself.

The Federal Republic of Germany was in some respects a more exasperating country. She was dependent entirely on imported tin; her economic system was closely related to the systems of the other five members of the E.E.C., who had all been members of the tin agreements from 1956. It might have been expected from German economic history that the mere whisper of a charge against the Tin Council of being a cartel would have at once attracted Germany into membership. In fact, there was a long drawn out internal fight. The Foreign Office wanted to rally itself with its fellow members of the E.E.C. in membership. The dominant political party for many years, and especially

[8] U.S.I.S., 8 August, 1961.
[9] *The American Metal Market,* 29 December, 1965.

Chancellor Erhardt, were tightly committed to the doctrine of fully competitive markets; the German tinplate industry was obsessed with the fear of high manipulated tin prices and was frightened of competition, especially from plastics. It was, however, not until 1971 that Federal Germany was lined up by the other E.E.C. members to be pushed into entering the tin agreement.

The biggest fish was not to be caught yet. The United States had played the dominant role in drawing up the Havana Charter; it had supported the International Tin Study Group as a member from 1948 to 1956; it had produced in Washington in 1948 one of the draft international tin agreements with which the Study Group was so prolific. It took a part, sometimes a major part, in the United Nations Tin Conferences of 1950, 1953, 1960, 1965 and 1970. It went so far as to say, unnecessarily, in 1954 that it would not object to other countries entering the first tin agreement and thereby gave the Tin Council officials a golden, and perhaps misleading, opportunity to label the U.S. attitude as one of "benevolent neutrality". The U.S.A. was perhaps moving towards membership in the early 1960s when the Kennedy administration was pressing its policy of "good neighbours" in South America and was pouring money into the rehabilitation of the Bolivian tin mines. At the meeting of the Inter-American Economic and Social Council in Punta del Este in August, 1961 Douglas Dillon, the U.S. Secretary to the Treasury, went so far as to say: "Tin is another commodity of importance to this hemisphere. In order to strengthen and support the International Tin Agreement we plan to discuss with the International Tin Council at an early date the terms of possible United States accession to the agreement",[8] and again in November, 1965, at the meeting of the Organisation of American States, Secretary of State Rusk said: "We have now under active consideration the question of signature by the United States (to the agreement)."[9]

But the administration was far ahead of U.S. business thinking. The hard core of business thought was never persuaded. In 1956 the National Foreign Trade Council had wholeheartedly damned the stabilisation activities of international commodity organisations as leading inevitably to a regimentation extending over the whole range of the economies of participating countries. The Senate Committee of Commerce by 1961 was more accommodating and asked for an open mind.[10] The tinplate producers, although grateful for some degree of stability in the tin price, were unequivocally opposed to any other regulation of the tin market. The metal traders, as represented by the American Tin Trade Association, naturally never saw any reason for a price stabilisation which would take the bread out of their speculative mouths. In 1961 a meeting of tinplaters, can-making firms and the traders made clear to the State Department their opposition, an opposition repeated by another similar meeting held in late 1965.

It was perhaps fortunate for the State Department that in 1965, when its officials were believed to be seriously considering signing the third agreement, the sudden, temporary, refusal of the largest producer – Malaysia – to sign the agreement gave the U.S. administration time to reconsider and to reject. At the Geneva conference on the fourth agreement in 1970 the U.S.A. was present but unobtrusive.

Underlying the U.S. attitude was the strategic stockpile. At the beginning of 1973, this still contained over 218,000 tons of tin. The State Department, however sympathetic to the Council it showed itself in the disposals from that stockpile during the 1960s, was none too willing to embark on the task of persuading Congress to hand over to any external international organisation rights in a strategic stockpile which had been acquired by American money for American defence; and the State Department was equally well aware that there existed no tin lobby for (but every business lobby against) tin in the U.S.A. Even the prospect of the U.S.S.R. as a member of the Tin Council did not tempt the U.S.A. to counter-membership.

[10]*The United States and World Trade.* Final report of the Committee on Commerce, U.S. Senate, 1961.

[11]In the first agreement the chairman of the Statistical Committee (H. Depage: Congo) and the chairman of the Statistical Working Party (Sir Vincent del Tufo: Malaya) were producers; in the second agreement the Committee chairman was Malcolm Brooke, a delegate of Australia, a consuming country, but himself with a very strong producing background; in the third agreement the Committee chairman was J. van Diest of the Netherlands, a consumer and a most conscientious and devoted worker.

The administration of the Council

The agreements operated on a fairly simple machinery. The Tin Council aimed at four meetings a year, but in fact held a total of 28 under the first, 23 under the second and 20 under the third agreement. Its headquarters were in London, its normal meeting place, but it held some Council meetings both in producing countries (Thailand, Nigeria, Congo-Zaire, Bolivia and Indonesia) and in consuming countries (Japan, France, Turkey, etc.), meetings which were sometimes as useful to the international prestige of the inviting governments as they were informative to the members of the Council.

The most immediately influential committee of the Council was the Statistical Committee[11] which gave to almost every Council meeting an assessment of the forthcoming supply and consumption position. The assessments were not conspicuously accurate, especially during the second agreement. This was not surprising. The producing countries normally tended to overestimate future production since this in itself is the first line of justification for export control or the first line of defence in denying the need for releases from the American stockpile; the consumers tended to prolong into the future the particular optimism or pessimism current at the time of making the forecast. The Statistical Committee from a fairly early stage brought into attendance the U.S.A. and Federal Germany and, at a later stage, Brazil (although only the second of these countries had become a Council member by 1972). Behind the Committee stood the statistical and information services which the secretariat had built up since the days of the International Tin Study Group and which were generally held in high repute.

Other standing committees[12] discussed the problems of freedom (ultimately granted by the United Kingdom authorities) to hold all the cash resources of the buffer stock outside the sterling area, the detailed arithmetic (but not the principles) of export control, the long-term problems of production and consumption and also, every five years, the renewal of the agreement.

For the fourth agreement the Council established as early as December, 1968 a Preparatory Committee to which non-members of the current agreement were invited, appointed almost at once a small drafting committee for the agreement, and approved in January, 1970 a draft text to be sent to the UNCTAD and circulated by that body to the Geneva conference in April, 1970.

The main officer of the Council was the Chairman,[13] elected by the Council and holding no vote, but responsible to the Council for the administration of the agreement. The buffer stock Manager was responsible to him. There was a natural and human tendency for the Manager of 1959-65 to assume that his duties and obligations were determined primarily by the agreement, not by the Chairman. The possibility of these conflicting views resulted in the second agreement providing, on the one hand, that the Manager should act within the framework of the instructions of the Council and, on the other hand, that the Council could give instructions to the Chairman as to the manner in which the Manager should operate. These curiously tautological provisions were repeated in the third and fourth agreements. The Chairman himself had some other important, if temporary, functions in the buffer stock. The first agreement had made no provision for suspending the operations of the buffer stock. When, therefore, the buffer stock was bought out by speculative interests only eleven days before the end of the first

[12]The more important of the committees and working parties with the chairmen longest holding office were the Finance Committee (J. A. Turpin of the United Kingdom and J. M. Rochon of Canada), divided in 1969 into an Administrative Committee (C. Bué, France) and Buffer Finance (M. S. R. Burhanuddin, Malaysia, and Pandit Bunyapana of Thailand), the Credentials Committee (J. Gregersen of Denmark), the Drafting Committee for the Fourth Agreement (R. van Achter of Congo-Zaïre), the Working Party on Consumption (Miloslav Had of Czechoslovakia), the Working Party on the long-term tin position (M. A. Brooke of Australia) and the Standing Committee on Production (H. W. Allen). Other ad hoc bodies were created from time to time (e.g. a committee on contact with the U.S.S.R. in 1958 and a committee on barter).

[13]In the fourth agreement re-christened the Executive Chairman. The holders were G. Péter, July, 1956-July, 1962, who then went on to United Nations work in South America; Don Manuel Barrau (Bolivia) as Acting-chairman July-November, 1962; and H. W. Allen (Australia) who was elected in November, 1962, after a hotly-disputed contest between G. C. Monture (Canada) and Y. Coppieters 't Wallant (Belgium) when the producers' vote shifted to him.

[14]W. K. Davey, July, 1956 to April, 1959; J. B. M. Lochtenberg, May, 1959 to October, 1965; and R. T. Adnan from January, 1966.

agreement, neither Chairman nor Manager had the legal authority or was prepared to stop selling. The Council realised the folly of leaving itself in so exposed a position. The third agreement accepted the right of the Chairman to suspend the operations of the buffer stock at the floor or ceiling price – an authority used when the price shot through the ceiling at the end of 1969. The Chairman had a further authority for short-term action. In all the agreements he was empowered to suspend the operations of the buffer stock if currency changes made a review of the existing price scale of the Council necessary. The collapse of sterling (in which the prices in the first three agreements were expressed) made him temporarily suspend buffer stock action in November, 1967 for a couple of days until the Council determined new prices.

The first three buffer stock Managers of the Council[14] were men of diverse character. The first one, W. K. Davey, was English, an ex-chairman of the London Metal Exchange and a man of remarkable integrity. He was almost overwhelmed by the flood of tin loosed on the market in 1957 and the inadequacy of his financial resources; but he handled the acquisition of some additional 10,000 tons of tin under the special fund of 1958 admirably, and his sale from the special fund on a depressed market in 1958-59 was perhaps the most skilful piece of tin marketing ever seen. The second, J. B. M. Lochtenberg, a Dutchman with Indonesian connections, had been appointed initially as deputy manager in 1956 to ensure appropriate producers' influence in the buffer stock. As Manager he sold in 1959-60 the United Kingdom strategic stockpile with such a tight hand as to give an almost flat fixed price on the tin market and thus to provoke a very strong and successful revolt by the London Metal Exchange; his stock of 9,000 tons was taken from him by speculators in a few days in June, 1961; he was bought out again in 1963. R. T. Adnan, an Indonesian, had been with the Indonesian Tin Sales Organisation in London and was for many years a loquacious member of the Indonesian delegation to the Tin Council. He built up a buffer stock of some 11,000 tons in 1967-68; he strongly advocated the introduction of export control in 1968-69; he reduced his stock to under 1,000 tons in 1970 but raised it again to over 6,000 metric tons by the end of 1971 and 12,500 tons by the beginning of 1973. As a producers' representative on the Council he had often advocated the extension of buffer stock action into the middle sector of the price range.

The third official –the Secretary of the Council – under the first three agreements was in control of the statistical work of the Council, of the memoranda and information issued to and by the Council, of the organisation of its meetings and of its conferences on production and all other paper work.[15]

Voting on the Council

The framers of the first post-war agreement were bitterly aware of the harassing and undignified disputes under the prewar control schemes about the standard tonnages of producing members; they were also aware of the argument that any tonnages fixed initially at the beginning of an agreement should be subject to some adjustment within the life of that agreement so as to reflect changes in the actual production of members. It was agreed early on that the basis of producers' voting should correspond, with modifications, to the agreed tonnage of production in an agreed period. These agreed tonnages were the result of compromises and the usual blackmail to be expected at international conferences. Generally, but not invariably, the agreed figures were at the expense of Malaysia who, as the biggest producer, had the most fat to surrender. Annual adjustment was applied, but naturally not in such a way as to impose too great a hardship on any individual country. The agreements did, however, avoid any acceptance of the idea of production capacity (as

[15]The Secretary from 1956 to 1971 was William Fox. The information issued included the *Statistical Year Book: Tin, Tinplate and Canning* (issued at approximately two-yearly intervals since 1949), the annual reports (from 1956-57), *Notes on Tin* (duplicated) monthly from 1951, *A Technical Conference on Tin, London, 1967* (2 volumes), *A Second Technical Conference on Tin, Bangkok, 1969* (3 volumes), etc.

[16]The terms granted to Australia under the fourth agreement – a tonnage based on exports and not on production –were not exceptional. Australia was the only producer which had a substantial home consumption; with the other producer members exports and production were almost the same.

distinct from performance) as the basis for assessing importance and at no time were they prepared to revert to the mistaken pre-1939 practice of bribing new members into the agreement with the grant of tonnages exempt from the provisions of export control.[16]

The need for flexibility

The system in the agreement had to be flexible (both for producers and consumers). In the first agreement, this flexibility for producers was obtained by pooling each year one-twentieth of each percentage and then re-allocating this pool. No principles were laid down for the redistribution and in consequence each year saw an indecent squabble amongst producers for disproportionate shares in the pool. The second and later agreements sensibly shifted to an adjusting formula, invented by Sir Vincent de Tufo (Malaysia), on an arithmetical basis. The severity of adjustment was also smoothed. The maximum amount of reduction that could be made in any percentage was limited and the Council was given power to accept the evidence of exceptional circumstances and even not to apply any reduction at all. Thus, in 1961 and 1962 Congo-Zaïre successfully claimed exceptional circumstances, and therefore suffered no reduction in percentage.

The system of annual adjustment, certainly over the four agreements, has kept the percentages roughly in harmony with real production. Between 1957 and 1972 (see Table 12) the share of Malaysia in the percentages rose by one-quarter and the share of Thailand doubled; the areas in relative productive decline drifted down, sometimes sharply – Congo-Zaïre and Indonesia by one-half, Bolivia by one-seventh and Nigeria by one-tenth. Since the percentages very largely determined voting power, the result was a steady increase in the wyight of Malaysia and, particularly, in the joint weight of Malaysia and Thailand. By 1972 these two countries held in total 573 out of the total 1,000 producing votes (against 426 in 1957). In practice, the massive Malaysian vote was used, so far as I saw, throughout the three agreements with a surprising degree of reasonable care for the interests of the other producers.

The voting of consumers

The voting system for consumers was simpler, since the issues at stake were much less sharp. In the first agreement the consumers' tonnages, which largely determined voting power, were based on the average of net imports and consumption.[17] This clearly gave an undue importance to the trading interests of the United Kingdom and the Netherlands, which might be handling buffer stock sales or purchases. In the second and later agreements a new and simpler formula on the average consumption of tin over the three lastest past years was adopted. One esception was made in the fourth agreement. The U.S.S.R., although a consuming member, was allowed to select as its basis for voting not consumption (which the Russians were unwilling or unable by law to make public) but imports of tin. This concession understated substantially the Russian vote in relation to other consumers.

The United Kingdom and France, with over half the total, dominated the consumers' vote in the first agreement, but by the third agreement their joint share had fallen to about one-third.

---

[17]Consumers, as did producers, also began with a basic vote of five.

[18]It is regrettable that the Tin Council has never seriously examined any other basic approach to the solution of the tin problem, perhaps one which would have taken into account that the differences between the wealth of individual producing countries (e.g. between Bolivia and Malaysia) are so wide that it is a misnomer to put them all under the same general heading of "developing" countries and that there is a case for self-help between the producers. I remember discussing with Sir Vincent del Tufo of Malaysia the possibilities of a plan for governments creaming off profits by increasing their duty or levy – when the tin price was above the ceiling in the agreement – into an excess profits pool. This pool, under the control of the Council or of the producers as a group, would be the basis of a fund to buy up and hold, or perhaps sterilise, unwanted production and to provide compensation for mines killed off. The fund could also help to diversify employment in areas where mining proved surplus. He saw very practical objections (the objection of governments to local levies being used for action elsewhere, the difficulties, especially in Bolivia, of finding anything in the mining areas into which to diversify redundant labour, the problem of deciding when capacity could be sterilised), but I now regret that this suggestion – or, indeed, any other suggestion – has not gone further.

This was due in part to the entry into membership of Japan in 1961; and by 1972 Japan held the largest single consumer vote (one-fifth of the whole). In general, the consumers on the Council did not usually show the same community of interest as did the producers. The Netherlands, Belgium and Spain could normally be expected to vote with the producers on major points of importance; the United Kingdom tended to hedge; France poured oil on the troubled waters; Japan and, to a lesser extent, Canada stood firmly in resistance to proposals for increases in the price scale in the agreement.

At the beginning of the fourth agreement in 1972 the six members of the European Economic Community held one-third of the total consumers' votes and, with the United Kingdom, nearly one-half. The E.E.C. was responsible for the general policy of its members on inter-governmental commodity agreements (including those members who hitherto might normally have voted with producers) and it is possible that this importance of the E.E.C. would show itself in a greater rigidity of position within the Council on price questions.

The control of exports

The members of the agreement, whether consuming or producing, were committed from the beginning to the doctrine that the only solution to long-term surplus in the tin industry was through the use of export limitation.[18]

The pre-1939 system of export control had never been able to discipline itself properly. There had been no penalties provided for over-export or under-export. There had been much under-export (in 1936 one-eighth of the total allocated for export had not been exported) but there had been little over-export. Any departure from the norm of the stated quota tonnage is an offence against stability. But under-export from one producer in a period of export control is in itself likely to be a stimulus to the price for all producers. The producers in the post-war agreements were not willing to penalise under-export. Over-export, which might cut incomes of other countries already reduced by control, was a different matter and attracted from the first agreement as severe penalties as the agreement could persuade members to accept. The penalties became more severe with accumulated offences. For the first breach the Council might impose an additional contribution to the buffer stock, thus permitting the Council machinery to take off the market the particular excess which one member was throwing on to the market. For an accumulation of excesses the Council was obliged to reduce the later export allocations to the offender; for a further cumulation of excesses it might take additional and stronger action – the forfeiture of part of the offender's rights to the buffer stock.

It is regrettable that the Tin Council has never seriously examined any other basic approach to the solution of the tin problem, perhaps one which would have taken into account that the differences between the wealth of individual producing countries (e.g. between Bolivia and Malaysia) are so wide that it is a misnomer to put them all under the same general heading of "developing" countries and that there is a case for self-help between the producers.

The observance of quotas

In general during the very serious export crisis of 1958-60 the exporting members observed their export quotas reasonably firmly. In the eight control periods to the end of 1959 a total quota allocation of 191,000 tons produced an actual export of 190,100 tons, a quite remarkable closeness to the target. There were, however, two serious breaks. In 1960, as the severity of the quotas was decreased, Bolivia found it difficult to adjust herself to a rising target and in the first three quarters of that year Bolivia managed to export only 14,000 tons out of her total allocation of about 21,000 tons. This under-export was no offence under the agreement. The second case was more unusual. A substantial tonnage was shipped out of Thailand in 1959 in excess of the permitted quota. The tonnage was not recorded, possibly because of bribery or directions from high Thai authority to the customs officials. Evidence from Malaysia, from shipping lines, and from the recording import authorities as to the destination of the tin concentrates in the United States indicated an amount of the order of 1,200 tons of tin content. The problem was delicate.

Thailand maintained the non-existence of the excess shipments or of the highly placed personnages who were rumoured to be behind the evasion, but ultimately agreed to make a voluntary contribution of £400,000 in 1960 to the buffer stock. The settlement had established a principle of liability on a sovereign government, even if the penalty was scarcely crushing since Thailand was to receive repayment of its voluntary contribution in the middle of the year 1961.

The second period of export control in 1968-69 told another story. That period was brief, was entered into in haste, was relatively light in its total degree of limitation, and soon became confused in practice. It at least secured the aggregate amount of tin aimed at. But to secure that aggregate in the second quarter of 1969 a shortfall declared by Nigeria had to be shifted as increases to Indonesia and Thailand; in the third quarter a further shortfall declared by Nigeria had to be shifted as increases to Bolivia, Indonesia and Thailand, and in the fourth quarter shortfalls had to be shifted from Malaysia and Nigeria to Bolivia and Indonesia. Before that fourth quarter had ended the Council called upon the producing countries to increase their exports as much as possible and agreed not to impose any penalties under the agreement on over-exports. In that quarter, in spite of the appeal, Congo-Zaïre, Malaysia, Nigeria and Thailand exported below their quota.

The use of the buffer stock

In the three agreements from 1956 to 1971 the buffer stock was the lesser weapon in the Council's armour against price fluctuations. The principle of the buffer stock was very simple, whether permissive or compulsory. When the price of tin was falling in a lower sector of the price range fixed by the Council tin *might* be bought; when it had fallen to the floor at the bottom of that range tin *must* be bought. When the price of tin was rising through the upper sector of the price range tin *might* be sold to ease the market price for consumers; when the price reached the ceiling of the range tin *must* be sold. The middle sector of the price range was to be regarded generally as the area within which the buffer stock should have no desire, need or power to operate. The specific expression of this simple principle was more difficult; and the practice of using money (with which the buffer stock was usually fairly well supplied) to buy tin which the market was anxious to sell to the Manager was a much easier task than was the practice of using metal (which the Manager very often did not have) to fill a market that in conditions of rising prices could swallow everything the buffer stock could offer. The principle of the buffer stock remained unchanged through the four agreements; but important changes were made in its permitted practices. The fourth agreement saw, partly by implication, an important upward change in the status of the buffer stock relative to export control.

The first agreement provided (and the next three agreements maintained) that at the floor price in the agreement the buffer stock was obliged to buy tin metal until his funds were exhausted. The floor price was broken for a few days in September, 1958, but not on any other occasion in the three agreements, although for a large part of 1957-58 the price hovered very near to the floor and again in the summer of 1968 it was only a little above floor.

In the lower sector the first agreement gave the Manager authority to buy at the market price "if he considers it necessary to prevent the market price from falling too steeply". This provision in the agreements gave the Manager in the falling lower sector and in the rising upper sector – covering between them about two-thirds of the whole range between floor and ceiling – a very wide nominal freedom of action in which he was fettered solely by his own opinion. But, in practice, the exercise of his authority was subject to the supervision of a Council which met normally once a quarter and which could be very easily convened at more frequent intervals. Most of the members of the Council could draw on advisers who knew market behaviour, market practices and market reactions or who could provide often extremely accurate estimates of buffer stock transactions or holdings. The Council could call upon the Manager for explanations or even (as it often did) for up-to-date returns on the tonnages of tin that he held.

The exercise of his buying judgment by the Manager showed in the later agreements a strong tendency towards buying fairly soon in the higher bracket of the permitted area. In the first spasm of buffer stock buying in April-June, 1957 the bulk of his buying was in the higher part of the lower range; but in the later part of that year the flood of tin into the buffer stock was too sudden for the Manager to consider a policy of gradual defence. In July-August, 1962 when the break in price under the threat of stockpile disposals from the U.S.A. surprised the Manager and where there had been a long previous period of decline without provoking buffer stock activity, the support, when given in the autumn, was at the very top of the lower sector. In January, 1967 buffer stock buying was made at practically the top of the lower sector. In the first half of 1969 the buying undertaken was towards the bottom part of the lower sector and very little above the floor price. In 1971 buying was tried, first, high up in the lower sector and, later, about half-way down the sector. Buying in the second half of 1972 was again towards the top end of the lower sector.

In the six months August, 1971-February, 1972 buffer stock activity (which, under the fourth agreement, could cover both buying and selling in any sector except the middle range) had produced a tin price which was virtually static. In those months, the variation in the monthly average cash price had been only £18 or 1.3 per cent as compared with £48 or 3.3 per cent in the previous six months. Thus the Manager was exercising the control of nearby supplies, which meant that he was manipulating the market like any large scale speculator, although from different motives. The market had, however, an outlet provided by the agreement – namely, that if the price were forced up into the middle range the Manager lost his power to operate. In March, 1972, the outlet came; the price, so released, soared temporarily by about £60 a ton.

The question of the quality of reaction in the exercise of the Manager's discretion in selling metal in the upper sector cannot be judged. Where selling took place the Manager seemed merely to follow (and probably could do little else) a rising price which was being dictated to him by market circumstances; there seems no proof that he was even an influence in retarding the speed of the rise. The provisions of the fourth agreement which enable the Manager to sell tin at above the ceiling price may tempt him to hold back metal for more profitable sale above that price instead of selling it on the way up.

The size and financing of the buffer stock

Each agreement has provided for a buffer stock. Contributions to this buffer stock are compulsory on producing member countries. Voluntary contributions may be made by producing or consuming members. The compulsory contributions may be in metal or cash as the Council determines. Provision is made for default in payment, penalising a country which failed to make its payments; but when that default happened, as with the Congo in 1966, the Council naturally leaned over backwards to prevent any member being tarred with the brush of failure to meet its legal obligations.[19]

TABLE 12
Buffer stock finance, 1956-71
in £ millions

	First agreement, 1956-61	Second agreement, 1961-66	Third agreement, 1966-71
Contributions	16·9	14·6	20·0
End position	20·3	19·3	27·6
Surplus:			
from interest	1·4	4·7	4·8
from trade	3·0	0·1	2·9
total	4·4	4·8	7·6

In the discussions through the 1950s, preceding the tin agreements, there was no challenge to the principle of a buffer stock as an essential element in stabilising prices. But there was very considerable discussion on the appropriate size for that stock and on the question as to who would finance the stock. So long as that financing was to be done only by the producing members, many of whom had just left colonial status and all of

[19]The Congo failed in July, 1966 to pay its due contribution to the buffer stock. The Council permitted a voluntary temporary contribution of £456,000 to be made by two banks, backed by the Belgian tin mining companies, in the name of the Congo government.

whom were short of capital, these countries wanted the minimum tonnage on which an effective buffer stock would scrape by (a figure which would certainly have been higher than the 15,000 tons which they offered in the 1953 discussions). The Americans were, in the light of experience, proved right when in the same discussions they urged 35,000 tons or about three months' world production. But neither the U.S.A. then nor any other consumer until 1970 showed any indications of being willing to put money into the buffer stock, and the consumers could scarcely press any high figure as its size. The producers were therefore under the three agreements after 1956 in happy possession of a constant grievance.

But the passion of the producers' stress on their sole liability to finance the buffer stock sometimes seemed a little unreal. The contributions to the buffer stock were, except in the nationalised industries in Bolivia and Indonesia, raised by the local governments from the private miners; this meant that the British-owned mines in Malaya and Nigeria and the Belgian-owned mines in the Congo for many years helped in the capital contribution.

The buffer stock of the first agreement was £16·0 millions (the equivalent of 25,000 tons of tin). This payment was no more than 16 per cent of the total production in 1955 of the six producing members. The second agreement had a buffer stock of £14·6 millions or 20,000 tons, the third a stock of £20 millions or 20,000 tons and the fourth a stock also of 20,000 tons or £27 millions.

The buffer stock was expressed in terms of tonnage equivalent, converted into cash at the floor price per ton expressed in the agreement. As the floor price was steadily moved upwards during the agreements, the nominal buying power of the Manager was constantly being eroded. Thus, the buying power of the first stock was reduced to 23,000 tons even before the great crisis of 1958 broke out; and the third buffer stock was reduced from its original 20,000 tons to the order of 16,000/17,000 tons by the end of 1968. On the other hand the buying power of the fourth stock was considerably increased by the voluntary contributions in 1971-72, approaching £1·9 millions, from the Netherlands and France.

The buffer stock had three functions – to intervene actively in holding or stabilising the price, to exist as an insurance policy against the contingencies of sharp price falls or rises and to invest the very substantial funds often left unused in buffer stock hands. In its first role, it was bound in the circumstances of the first agreement (a price at the floor in the early stages and at the top in the late stages) and in the circumstances of the third agreement (steady buying in the first year at low prices, steady selling in the second and third years at higher prices) to make a profit. The buffer stock has not yet faced an agreement where the prices have opened at a high level but have persistently moved down through the next five years, and it is perhaps too optimistic to assume that it will be – even over more than one agreement – always profitable.

The buffer stock retains its insurance aspect – whose effect on price sentiment is very important, if incalculable – even when, as under the second agreement, it has had very little operative work.

The investment aspect of the buffer stock was seen most clearly under the second agreement.[20] Contributions to the buffer stock had been called up in full even when the funds so collected were unlikely to be required in the reasonably near future for supporting the market. In that agreement buying and selling actively covered only twelve months and the maximum amount of tin held at any time was only 3,270 tons. The buffer stock became in effect an investment trust investing short-term in the United Kingdom (in bank accounts, in loans to local authorities and sometimes as widely as finance companies). In the third agreement, although the buffer stock held tin metal for most of its life, its maximum holding was only 11,290 tons for nine months and its average quarterly holding was about 4,500 tons. Half-way through the third agreement the buffer stock held £8·0 millions in cash and at the end £23·8 millions. That buffer

[20]The cash resources of the buffer stock rose from £15·4 millions in June, 1962 to £19·26 millions in June, 1966.

[21]The changes in the floor price in the third agreement meant that, if used, the bank facility would have bought only about 7,000 tons of tin.

[22]It is estimated for the I.M.F. and the World Bank that the maximum tonnage requisite for the buffer stock is around 40–45,000 tons of tin. This figure might rise as, and if, world total consumption moves upwards. It is under three months of world consumption.

stock had in one sense a wider range of investment, since it was permitted by the United Kingdom to hold a proportion of its cash resources outside the sterling area. It also enjoyed a long run of high interest rates for its money.

One additional source of finance was created but not utilised by the buffer stock. In the first agreement there was no authority to Council to borrow money for the buffer stock, and the creation of the special fund in 1958 was an exceptional measure, accepted by a number of members, with the buffer stock Manager operating the Council machinery as a matter of convenience. The second agreement legalised the principle of the borrowing of money for the purposes of the buffer stock and upon the security of the tin metal warrants held by the Manager. No use was needed for this provision under the second agreement since the buffer stock itself was flooded with cash. The third agreement repeated the authority (as did the fourth agreement) and in 1966 the Council paid a commitment fee of £112,500 for the right to draw up to £10 millions during the life of the agreement from a consortium of sterling banks in London headed by Hambros Bank. The arrangement was a purely commercial one with interest rates on the overdraft tied to standard commercial terms and the standard commercial deposit of warrants as the security. The overdraft was not drawn upon.[21] Interest rates under the third agreement were high and the Council preferred in 1968 to meet the problem of over-production by export control rather than by further buffer stock buying financed from its own funds or from the overdraft. In 1972, however, in spite of its reluctance to use the banking facility under the previous agreement, the Council agreed again to make arrangements for £8 millions.

The problem of financing the buffer stock entered a new and very important phase in the discussions on the fourth agreement. The possibility of such financing being shifted in one way or another on to the shoulders of the international agencies of the United Nations, in particular the World Bank (International Bank for Reconstruction and Development) and the International Monetary Fund, had been in mind for a number of years. In the early 1960s, informal talks had taken place with the World Bank,[22] which had replied that the possibilitiy then lay outside the charter of the Bank but that the matter could be reviewed again in five years' time.

Pressure was mounting at UNCTAD for the international financing of approved international commodity schemes with an approved buffer stock. In 1969, the International Monetary Fund accepted that a member with a balance of payments problem could, on certain conditions, use his drawing rights on the Fund to pay his contributions to a buffer stock comprised in an accredited inter-governmental commodity scheme conforming to United Nations principles. At that point the only commodity agreement which qualified was the tin agreement; and some of its producing members had balance of payments difficulties. During the fourth agreement many producing members used their rights on the I.M.F. to cover buffer stock contributions.

Buffer stock action in the middle sector

In the basic, but often challenged, theory of the agreement, the middle sector of the price range represents the zone where normally the buffer stock would be quiescent and where the price would be left to find its own level, under the influence of uncontrolled production, unrestrained consumption and other market forces. It was perhaps almost naïve to establish an international body to stabilise prices and then to expect that body, many of whose members have the strongest interest in pushing stabilisation further, not to be constantly itching for entry into that banned sector.

The first agreement provided that the Manager of the buffer stock could neither buy nor sell in the middle sector unless the Council decided otherwise; in the third and fourth agreements he was barred except by the special authorisation of the Council.

Some producers (Indonesia and Bolivia, for example) hankered persistently for operations by the buffer stock in the middle range. The reason was obvious. The higher the point within the

[23]Export control continued to September, 1960 and permission to operate in the middle sector until June, 1960.

[24]The amount of tin then in the buffer stock was 11,290 tons. The liquidation of the agreement and therefore the distribution of the buffer stock was only two years ahead.

price range at which the Manager could operate the higher they hoped he would tend to peg the price; and in any case the nearer he would be to the principle, which many producers and particularly Bolivia approved, that the object of the agreement was to obtain an actual market price around the mid-point between floor and ceiling prices. The Malaysian producers were more conscious of their lower costs and, therefore, of the need for a lower price level, and were also closer to the actual free operations of the Malaysian market.

The attitude of the consumers, generally (but not invariably) in favour of maintaining free markets in existence or, at least, limited existence, depended on the application of export control. Export control might be misjudged and might clamp down too severely; or individual export quotas might not be fulfilled; or the buffer stock Manager might buy too lavishly in the lower sector. The price might be forced too rapidly from the lower into and then through the middle sector. The consumers were prepared for action in the middle range (that is, for selling) as many producers were also prepared for action (that is, buying). The consumers had a better bargaining position, since by their own approval to export control they could ensure the producers' support of action in the middle sector.

The introduction of export control in December, 1957 was accompanied by the acceptance of buffer stock operations in the middle sector.[23] The acceptance was to ensure that there would be no undue rise in the tin price during export control and to enable the Manager to maintain and achieve a proper balance between tin and cash in the buffer stock. Both consumers and producers realised that there was sense in allowing some part of a swollen metal stock being disposed of in the the middle range (as was done when the buffer stock was sold down from 23,000 tons to 10,000 tons during 1959-60 when export control and selling in the middle range coincided in time).

In March, 1969, after control had again been applied, the Manager was again given authority to buy and sell in the middle range, but with a tighter hand. His duty was to reduce the quantity of tin in the buffer stock without disrupting the market[24]; he was directed to aim at a balance between cash and metal in the stock; and he should be at all times a net seller of tin. Equally important were the negative directions: the Manager was not to lead the market and, in particular, he must not try to maintain the price in the middle sector. The repetition of his authority in September and again in December, 1969 referred only to his aim of mitigating where possible sharp price fluctuations, omitted the reference to net selling, but retained his other obligations. The December authority was granted even though export control had terminated and the market price position had changed very rapidly upwards. In January, 1973, with export control re-imposed, he was permitted to operate in the middle sector.

Did operations in the middle sector give a wider degree of stability to the price? Normally in the periods of conjoint authority, export control was the major activity in the sense that the export control quota was kept at an artificially low figure to allow more tonnages of the buffer stock to be sold. Through 1958 when no sales in the middle range were made, the price moved from £731 to £756; through 1959 when selling was active (to the extent of 13,000 tons in a year) the price moved between £759 and £789; in the first half of 1960 when no sales were made (but the export quotas were being increased) and in the second half of 1960 (when both export control and action in the middle sector were formally abandoned) the price was roughly a straight line. In the second export control period 1968-69, although the buffer stock sold nearly 7,000 tons whilst export control was operative, the price soared by no less than £250 a ton from the beginning to the end of the year.

The balance of metal and cash in the buffer stock varied. It was made better in 1959 by selling in the middle range (40 per cent of the nominal buffer stock total being in cash by the end of that year); but the metal element (fairly large, about one-half) at the beginning of export control in the autumn of 1968 was reduced during 1969 by buffer stock sales to a negligible proportion (only about one-fifth). In that latter period, the operative force in the Council's real thinking may have been the desire to end the tin agreement in June, 1971, not with a balanced buffer stock but with a buffer stock full of cash available for distribution to the buffer stock contributors, a concept which is understandable enough where each agreement is a new agreement, containing perhaps new obligations, but which makes the applications of general principles to practical buffer stock policy difficult.

The evidence, so far as it is relevant, does not seem to suggest that operations in the middle range have, in fact, provided any greater degree of stability to the market price. The general theory of the market might assume that – so long as the free market is allowed to exist at all – the limitation of its activities to the middle sector might merely compress it within that sphere in a more explosive form only too apt to force itself into pushing the price unreasonably into the sector below or the sector above. In this connection it should be noted that, given the long history of price instability in tin, the agreements have always accepted a surprisingly narrow belt for the middle sector (and, indeed, for all sectors). That belt was only £50 for most of the first eight years of the agreements, £100 or little more for the next six years and then even reduced again to £80 for the next two years. The narrowness of the belt, which was relatively much the same throughout, has always been a temptation for the market to break and a great difficulty for the buffer stock to maintain.

The timing of export control

The pre-1939 control schemes had been clear in their general concept of export control but remarkably shy of detailed argument as to the factors which inspired their decisions on export quotas. They had also been erratic in arriving at and in changing these decisions. The new agreements after 1956 were determined not to repeat either of those errors. They made clear and public the bases on which their decisions were to be made. "In the light of its examination of the (quarterly) estimates of production and consumption", said the fourth agreement, ". . . and taking account of the quantity of tin metal and cash held in the buffer stock, the quantity, availability and probable trend of other stocks, the trade in tin, the current price of tin metal and any other relevant factors, the Council may from time to time determine the quantities of tin which may be exported from producing countries . . . and may declare a control period and shall . . . fix a total permissible export tonnage for that control period. In fixing such tonnage, it shall be the duty of the Council to adjust supply to demand so as to maintain the price of tin metal between the floor and ceiling prices".

In general, in weighing so far as it could the varying importance of these factors, the Council paid great attention to the estimates of mine production and commercial consumption made by its statistical committee. But its two periods of export control had different histories for different reasons. In the first period, 1957-60, the over-supply of tin was made glaringly clear by the collapse of the tin price and the flood of tin imports from the U.S.S.R., but, in the later stages when the price crisis had been conquered, export control was maintained primarily to permit, first, the stock in the special fund and, then, a substantial part of the ordinary buffer stock to be sold. In the event the Council made a decision parallel to the decision for which it had been criticised in the working of the pre-war agreements; it subordinated the interests of the industry side of the scheme to the interests of the buffer stock side. This in itself perhaps did not matter much (the interests on both sides of the scheme were the same) but the longer-term consequence of the retention of export control for too long and the absence of almost any buffer stock at all from June, 1961, left a market wide open under the second agreement to the U.S. surplus strategic stockpile.

In the control period 1968-69 when the surplus of production was in no sense as sharp, there was no flood of tin from the U.S.S.R., disposals from the U.S. stockpile had stopped, consumption was later found to be rising through 1969 to almost its highest record, and the buffer stock still had ample cash in hand. It is likely that export control was entered into too hastily and perhaps unnecessarily and was kept on too long. It also illustrated, in the later stages of exploding prices, the consequences of miscalculating the right time to get out of export control.

XIII

The Agreement and Russian Tin, 1956-61

The first agreement came into effect in July, 1956. The moment was not very happy. The U.S.A. purchases of tin for the strategic stockpile had ended and this artificial prop, which had supported the price for a number of years, could no longer be relied upon. The industry now had to depend on purely commercial factors. But the figures over the three years 1954-56 showed an aggregate surplus of mine production over commercial consumption of no less than 78,000 tons and there was no indication that the commercial market could absorb anything like this over-stimulated figure in the first year or two of the agreement.

The potential danger of the situation was not, however, yet reflected in the tin price. Admittedly, that price had been moving down through the first half of 1956, but it had then been stimulated, first by the successful entry into force of the agreement, secondly, by the belief that the pressure from the producing members to raise the floor price in the agreement would soon be successful and, thirdly and most effectively, by the Suez crisis in October-November, 1956. During that period the price shot for a brief day through the new agreement's ceiling price of £880 a ton.

This last episode helped to emphasise that price movement in the first six months of the agreement was as volatile as it had been in the six months prior to the entry into force. It also emphasised one of the Council's recurrent weaknesses. While the market price was within the upper or middle sector of the price range in the agreement the buffer stock Manager had ample cash (he had raised an initial contribution from producers of £9·6 millions), but he had no metal to sell within the upper sector and therefore no means of influencing the price; nor had he the authorisation required from the Council to permit him to operate in the middle sector.

In January-February, 1957 the reaction against the Suez boom helped to bring the price down into the current middle sector. The Council had little sense of unease. It was involved in considering an increase in its price range. That range had been arrived at in the discussions at Geneva in December, 1953. In the light of current circumstances, with a consumption higher now than it had been in 1952-53 and with the production showing signs of slackening, there seemed to be a clear case for price adjustment. In December, 1956 and in March, 1957 the producers on the Council requested an increase in both the floor of £640 and the ceiling price of £880.[1] They pointed out, *inter alia,* that on the current lower sector the buffer stock was unlikely to be buying; that, without prior buying of metal, the buffer stock would (as in the Suez crisis) not have the metal to stop a price rise; and that an increase in the price range, by increasing also the lower sector, would enable the buffer stock to accumulate metal and thus to fulfil the requirement that "the Council shall also aim to maintain available in the buffer stock tin and cash adequate to rectify any discrepancies between supply and demand which may arise through unforeseen circumstances".

The combination of very diverse but not unreasonable arguments was successful only in part. In March, 1957 the floor price was raised from £640 to £730. This was the largest proportionate change in the floor price made (except on currency devaluation grounds) during the three agreements. There was no change in the ceiling price of £880 but the effect of the divisions of the range was to give a middle sector of £780-830 with a mid-point of £805 against a previous mid-point of £760. The widths in the

[1] The Congo wanted a floor of £720 and a ceiling of £920, Bolivia £720 to £960, Indonesia £720 to £990 and Nigeria a range of £150 around a medial point based on the actual average of a past period and the buffer stock Manager's estimate of future prices.

[2] At the July meeting, the Chairman of the Council suggested that the one-twentieth be divided according to the proportions of the 1954-56 outputs held by the members, that the resultant Bolivian figure be rounded up to the nearest 0·5 per cent, that the offer of the Congo and Malaya for a small reduction for each be accepted and that Nigeria be lifted to 5·3 per cent and Thailand to 7·10 per cent. Neither Indonesia nor Thailand would accept.

[3] To ensure Bolivia and Indonesia a minimum of 95 per cent of their current basic tonnages and then to share the degree of restriction equally. In result, Thailand would get 8·0 per cent but was prepared to accept 7·35 per cent.

upper and lower sectors of defence were reduced to £50 (a lower of £730-780 and an upper of £830-880). The effect of lifting the lower sector was to bring within its scope the current market price of tin and to make the buffer stock a force in the market. The Manager started promptly to buy. In the next three months April-June, 1957 he bought 3,915 tons of tin. This buying was done when the price was between about £760 and £780 a ton, that is, in the top bracket of the lower sector. It was successful in holding the price, and in particular the forward price, without too great a fall. The tonnage bought was perhaps ominously higher than statistical estimates showed the rate of surplus to be but it was not in itself abnormal in the light of later buying practice.

The producers' percentages

With no obvious sign of serious trouble ahead, the Council could turn to an internal problem. This related to the producers' percentages. The Council had to proceed in July, 1957 to its first re-allocation of the pool of one-twentieth of the percentages of the producers. The producers argued, mainly among themselves, not only for the re-allocation of the one-twentieth, which was relevant, but also for the adjustment of the basic percentages, which was outside the power of the Council. Thailand claimed that its basic percentage had been arrived at for a period when its post-war rehabilitation was still lagging; its basic tonnage should take account of its likely production in 1957; and on that approach its basic percentage should be raised from 6·29 to 9·20 per cent. The Congo claimed that, if the years 1954-56 were used as the basis, its percentage would be raised to well over 9 per cent. The Indonesians, whose production in 1954-56 had been lower than in the basic period, still wanted to increase their percentage. The consumers read patiently very detailed mathematical suggestions by M. Depage (Belgian Congo). Proposals were put forward and rejected.[2] In the October meeting, Thailand made a fresh proposal.[3] It was necessary for the Council in October to force on Indonesia the acceptance of a compromise from which the percentages for 1957-58 emerged with the full reduction for Bolivia and Indonesia, a substantial

increase for Thailand and a smaller one for Malaya[4]. The principle had been made clear that the annual re-allocation was confined to the one-twentieth only and did not apply to the whole of
the percentages.

The statistical position and its assessment

In the colder, outside world the Statistical Working Party was
still misleading itself. In July it was convinced that its earlier
estimates of 161,000 tons as mine production and of 158,000 tons
as commercial consumption for 1957 should be left unchanged.
It looked at the movement of imports from the U.S.S.R. and
China and concluded these would be around 2,500 tons. All told,
the surplus of tin metal available in 1957 would be no more than
6,300 tons.

No other stock disposals were likely to upset these calculations. The two stocks known to be coming forward for disposal –
but at future dates – were small. At the end of 1956, the United
Kingdom had given notice of its intention to sell 2,500 tons of its
non-commercial stocks at a future date. In July, 1957 it
emphasised that its government held no mandate for retaining
stocks once the defence necessity had gone; it was obliged to
liquidate the 2,500 tons as soon as possible but it would sell at
prices which would not depress the market. It would not sell
below the floor price of £730 a ton and, in fact, it would also
accept a limitation on sales when the market price was in the
lower sector of the price range. The producers unanimously
recommended that any member disposing of strategic stocks
should sell only at a price above the lower sector (that is, above
£780 a ton). At the moment the discussion was academic since
no tin was sold from the British stockpile until the middle of
1959. The Canadian strategic stock had been built up during the
Korean panic. It amounted to some 3,000 tons. In June, 1957 the
Canadians gave notice of their intention to dispose. The gov-

[4] The voting was of some significance. Five producers voted for the settlement,
one (Indonesia) against (787 votes against 213). The consumers were less
united. One (the Netherlands) supported Indonesia by voting against; four
with about one-third of the votes abstained; only seven (including the United
Kingdom) with a total of 600 votes were in favour.

ernment did not visualise offerings at prices below the upper sector (that is, £780 to £830 a ton) in the agreement price range and sales would, in effect, be limited to Canadian internal consumption. Here again no disposals took place for a couple of years (and then only gradually). Canada, although a consumer member of the agreement, was a very important producer of many other non-ferrous metals which were held to overflowing in the U.S.A. strategic stockpile and her commitments to the Tin Council on her own tin stock made her keep a careful eye on the precedent likely to be created for U.S. disposals in those other metals.

The statistical position was now changing sharply, but in June and July the Council committed itself publicly and unwisely to the position that the examination "indicated a small surplus during the year 1957" and its Statistical Committee did not meet again for three months.

Through early July the London cash and forward prices fell towards the bottom end of the lower sector, but it took relatively little buying by the buffer stock (no more than 400 tons in the third quarter of 1957) to hold it off the floor.

The Council had made two serious – indeed almost fatal – errors in its statistical calculations. In the first place, it had consistently under-estimated the tonnages likely to come to Western Europe from the U.S.S.R. That country, in the later 1950s, had entered into long-term arrangements to take supplies of tin from Mainland China. The extent of these supplies was unknown to the Council, which otherwise might have been forewarned, but they were on a massive scale well beyond the comprehension of Western observers. In the three years 1956-58 alone the U.S.S.R. took 56,000 tons of tin from China; in the next two years it was to take another 38,000 tons. It began exporting small quantities of this tin during 1956, but this trade was mainly to the other socialist centrally-planned countries in Eastern Europe and made no impact on the western market. Russian policy changed in 1957 and tin from the U.S.S.R. began to move into the West.

At its meeting in October, 1957, however, in the light of the figures of imports and transit trade from the U.S.S.R. and China available to it (2,500 tons for the first half-year), the Statistical Working Party seemed not unjustified in assessing the likely U.S.S.R.-Chinese tonnages at no more than 4,000 tons a year for the whole of 1957. The result was a most unfortunate underestimate.

The other error lay in the consumption figure. Earlier in the year it had been assumed that the movement of tin consumption would continue to follow the rising curve which it had shown for the last four years; on that assumption the U.S.A. consumption for 1957, for example, would be 64,000 tons. But through 1957 actual consumption in the U.S.A. and also in the United Kingdom was sliding downwards, and the Working Party in October now accepted a reduction in the U.S. figure to 57,500 tons (a figure which still proved in the result to be too optimistic). There was no such drastic change in the estimates of consumers elsewhere, but the Statistical Committee found itself now with an estimated surplus for the year of 12,800 tons.

In a third field – the production of tin – the Council was, on the other hand, remarkably correct in its judgement. In June, 1957 it was estimating for the year 1957 a mine production of 161,000 tons and a metal production of 162,600 tons; the ultimate actual figures proved to be 163,000 tons and 161,000 tons. But timing was not on its side. There is in tin a fairly general cyclical swing in total mine production which throws up a slightly higher output in the later quarters of the year and a fairly general seasonal swing downwards in consumption, especially in the U.S.A., in the last quarter of the year. For 1957, effect of this scissors movement was aggravated in that quarter by an abnormally high export of concentrates from Bolivia and an abnormally low metal consumption in the U.S.A.

[5] This figure was probably an under-estimate. The intake into Western Europe from the U.S.S.R. and China during 1957 was believed by the Council to be about 7,600 tons. Actual exports from the U.S.S.R. alone were later known to be at least 9,900 tons.

[6] The subsequent contributions were assessed at the new floor price operative in the agreement since March, 1957 (that is, at £730 a ton instead of £640); this was unlike the practice of later agreements.

[7] In doing so it was still behind some market opinion. As early as September-October, 1957, Vivian, Younger and Bond Ltd. (in their *Tin Review*) had estimated imports from the U.S.S.R. alone at 7,000 tons for 1957.

The tin market was quicker on the uptake than was the Council. In early October the tin price fell and reached the floor price of £730. The buffer stock was compelled to buy very heavily. In the seven weeks from 1 October to 21 November it bought 5,400 tons or more than it had bought in the whole of its previous buying. Tin from the U.S.S.R. and China had become a flood, especially through the Netherlands, and in the fourth quarter of the year the reported imports were known to be over 4,000 tons.[5]

On 21 November, 1957 the buffer stock held 10,000 tons of tin metal. This meant that the initial cash contribution of £9·6 millions would soon be exhausted. The agreement provided that, when the 10,000 tons holding was reached, there should be called up for the buffer stock a further and first subsequent contribution of the equivalent of 5,000 tons (£3·65 millions). On 21 November this amount was called up for payment within three months. The producers agreed at once to waive their right to make payment in metal of part of this amount and to contribute wholly in cash (by 6 December, if possible). They also agreed that the second subsequent contribution of 5,000 tons (£3·65 millions),[6] which was due to be called up when the buffer stock held 15,000 tons of tin, should be paid as soon as possible after that due date, and some producers stated very optimistically that they might make payment even before the due date.

Through December the buffer stock was buying at the alarming rate of 1,000 tons of tin a week and by the end of the month had accumulated no less than 15,300 tons. The buffer stock was reduced to £2·3 millions in cash. The second subsequent contributions were thereupon formally called up for payment no later than 30 March, 1958. Almost all of this was paid in cash (one-quarter immediately).

At the end of November, the Statistical Working Party at last provided a much more realistic picture of the situation. It accepted that the imports from China and the U.S.S.R. in 1957 would be as much as 9,000 tons;[7] that commercial consumption would fall as low as 147,500; and that, in consequence, the surplus for the year would be no less than 19,500 tons. This figure would swallow almost all the total resources of the buffer stock which, as the result of the change in the floor price, had

been reduced to a buying value of about 23,500 tons. The surplus
for 1958 might be considerably higher. For the first quarter of
that year the Working Party foresaw imports of 3,200 tons from
the U.S.S.R. and a net effective demand[8] of only 28,000 tons
against a normal production from the producing members of
36,000 tons. The producing members of the Council asked for
the introduction of export control almost immediately.

The introduction of export control, 1957

On the statistical assessment, on the actual market prices and
on the position of the buffer stock, there could be no argument as
to the need for export control. The only discussion could be on
the timing and tonnage. On the first point, the producing countries could scarcely be expected, in view of the administrative
machinery needed, to apply control backwards or immediately
(that is, on 4 December, the date of the Council meeting), but
they accepted 15 December. This was very ready acceptance, but
even this short delay aggravated one problem which arises at the
time export control is first introduced. There is, between the
point of export, however this is defined, in a producing country
and the consumer of metal a pipe-line of material, sometimes
unsmelted concentrates, sometimes metal. Tin concentrates
produced in Thailand and Malaya may be smelted at the
Malayan smelter within at the most a fortnight; the resultant
metal may take two months for delivery to the consumer in
Europe or the U.S.A. Bolivian concentrates may not reach the
smelter in Britain until perhaps three or four months after mining and may then wait for a couple of months for smelting;
Congo concentrates (whether shipped to Belgium for smelting
or smelted locally) may reach the market as metal only after six

[8] That is, consumption 34,000 tons, mine production from other than the producer members 2,800 tons, imports from the U.S.S.R. 3,250 tons, therefore net
effective demand upon producing members 28,000 tons.

[9] Calculation of the reductions is on the total mine production of the six members over the twelve months ending on 30 September, 1957, a quarterly average
of 37,777 tons.

[10]On a total tonnage of 27,000 tons the percentage degree of limitation worked
out at the Congo 29, Bolivia 18, Malaya 32, Nigeria 40, Indonesia 21 and
Thailand 39.

months, sometimes more. This pipe-line never runs dry but clearly, if the total tonnage of tin allowed to be produced, and therefore consumed, is reduced the necessary tonnage in the pipe-line may be reduced. But at the time of the initial introduction of export control there will be an excess in the pipe-line which has already passed the point of export and whose flow cannot therefore be stopped. This excess the Council probably over-estimated at the order of 5,000 to 7,000 tons. Between 4 December, when control was agreed, and 15 December, when control started, each producer would naturally try to throw into the pipe-line as much tin as he could lay hands on; there might be another 1,000-2,000 tons of tin in concentrates which would thereby escape control but which would become marketable tin metal during the first control period.

The statistics in front of the Council indicated a figure of 28,000 tons as the export quota for the first control period (from 15 December, 1957 to 14 March, 1958). The producers were split on the figure (Malaya, the Congo and Indonesia for 26,000 tons, the three others for a higher level); Thailand wanted 32,000 tons, Nigeria would accept 28,000 tons. The consumers agreed on 26,000 tons after considering a higher figure. A compromise by the Netherlands of 27,000 tons was accepted.

This permissible export amount represented an overall reduction of 28 per cent in the exports of the participating countries. In the individual countries the tonnage allocation depended on the percentage held as this had been re-allocated in October, 1957; individual countries suffered therefore very different degrees of limitation.[10]

Most, but not all, members of the Council saw one danger in the introduction of export control, especially on the severe rate now being imposed. If the pressure of restriction were excessive, the price might be forced from its current floor price level through the lower sector and then into the middle sector. No check could be imposed on that price rise until the price reached the upper sector. The second part of the argument was that sales in the middle sector would restore the balance, now so badly distorted, between cash and metal in the buffer stock. With questionable logic, since it seemed to be an essence of the agreement that a free middle sector was desirable, the consumers pressed for authority to the buffer stock to operate in the middle range whilst export control was operative. The producers were less strongly in favour of the proposal but the consumers had tied the question to the acceptance of export control. With two of the producers (Indonesia and Thailand) voting against, the authority to act in the middle range was granted. The same arguments were to be repeated and the same or a similar resolution passed for the next 2½ years, but in no case was there any serious danger of a price rise substantially into, let alone through, the middle range.

Backed by export control and with his funds replenished by the subsequent contributions, the buffer stock Manager began another round of massive buying (4,690 tons in January, 1958 against 4,160 tons in December and 5,480 tons in November, 1957). He continued to be successful in his major aim of holding the floor price[11] but with all his effort it was only spasmodically that the cash price was moved a few pounds above the floor. He had become the waste bin for all the tin that the world could transfer to him. The imports of Russian-Chinese tin into Western Europe in the first quarter of 1958 were to reach about 4,000 tons. The cash from subsequent contributions to the buffer stock from the producers passed through his hands almost as soon as received. There was no help from consumption which, in the world as a whole in the first quarter of 1958, was to be about one-seventh below what it had been a year previously. Many producers naturally tended to deliver more of their permitted quarterly quota to the smelters early in the quarter while the buffer stock was seen to be still buying.

By late January, 1958 the producers met under V. Paz Estenssoro (Bolivia) and reviewed a gloomy future. The amount of tin in the pipe-line and its rate of flow in December and January had

[11]He had bought forward tin (fairly substantially, about a fifth of the whole) in the third quarter of 1957 at £751 against £738 for cash tin, but he practically stopped buying forward tin in the fourth quarter and bought no more forward until he both bought and sold forward in the third quarter of 1959.

[12]The Congo contributed £120,000, Malaya £506,000 (in the form of a bank guarantee), Bolivia £276,000, Nigeria £72,000, Thailand £99,000 and Indonesia nothing.

been greater than anticipated and during March the resources of the buffer stock could fall 5,000 tons short of the amount necessary to stop the flood.

The "Special Fund"

At this point the producers agreed to recommend their governments to create a special fund as a voluntary contribution to the buffer stock, the amounts to be in accordance with their percentages under the agreement, payment to be spread over three months and the total not to exceed 10,000 tons equivalent. This bold decision, for which Sir Vincent del Tufo (Malaya) was largely responsible, was to save the Council.

The consumers (who bore no responsibility for the proposed special fund) agreed to a further tightening of control, and the control period (15 December, 1957 to 14 March, 1958) was extended by another fortnight to 31 March without any change in the total export amount of 27,000 tons. That reduced tonnage rate (23,000 tons quarterly) was, in effect, continued for the second quarter of 1958. Four of the producers and most of the consumers (including the United Kingdom and France) voted for 23,000 tons (Thailand had wanted 26,000 or 27,000 tons and voted against, Bolivia abstained).

The Council solved easily enough (by ignoring them) any legal problems likely to arise from the creation of the special fund. By resolution it noted that the special fund was to be put at the disposal of the buffer stock Manager; the Chairman of the Council might, as soon as the present emergency was no longer in existence, authorise the Manager to liquidate the fund by selling in a manner calculated not to disturb the market and to refund the proceeds to the contributing countries. The Council would not be told of the fund's operations. Contributions of £1,075,000 were made to the fund.[12]

The cash contributions and guarantees to the special fund alone would have bought only some 1,500 tons of tin and would have been quite inadequate to meet the situation, since by the end of March, 1958 the buffer stock had bought 22,200 tons out of its total buying capacity of 23,350 tons and had then only £880,000 in cash left. For the next six months the buffer stock was out of the market and the defence of the floor price fell on the shoulders of the special fund and on the continuation (and later tightening) of export control.

The Manager approached banking sources for the remainder of his special fund requirements. The Chartered Bank agreed to make overdraft facilities available (a figure of up to £6 millions was mentioned); at a later stage Hambros Bank provided drawing facilities (over £500,000); a short term loan of £750,000 came from the metal dealing firm of Vivian Younger and Bond Ltd. The Bank of British West Africa was prepared to help; the Bank of London and South America (Bolsa) declined.

The first purchases on the Chartered facility were made in mid-February, 1958 (1,770 tons in March, 910 tons in April, 1,000 tons in May and 2,400 tons in June); by the end of June purchases on that facility were 6,990 tons and by end-September, 9,490 tons when purchases ceased. In order to disguise the operations of the fund there were some sales through the Chartered facility in March-April (about 1,750 tons). Sales began again in October, 1958; were very heavy in January, 1959 (1,480 tons), in February (2,800 tons) and in March (1,820 tons); they were completed (total 7,730 tons) in April, 1959. The Hambros facilities were used to buy over 800 tons of metal in September, 1958[13] and to sell this metal over the next six months to February, 1959. (See chart on page 310).

In total the purchases made by the special fund were 10,262 tons and sales 10,220 tons. The amount raised from producers was £1,075,000, the loan from Chartered was £4·6 millions and from Hambros £523,000, the gross profit about £235,000 and the net profit £40,000.

In some six months the special fund sold net over 8,500 tons of metal or a figure equal to about one-seventh of the total world metal production in that period and in a time of intense restriction of exports from the producing countries. In that period the tin price moved up from £741 to £779 a ton, a change of only 5 per cent. This was a remarkable result.

The battle was long. World consumption of tin metal was not to move upwards – and even then not substantially – until the last

[13]Apparently at about £673 a ton, therefore well below the floor price.

quarter of 1958. The imports from the U.S.S.R. and China showed no signs of stopping (they were estimated at 4,500 tons for the first quarter of 1958 against 4,700 tons in the last quarter of 1957); they even increased and were estimated for April-May, 1958 alone at 3,900 tons. There was only one consolation – world production of metal, whose high level had reflected the using up of tonnages in the pipe-line, moved downwards in the second quarter of the year and became more closely in line with the limited mine output. Between April and August the cash price averaged a shade over the floor of £730; the forward price – more volatile – ranged between £722 and £740.

The discussions with the U.S.S.R.

It was clear that the Russians would have to be tackled. The Council adopted a two-fold approach. It first considered the possibility of consuming members limiting imports of tin metal from non-member countries (that is, from the U.S.S.R.). When export limitation was in effect for producing members, the consuming members should limit their imports from non-members (perhaps in the same degree as export control was limited). Some consumers had their doubts about the proposal. Mr. Herinckx (The Netherlands) pointed out that such an extension of the authority of the agreement would be a breach of general trade policy and, in any case, might merely shift Russian tin from the consuming countries into the world of non-members.

A committee on imports from outside countries in June, 1958 also thought that the proposed import controls might conflict with the other obligations of member countries under the General Agreement for Tariffs and Trade; that such important non-members as the U.S.A. and Federal Germany might regard the action as discriminatory; that, even if the trade were controlled, the powers would not be effective in view of the ease with which tin moved in international trade; and that, in any case, the Council had no power to do what the agreement did not specifically allow it to do. But the Council, which now was finding itself cutting the export quota for the fourth quarter of 1958 to a figure of only 20,000 tons, or little more than half normal production, felt itself obliged to do something. In June, 1958 it drew attention by resolution to the very heavy burden on the producing countries of their contributions to the buffer stock, their drastic limitations of production, the closing down of mines and the serious unemployment in the industry and declared that it was essential to secure some limitations on the exports of tin

TABLE 13
U.S.S.R. trade in tin, 1955-72
(long tons)

	from China	Imports from others	total	Exports
1955	16,663	—	16,663	2,067
1956	15,452	—	15,452	3,248
1957	21,653	—	21,653	18,011
1958	18,995	99	19,094	21,948
1959	20,471	35	20,506	17,816
1960	17,420	44	17,464	11,318
1961	11,023	39	11,062	5,610
1962	8,563	1,181	9,744	492
1963	4,232	3,445	7,677	689
1964	1,083	4,330	5,413	11
1965	492	5,216	5,708	7
1966	492	4,232	4,724	7
1967	98	5,512	5,610	6
1968	295	6,693	6,988	—
1969★	30	6,659	6,689	—
1970★	197	7,967	8,164	—
1971★	492	3,815	4,307	—
1972	784	3,425	4,209	—

★In addition, the U.S.S.R. reported the import, as tin-in-concentrates, of nil tons in 1969, 1,377 tons in 1970 and 3,836 tons in 1971. The table makes clear (i) the disappearance after 1963 and the re-emergence on the smallest scale in 1971 of tin metal imports into the U.S.S.R. from China, (i) the general relationship between 1955 and 1961 of the exports of metal from the U.S.S.R. (including exports to western Europe) and (iii) the steady maintenance after 1963 of imports by the U.S.S.R. Mine or smelter production in the U.S.S.R. is never published.

[14] H. Depage (delegate for the Belgian Congo) said: "The U.S.S.R. action had helped to aggravate the general tendency of the market. They are clearly within the framework of the offensive, both political and economic, of the U.S.S.R. against the free world". *Agence économique et financière*, Brussels, 23 September, 1958.

[15] According to the official Russian trade returns exports were at a price per ton only slightly below the prevailing London market price.

from the U.S.S.R. It asked the Chairman of the Council to stress to the Russian authorities the impact of these exports on the economies of the under-developed countries. It asked him formally to invite Russia to become a producer participant in the agreement and, in any event, to limit as soon as practicable its tin exports to countries outside the Sino-Soviet bloc. That limitation should be in the same proportion as the limitation imposed by the Council on its own members. Delegates were asked to transmit the resolution to their governments with a view to such action as might be possible to limit this Russian threat to the continued effective operation of the agreement. The Chairman was also asked to approach the most important other non-members – Federal Germany, the U.S.A. and Japan – to seek their co-operation in the solution of this problem.

The approach to the Russians needed to be extremely delicate. It was notorious that the U.S.S.R. government resented very strongly any pressure from outside which could in any way be regarded as infringing its sovereign rights. No one knew whether its sales of tin at a time of a world crisis in the industry had been actuated by malice to destroy economically an industry of a flagrant colonial character[14] or had arisen from the stupidity of a bureaucracy which, told to sell massive accumulated stocks of tin, sold them at the worst time and price for itself. There seems to have been little, if any, actual price cutting, in spite of the market rumours.[15] There certainly was confusion amongst the Russians as to the results of the sales. J. E. Chadaev, Deputy Chairman of the Soviet Planning Committee, whipping up Malayan goodwill in February, 1958, still stated that the U.S.S.R. needed all the tin that Malaya could produce.

The first approaches to the U.S.S.R. were made through a sympathetic party (through the delegate of India) with a suggestion for a discussion on the importance of stability and on the rate of Russian tin sales. In June, the Chairman of the Council (G. Péter) wrote to the Russian Ambassador in London. He pointed out that the substantial sales by the U.S.S.R. had weakened the stabilising objective of the agreement, had depressed the price of tin and had in consequence resulted in a still tighter degree of export control and more unemployment, consequences which might have escaped the notice of the U.S.S.R. government. He invited the U.S.S.R. to consider membership of the agreement as a producer and to limit its exports outside the Sino-Soviet bloc.

Indonesia and Malaya made diplomatic representations. But the attitude of the importing countries was far more significant. The Netherlands, the main trader in Russian tin, was doubtful; but the United Kingdom agreed that its ambassador in Moscow should approach the Russians.

The Russians replied to the Chairman of the Council late in July, 1958 that: "Since the (appropriate Soviet) organisations are not well acquainted with the functioning of the International Tin Agreement and since the U.S.S.R. is a country exporting and importing tin at the same time it has been found expedient to have the U.S.S.R. represented on the International Tin Council for a certain time unless the ... Council has objections".

This delaying letter in no way checked the movement of tin from the U.S.S.R. In June, in August and again in the first week of September there were selling waves of cash tin on the London Metal Exchange. The buffer stock Manager (with 23,350 tons in his stock and 8,430 tons in the special fund at the end of September) had almost exhausted his resources and on 18 September, 1958 stopped supporting tin at the floor price of £730. Prices collapsed at once to £640 a ton. On the same day the Council Chairman met V. A. Kamenskij, the Trade Commissioner of the U.S.S.R. in London, who was by now well aware of the political and financial implications of his sales. He promptly stopped sales of U.S.S.R. cash tin on the London market and he agreed in discussions that he did not consider it in the interests of the U.S.S.R. to sell tin at low prices.

Government action against Russian imports had already started. The lead was taken by the United Kingdom. From 30 August imports of tin and tin alloys into the United Kingdom were restricted on license. The three months' quota was only 750 tons or little over one-third of what actual imports had been in

---

[16]Sweden and Switzerland were sympathetic. The U.S.A. also said the Russian tin was classified as Chinese tin and therefore excluded from the U.S.A. France, as a member of the agreement, thought that the proposed restrictions of Russian tin might show an unfriendly attitude; it was not unfavourable to an observer from the U.S.S.R. on the Tin Council.

the second quarter of 1958. Other countries on the Council followed suit, although not necessarily with the same tight degree of restriction. In October, Canada agreed that tin metal should not be imported from countries not members of the agreement. In France no new contracts were entered into for Russian tin. Denmark agreed to limit, but not publicly. Perhaps most important, the Netherlands (in spite of the danger to its trade with the U.S.S.R.) accepted limitations. The major entrances for Russian tin had therefore been abruptly narrowed.

In the part of its campaign to attract new consumer members the Council was less successful. Japan did not find it possible to join the agreement and hedged on the possibility of restricting imports of tin from the U.S.S.R. Federal Germany continued to be favourably disposed to the activities of the Council, would continue to comply with the Council's request for a policy of restricting imports but would not join the Council. The U.S.A. said that a reply to the invitation to join would be negative.[16]

The committee on the Russian observership talked with Kamenskij in October and felt that Russia was unlikely to be interested in membership as a producer. The Russians had already the specific intention of limiting their export level in 1959, perhaps in the degree of one-fifth. That limitation could not affect supplies to the West from China, which was a sovereign power in its own right. In December, Kamenskij was prepared to agree a limitation of exports from the U.S.S.R. to 13,500 tons for 1959. The producers regarded this proposed limit as far too high and wanted the Russians to enforce restrictions on the re-export of Russian tin from the other East European socialist countries. The unwise pressure of the producers for a tighter bargain with the Russians was continued by M. Depage; he felt that any failure to obtain a commitment on re-exports left the future dangerously open. But Nigeria (Mr. Goble) attached the very greatest importance to an agreement with the Russians as an insurance policy against any further raids on the tin market from behind one section of the iron curtain (if not from China); Sr. V. Paz Estenssoro (Bolivia), the later President, thought the tonnage offered too high, but the Russians had better cards and would hold more if the consumers abandoned their restrictions on imports from Russia; and Sir Vincent del Tufo (Malaya) stressed the need to create a good atmosphere with the Russians.

The decision of the Council was determined by the Netherlands, the country which had handled most of the tin from the U.S.S.R. The Netherlands had strong reasons for trying to maintain its old connections with tin-producing Indonesia, but it was also heavily dependent on Russian timber for housing and on its carrying trade with the U.S.S.R. It had entered into the restrictions on Russian tin reluctantly and had no desire to embitter its trade relations to the possible extent of Russian retaliations. It had adopted its restrictions to cover an emergency situation, to provide a reasonable safeguard for the prosperity of the producers and to give time for co-operation between the Council and the U.S.S.R. In mid-December it decided firmly that the tin position, measured in terms of the tin price, had moved out of the emergency and that other countries were giving clear indication of increasing their tin trade with the U.S.S.R.; it decided to lift its restrictions almost at once.

The consumer members of the Council voted almost unanimously[17] in favour of a settlement with the U.S.S.R. on the basis of the 13,500 tons export rate and a heavy majority (in voting power) of the producers followed suit.[18]

The settlement was embodied in two letters in which the Russians agreed their intention of exporting less in 1959 than in 1958 to the non-socialist countries (less than 13,500 tons with not more than 750 tons a quarter to the United Kingdom); they would request the Socialist countries not to re-export the tin they had obtained from the U.S.S.R. and the Council would secure the suspension of the import restrictions.

In the meantime the Russian request for an observer had been dealt with. It is difficult to believe in the seriousness of the Russian request to sit in on the discussions of an international

---

[17]Korea voted against, Belgium abstained.

[18]Bolivia, Indonesia and Malaya for, the Belgian Congo and Nigeria abstained and Thailand against.

[19]The fall in 1959 was mainly to Britain and in 1960 mainly to the Netherlands. The reasons for the very sharp overall drop in 1960 are not known; it may have been in anticipation of the forthcoming drop in the Russian intake from China in 1961.

body where many, perhaps almost all, members believed that the
result of current Russian sales was, by ignorance or malice, to
help to destroy the price structure which the body was trying, at
great cost, to maintain. Fortunately, there was a sufficient reason
to justify a refusal. The agreement made no provision for observers (although it contained no prohibition against observers
either) and the Council was able diplomatically to inform the
Russians:

"The Council has given full consideration to the request made
by the U.S.S.R. to become an observer at the Council. The
Council was not able to accede to this request; there is no
provision in the Tin Agreement which would permit countries
to become observers. The Tin Council was glad to be informed,
however, that the U.S.S.R. is willing to co-operate with it".

The arrangement, once arrived at, was firmly adhered to by the
Russians. Exports of tin from the U.S.S.R. to the four major free
markets (the Netherlands, United Kingdom, Federal Germany
and Japan) were 16,000 tons in 1958 and were reduced to 10,500
tons for 1959. They dropped in 1960 to only 5,300 tons.[19]

It was a major, but happily not an overwhelming, victory for
the Council. It had been won by consumers. Major importers,
especially the United Kingdom and the Netherlands, had risked
their trading interests with the U.S.S.R. for the sake of the
producers and few had wanted to push the Russians into a
humiliating retreat. The chairman of the Council (G. Péter) had
been excellent in diplomacy. There were, of course, disadvantages. If there had been any real likelihood of the U.S.S.R.
joining the Council the struggle must have hardened Russian
unwillingness; but it was probably desirable for the internal
harmony of the Council that, during the 1960s when its relations
with the U.S.A. on disposals from the U.S. strategic stockpile
were at the most acute, the U.S.S.R. was not present as a member
to make them more acute.

In a wider field, the case was one of the few examples of a
successful struggle yet recorded by an international commodity
organisation against a massive non-member.

The easing of export control, 1959-60

The break in price in September, 1958 was very short-lived. The Council, although it did not as yet know it, had turned the corner. The permissible export amount for the fourth quarter of 1958 had been put forward by the Chairman in July at 20,000 tons, the lowest quarterly tonnage yet. This was agreed, with Bolivia and Thailand objecting and wanting a higher figure. Coupled with the restrictions imposed on Russian tin, this tight measure succeeded in pushing up the average price for cash tin during October above the floor of £730 for the first time in any substantial degree in twelve months.

The consumers and even some producers smelled the spring. In November, 1958 when the quota for the first quarter of 1959 was discussed, Italy, one of the pure consumers on the Council, thought the retention of a quota as low as 20,000 tons was a desperate remedy and even France and the Netherlands voted against continuing at that level. The tin prices, which averaged £741 for October and £758 for November even while the special fund was being unloaded, showed the real fever was over. But the shift of consumption upwards was still very sedate and the negotiations with the U.S.S.R. had not yet been finalised.

The United Kingdom could scarcely agree to ease the severity of export control on the producers while it was imposing limitations on imports from the U.S.S.R. because of the severity of the producers' sacrifices. In any case the special fund required to be sold and its disposal would, it was hoped, prevent prices rising too sharply. With United Kingdom support the producers secured a repetition of the quota at 20,000 tons.

The size of the quota was now being decided on issues not relevant to the immediate supply-demand-price position. By February, 1959 the Statistical Working Party could at last look forward to an increase in consumption for the second quarter of the year, even although it under-estimated that improvement. On that under-estimate it arrived at a net balance to be provided from the participating producing countries of 28,500 tons. The Council was aware that, in spite of substantial sales from the special fund, the price was approaching the top of the lower sector and might be expected to enter the middle sector. The importers' restrictions on Russian tin had ended and the flow from the U.S.S.R. was now subject in turn to curtailment under an understanding. The producers were prepared to go up to 22,000 tons but the pressure from consumers for more tin was overwhelming and 23,000 tons was agreed (with Indonesia abstaining on the vote). The quota was now back to where it had been twelve months earlier.

In fact, the Manager of the buffer stock was now beginning to pour tin back on to the market almost as fast as he had withdrawn it a year earlier. He had begun selling from the special fund in November, 1958; he sold 1,290 tons in December, and a massive 6,600 tons in the first quarter of 1959. The fund was sold out in April. The ordinary buffer stock started to sell in March (nearly 1,500 tons). From the two sources he supplied the market with over 9,400 tons in five months.

The flow from the buffer stock was to continue almost as a flood. Heavy selling (including selling of the U.K. stock on that government's account) brought the buffer stock holding down from 23,200 tons in February to only 6,550 tons in November, and the Manager found himself obliged to buy, not only to hide his selling transactions but also to bring his holdings back nearer to the 10,000 tons on which the continuance of export control depended. It was also necessary for him to continue fairly heavy net buying through the first half of 1960 so that he could reach his 10,000 tons target by the middle of the year.

By May, 1959, with the price pushed up into the middle sector, the United Kingdom felt the basic position promising enough to justify it in giving the Council notice that it would be selling 2,500 tons of tin from its stocks from the middle of 1959. Even more important as an indication of the change was a challenge to the floor price in the agreement itself. The current range with its floor of £730 had been established certainly at the wrong time in March, 1957. That change had reduced the power of the Manager to buy and had brought forward the date at which he had acquired the 10,000 tons of tin and from which export control had flowed. Once the original error had been made, it would have been disastrous and demoralising to have attempted to change it in the intensity of the battle to save the floor. But by May, 1959 the market price was running £50 a ton above the floor and was in the middle sector.

The United Kingdom proposed to shift the Council's range downwards to a floor of £700 and a ceiling of £835. It maintained that the present floor would not have been lifted to £730 in 1957 if there had been more accurate forecasting or if the Council had been aware of the forthcoming flood of Russian tin. The present market price of around £785 was being supported (by export control) at a level above the real price of tin. In December, 1959 the United Kingdom repeated its argument and France (P. Legoux) in half-hearted support went so far as to suggest a reduction in the ceiling (to £830) but not in the floor. Most consumers doubted whether, since the whole matter of a new agreement with new prices would be before the United Nations Tin Conference during 1960, the time was ripe to discuss the matter now. The United Kingdom pressure was so mild as to suggest that their action might have been no more than a warning to the producers. The matter of the price range was not raised again for Council action until in October, 1961 in very different circumstances the producing countries embarked on a much more vigorous campaign for sharp increases in all the elements in the price range.

On the quota itself the problem had now become one of preventing it going up too fast rather than of forcing it down too much. In May, all the factors – including the producers themselves – were favouring an increased tonnage. The price had been for two months in the middle sector where, in principle, the

[20]He told the Council that there was likely to be 10,000 tons on 31 December, including the 2,417 tons of the United Kingdom strategic stocks which he would then hold not for the buffer stock but for disposal as agent for the U.K. At the end of December, he held only 7,500 tons of Council tin.

[21]The argument was remembered by the Council. Later agreements provided that the Council could not declare a period of export control unless 10,000 tons of tin metal were likely to be held in the buffer stock at the beginning of that period. There was one minor qualification (in declaring export control for the first time the tonnage could be 5,000); but there was a major qualification (that the Council by a special majority could in respect of any period fix a figure below either the 10,000 or the 5,000). That qualification gave the Council full freedom, if it so desired, to divorce the problem of export control from the problem of the buffer stock.

normal market forces should freely operate. Consumption was rising sharply outside the U.S.A. and, even in that country, the decline was beginning to flatten out. There was some reason to believe that in total the world might take in 1959 as much tin as it had taken in 1956. The buffer stock was selling tin very fast. (By the end of June its holdings would prove to be down to 13,990 tons). In some producing countries the burden of stocks had been eased by barter deals with the United States. Without the need for a vote the quota of 25,000 tons was approved for the third quarter of 1959.

The position was still brighter in September, 1959. The producers hoped to plan for a systematic steady rise in quarterly tonnages which would phase them out of export control within the next year. The United Kingdom was anxious to ensure that the market would still be able to absorb the 5,000 tons of strategic stocks of which it was now disposing. Most producers were prepared to accept a quota for the fourth quarter of 1959 of 28,000 tons; the consumers inevitably requested more; and without a vote, but with an understanding, there was agreed a figure of 30,000 tons. The understanding was that, if this quota proved insufficient and if, in consequence, the buffer stock had to sell tin and therefore fell below 10,000 tons of tin before the end of the quota period, a special meeting of the Council should be convened.

In December, 1959 the Council had to overcome a legal problem on export control. Under the agreement no permissible export amount could be fixed by the Council unless at least 10,000 tons of tin were in the buffer stock or unless the Council found that 10,000 tons was likely to be held before the end of the current control period. The buffer stock was now selling so fast that a holding of 10,000 tons was endangered. On the assurance of the buffer stock Manager[20] the Council passed the necessary resolution.[21]

But more important than the legality of the continuance of export control were the signs that long-continued restrictions had begun to sap the productive power of some producers. The permissible export amounts laid down by the Council had (with the exception of the illegal trade from Thailand) been observed by members with remarkable scrupulousness; and an initial over-export by Malaya in the first control period had been promptly corrected by a corresponding under-export in the secone period.

In the first seven control periods passing from 15 December, 1957 to 30 September, 1959 the producer participants had exported 160,900 tons against a total permissible tonnage of 161,000 tons.

In November, 1959 the producers made a proposal which indicated their expectations of trouble in obtaining the full quota from all members. They suggested that the quota for the current quarter might be increased, provided that no producer should be penalised for under-exporting his share in the increase and that any balances not exported should be carried forward into the first quarter of 1960, or that the producer should make a voluntary contribution to the buffer stock based on current percentages and payable by March, 1960. These ingenious ideas which, on the one hand, could ensure the fulfilment of an aggregate quota even if individual countries failed or, on the other hand, could ensure that the buffer stock continued to hold over 10,000 tons so that export control was also continued, did not go further but the underlying reason was made clear by Bolivia in December. Bolivia had economically the most rigid structure of all the producers. Nationalisation of the major mines in 1952, gross labour indiscipline and export control had succeeded in almost halving her total output within five years. The industry was disorganised and had made it clear that it had not the resilience to meet any increase in the export quota.

It was clear that the fixing of a quota would no longer bring the automatic guarantee of its fulfilment. The price was rising (for the last quarter of 1959 it was £790) and the Netherlands (Mr. Herinckx) rubbed at a sore point when he said that a rise above £800 a ton might bring sales from the U.S. stockpile. Some consumers wanted for the first quarter of 1960 a quota of 40,000 tons (that is, well above the pre-control tonnage) and even the

[22]The Manager in the discussion in March, 1960 on the quota for the coming quarter, unusually advocated 38,000 tons on the grounds that without this higher tonnage (and allowing for an under-export) the demand for tin might not be met, he might be forced to sell and the stock would then fall below 10,000 tons.

United Kingdom (with 2,400 tons of its own stock still waiting to be sold) wanted 38,000 tons. Bolivia was naturally opposed to either figure; the producers (Indonesia abstaining) offered 36,000 for the first quarter of 1960; and the consumers accepted this figure and a continuation of buffer stock power to act in the middle range. The buffer stock Manager said that his current operations were directed to keeping the price between £790 and £800, at retaining 10,000 tons in the stock and to make room for the U.K. stock sales[22].

In December, 1959 the Council agreed that if Bolivia under-exported in the ninth control period no penalty would be imposed on her. Bolivia fulfilled expectations by under-exporting some 700 tons in the eighth control period and by declaring in March, 1960 that she would be likely to under-export by 2,000 tons in the ninth control period. In the event, her under-export in that latter period proved to be as high as 3,100 tons; her actual exports were not much more than half her permitted quota. They were also below the actual Bolivian exports of any previous control period in spite of the growing easiness of the general situation. In the tenth quota period the continued under-export from Bolivia was 1,900 tons and in the eleventh and final period was 2,100 tons. In the nine months from January to September, 1960 the total under-export from Bolivia was 7,200 tons.

It was not surprising that, in the discussion in March, 1960 on the quota for the second quarter of 1960, Bolivia thought a total figure of 36,000 tons would be too high. The Australians, whose vested interest in production was energetically used, suggested a quota of 37,000 tons as being the figure most likely to produce an actual 35,000 tons. In the light of a tin market price which had been in the middle sector (even although very low down in that sector) for some twelve months, there could be little resistance from the producers and the agreed figure of 37,500 tons did not need a vote. At the same time, the permission to operate in the middle sector, which had been granted to the buffer stock Manager since the beginning of export control, ceased as at 30 June, 1960. Export control was about to die. In May, the Statistical Working Party calculated the net balance for the third quarter of 1960 at 32,000 tons. The economic recovery was pushing consumption well above the level of 1956. The price over the first half of the year 1960 was to average slightly more than the yearly average for 1956; it was to move in June, 1960, if only temporarily, to the mid-point of the whole price range. All the 4,900 tons from the British stockpile had been rapidly swallowed by the market. Some consumers in May pressed for a quota for the eleventh control period of well over 37,500 tons but were prepared to compromise on 38,500 tons if the buffer stock Manager were continued with authority to sell in the middle range. The Netherlands favoured 37,500 with perhaps the buffer stock operating only at a price above £805 a ton (the mid-point of the middle sector). The producers were anxious to abandon the buffer stock action in the middle sector now that prices showed signs of rising fast; they were willing to accept 37,500 tons if the buffer stock authority in the middle sector were abandoned.

In August, 1960 the producers tabled their last and not very strong arguments for an extension of control. Sir Vincent del Tufo stressed the stocks (from producers, from Canada and from Italy) which were available for disposal and Mr. Adnan (Indonesia) was dramatic, but rather uncertain, on the effect of the withdrawal of the loans of metal which had been made to the market by the buffer stock Manager and on the vacuum currently being created by the transfer of the smelting of Indonesian concentrates from the Netherlands to Malaya.

Export control was now well beyond justification. By the third quarter of 1960 the price had been set, apparently firmly, in the middle of the middle sector of the price range; the buffer stock was balanced with 10,000 tons; the British strategic stock had been sold; consumption had been higher than world production for two years; the flood of Russian tin was now shrinking to a mild stream; and the U.S.A. had made no mention of its stockpile. Export control had become nominal and was officially ended on 30 September.

Barter dealings with the U.S.A.

Inevitably, during the period of export control, some producers, for economic, social or political reasons, preferred to main-

tain their production at a rate higher than the limited export rate
and to carry as stock the concentrates which they were not
allowed to export. This had advantages from the point of view of
governments since it reduced the mass of unemployment which
evolved from export control, even though the stocks, so long as
they were not exported, added nothing to the revenues of the
governments. But these unexported stocks rapidly assumed
almost alarming proportions. At the end of 1956 the six exporting members of the agreement had held a little under 11,000 tons
of tin in concentrates in stock; at the end of 1957 the total was
nearly 20,000 tons and at the end of 1958 over 27,000 tons. It was
clear to the more intelligent producers that, when export control
was ended, this excess stock would be immediately available to
be thrown on the market with effects on the price that might
undermine much of the stabilising work of control and force the
buffer stock to take back from the market tin which it had
recently supplied to that market. This explained why they
insisted on the slow movement out of export control.

An unexpected source eased the position. In late 1958 the U.S.
Department of Agriculture offered to barter (through private
channels) surplus agricultural commodities against materials,
including tin, for the U.S. strategic stockpile. To the Bolivians,
short of dollars for essential imports and with an income from tin
almost halved by two years of export restriction, the proposal was
a godsend. Other producers – not Malaya – were eager.

In September, 1958 Bolivia had internal stocks of 3,336 tons of
tin; by December she was negotiating for barter deliveries of
5,000 tons of tin, perhaps against U.S. rice and tobacco. In
February, 1959 Thailand secured Council approval to barter
2,250 tons of tin in concentrates (for tobacco) and in May, 1959
the Congo for up to 1,000 tons. By the end of the year Bolivia had
bartered 5,291 tons of tin, the Congo 700 and Thailand 1,858
tons.

This total of nearly 8,000 tons going into the U.S. supplemental stockpile was a most welcome relief to the producers'
stock position; it was essential to the Bolivian economy, but it
was hard won. The Bolivian contracts with U.S. private interests
involved in practice a tin price paid to Bolivia of only about 90 per

cent of the world tin price, and diverted to the U.S. stock tin which, later, Bolivia could have used to fulfil her quotas under the tin agreement at a higher price.[23]

The Tin Council watched these arrangements with some natural suspicion. Canada, a tin importer but one of the world's great food exporters, was anxious to ensure that these barter transactions would not interfere with the normal marketing of food. The other producers thought that Bolivia might be tempted to build up her stocks during periods of export control so as to barter them to the U.S.A.; they gave little thought to the danger in the far future of swelling now the U.S. strategic stockpile or of the nearer danger of a Bolivian failure to fulfil her share in the Council's quarterly quota. But the case for ensuring that Bolivia was not starved to death for lack of dollars was overwhelmingly strong.

In December, 1958 the Tin Council approved the general principle that tonnages of tin bartered against agricultural U.S. commodities and destined for the U.S. supplemental stockpile should not be counted against a producer's permissible export amount, provided that the tonnages so shipped and the stocks in the producing country should not exceed the maximum stocks permitted by the agreement.

The buffer stock, 1959-60

The shortage of tin from the producing countries under export control during the first quarter of 1959 was made up for by the tonnages sold on the market from the special fund. On the

[23]Bolivia contracted to sell 1,000 tons of tin through Tennants and 4,000 tons through Philipp Bros. of New York. The price payable was determined largely by the price paid by these firms to the Commodity Credit Corporation for the agricultural commodity. The concentrates were to be smelted at Williams, Harvey in the United Kingdom. The Congo contract was with Philipp Bros. The Thai concentrates were to be smelted in the U.S.A.

[24]Partly to hide his transactions. But the Manager (1959-65) was a believer (probably correctly) in the greater importance of forward tin, instead of cash tin, as determining the longer-term market price.

[25]It is to be remembered that he did not like the speculative London Metal Exchange, that the Council knew his price was almost a straight line and that there was a deep-rooted belief in the Council (which was probably unjustified) that the Americans were insistent on the price not going above £800 a ton or 100 cents a pound.

cessation of those sales in April, 1959 it was necessary to choose whether the gap in the market should be filled from the sales from the ordinary buffer stock (including the forthcoming sale of the British stock) or by increases in the export quotas. The Council decided initially in favour of the first course and the buffer stock, which had been inactive since June, 1958 with 23,300 tons of metal and only some £10,000 in cash, was wakened. It sold 2,300 tons in the first quarter of 1959 when the sales from the special fund proved inadequate.

These sales became a flood in the second quarter of 1959, when it sold 7,200 tons of cash tin (but no forward tin) at a similar price, and by the middle of the year the Manager had reduced his metal holdings to 14,090 tons but had raised his cash holdings to £7.27 millions. In the third quarter the stock's policy seemed to shift (to heavy sales of cash tin and to heavier sales and purchases of forward tin[24]) but the result was an overall sale of nearly 3,000 tons. A similar policy in the fourth quarter brought his metal total down to 10,050 tons and his cash holdings up to £12.66 millions. In addition the Manager, acting as agent for the United Kingdom, had sold 2,500 tons of that country's stockpile in the third and fourth quarter and was to sell another 2,400 tons in the first quarter of 1960. The Council and the Manager had been imposing perhaps too tight a constraint on the market (his purchases of cash tin in the fourth quarter of 1959 had been at £792 and his sales of forward tin at £795) and the Manager had perhaps over-extended himself in the forward market (in the first and second quarters of 1960 he was providing the market heavily with tin in execution of forward contracts). This was in itself an obstacle to the too sudden termination of export control. But the Manager had to bear in mind that, under the agreement, export control could be operative only so long as the buffer held 10,000 tons of tin metal. His market operations now became lending operations to tide over the market (until the summer of 1960) but he maintained the totality of the tonnage at just over what he felt to be the necessary 10,000 tons and, perhaps more questionably, the price at almost a straight line.[25]

FIGURE 13

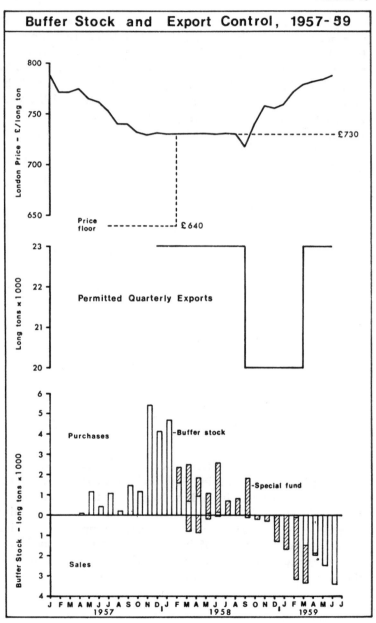

Buffer Stock and Export Control, 1957-59

Selling out the buffer stock, 1961

The buffer stock, with its holding of metal unchanged at 10,030 tons and its cash at over £11 millions, was inoperative for nearly a year after June, 1960. In the first quarter of 1961 the highest London cash price was £825, that is, within – although near to the top of – the neutral middle range of £780-830, in which the Manager could not now act. In April, 1961 the effect of a stronger market demand and a lagging production began to be felt more powerfully and the cash price moved into the upper range (£830-880) in the Agreement where the Manager had power to sell to prevent the market price from rising too steeply. Sales of tin (about 1,000 tons in two months) were made, but the price continued to rise through April and May, until it reached, early in June, the ceiling price of £880 in the Agreement. At that point the Manager was obliged under the Agreement to offer tin for sale until the cash price fell below the ceiling price or the tin at his disposal was exhausted.[26]

The market anticipated – correctly – a killing if on a rising market the second agreement were to come into force stripped of the metal stock which was the only defence against that rise. The buffer stock was, therefore, wide open to the strippers. At the end of May it held over 9,200 tons of metal, and the balance between metal and cash made the stock seem secure. On 12-16 June it sold over 2,200 tons of metal, on 19-20 June another 1,200 tons and then on 21 June it was overwhelmed with sales of over 4,200 tons. It was then left almost untouched until, on 29 June, it sold its last and useless 1,100 tons. In July, with nothing to stop it, the price rose to £914 and in August, as the new holders of the stock put the squeeze on the market, to £945.

[26]The Second Agreement was due to come into effect in a few days. Under that Agreement the Chairman of the Council had authority to suspend the operations of the buffer stock (at the floor and ceiling points) if, in its opinion, the discharge of the obligations laid upon the Manager would not achieve the purposes of the Agreement. But there was no such authority in the First Agreement. The Secretary urged that the Chairman should anticipate the operation of the Second Agreement and should suspend sales by the buffer stock which would leave the Second Agreement without any metal and therefore incapable of operating. The Chairman and the buffer stock Manager felt that they were bound by the terms of the current Agreement and that, under those terms, they could not even limit, let alone suspend, their daily sales.

XIV

New Prices and the Buffer Stock in the Second Agreement

AT the New York tin conference in 1960, the producers had argued strongly for the insertion in the proposed second agreement of price levels substantially higher than the range of £730-£880 then operative in the first agreement. They had not a strong case. The actual market price of tin, even with export control operating, had been kept only just above the floor price through 1958 and only a shade above the beginning of the middle sector through 1959 and the early part of 1960, but was still higher on the average over 1956-59 than it had been in 1955. A higher price in the agreement was regarded by many consumers as merely an attempt by the producers to recoup quickly some of the money which they had lost in the long battering of export restriction. Bolivia proposed a new floor of £760 and a new ceiling of £1,000 a ton and was supported by Nigeria and Indonesia, but not by Thailand. But even Australia would not support an increase of this order and the United Kingdom also felt that, on the evidence of the past actual swing of prices, the floor price might actually be reduced so as to give a mid-point in the price range of around £760 or £770 instead of the current £805. Against the obstinacy of the consumers, who pointed at length to the dangers to consumption of a higher price or to the political difficulties of putting now a new price into an agreement which could not operate until mid-1961, Bolivia put forward a modified price range of £750 to £900, a modification now acceptable to Thailand. The consumers (with M. Legoux their chairman) reinforced their objections with other arguments, including the point that increasing the price might lead to

[1]Wage rates in Nigeria were reported as follows:—

	1957	1960
	s - d	s - d
Day labourer, lowest per day	3 - 0	3 - 9
Grade I labourer, per day	4 - 3 to 5 - 3	4 - 11 to 5 - 11

over-stimulation of mine production and therefore to the surplus production which had overwhelmed the industry as recently as 1957 and which was now indicated as likely for 1961. No decision was taken by the conference.

The case for new prices, 1961

The producers renewed the attack under the second agreement in 1961. The market background of the industry had changed radically since the failure of the producers' claim in 1960. Export control had disappeared. In spite of the intake of metal from the U.S.S.R. and China, the year 1960 had ended with a small shortage. The first half of 1961 was to show the effective stoppage of the Russian flood and the clear emergence of a serious shortage. The buffer stock, so large in 1960, had been sold out entirely in the middle of 1961 and the market price of tin, which had broken into the higher bracket of the upper sector of the price range in May, 1961, had then burst through the ceiling price in June and by October was £85 a ton above that ceiling.

The producers, given now by circumstances a better case, still continued to base themselves in June and October, 1961 on the same arguments. Bolivia pointed out that the tin price had been kept too low for too long. The mid-point of £760 in the initial range of the first agreement had, in fact, been below the price actually averaged on the market over 1948-50. Its delegation quoted increases in the costs of equipment in recent years (as well as a 13 per cent increase in smelting charges in the United Kingdom and of 25 per cent in the ocean freight charges on concentrates from Antofagasta to Liverpool). They asked for a floor price of £840 and a ceiling price of £1,020 with a mid-point of £930, which was well below the actual market price of the first three months of the new agreement. Nigeria reported in some detail that production costs had moved up 12½ per cent in the years 1957-60; the average wages for daily labour had moved up by 20 per cent; equipment for draglines and mining materials by 7 per cent; the mining costs of three representative companies were 30.8 pence per cubic yard in 1960 against 27.0 pence in 1957.[1] The Congo Republic (now independent) stressed the importance of labour costs in its tin mining economy; in the three years since 1957 these costs, for skilled and unskilled labour, had risen by 30 per cent and transport costs had risen by 20 per cent. It asked for a floor price of £880 and a ceiling price of £1,000.

In June, the producers replied to the question which they knew would arise – how to fill the shortage which the Statistical Committee was now estimating for 1962 in the neighbourhood of 20,000 tons. They denied that there was any deliberate holding back of production in order to inflate the price, although they admitted that Indonesia, Bolivia and Thailand were subject to special difficulties. The industry had emerged dislocated from the export control crisis of 1958-60. Mines had necessarily been abandoned and labour dispersed; equipment and housing in the tropical areas had deteriorated; development work had been suspended; and the transition from surplus to shortage had been almost startlingly swift. The limiting factor on expansion and even on recovery of mine production was the price of tin. Continually increasing costs of production had reduced the volume of profitable tonnage. In an industry where the price structure had so recently been badly shaken and where the future was still uncertain with the threat of releases from the U.S. strategic stockpile, the difficulty of raising fresh capital for new production was great. The current market price was rising and if that rise were to continue many marginal mines would re-open. But a shift in the Council's price range was essential if increased output were to be broad-based.

[2] The Malaysians quoted other figures. The capital cost of a medium size gravel pump was $M120,000 to $M250,000. An 8-inch gravel pump with a through-put of 20,000 cubic yards of ground a month on land at 0.35 katis per cubic yard required a tin price of £1,000 a ton. The cost of transferring an existing dredge, capable after modernisation of digging 4 million cubic yards a year, was $M4.7 millions (including $M0.6 million for the property and prospecting and $M0.1 million expenditure on roads and bridges). Working 4 million cubic yards of ground a year at a recoverable value of 0.18 katis would mean a cash realisation over 15 years of $M31.5 millions. Dredge operating costs at 30 cts. a cubic yard would take $M18 millions, mine and head office expenses at 5 cts. a picul would take $M0.5 million, amortisation $M4.7 millions, leaving a net operating return over 15 years of $M553,000 a year or 11.77 per cent on the initial expenditure of $M4.7 millions.

The new price range, 1962

In October, 1961 each country produced to the Council more detailed supporting argument. Malaysia stressed that the gap now apparent between production and consumption would be largely uncorrected in the foreseeable future unless production were stimulated. To eat into existing reserves and to open up new reserves producers needed a higher price. In Malaysia, the average grade of ground (tin content per unit volume of ground) was declining and was likely to be ever-decreasing, although the margin of payability had also been reduced by technical mining improvements so that mines were now working ground of half pre-1939 quality. To treat this poorer ground and to work tailings and residual areas a heavier capital outlay was necessary, and the return on this outlay should be reflected in the price of tin. Two-thirds of Malaysian production came from dredges. The capital cost of reconstructing and transferring an existing dredge was $M4 to 5 millions and of building a new dredge $M10 millions. On a dredge with a life of 20 years, with a capacity of five million cubic yards of ground a year and on ground of 0.2 katis per cubic yard, working would be uneconomic at a tin price below £1,000 a ton, and even that price would give an annual return in profit of only 9.87 per cent[2]. A similar price was necessary for a new gravel pump enterprise. Indonesia pointed to the importance of the mid-point in the price range since it was around this mid-point that the actual prices would tend to settle when production and consumption were in balance. Since the market had become free of buffer stock influence in June and of export control influence in September, 1960, the market price had swung around to over £900 a ton and this should be the mid-point acceptable to the Council in a revised scale.

The producers put forward the following proposals (£ a long ton: Lower, Middle and Upper sectors).

From	Floor	Lower	Middle	Upper	Ceiling
Bolivia	840	840-900	900-960	960-1,020	1,020
Congo	850	850-900	900-950	950-1,000	1,000
Indonesia } Malaysia	800	800-870	870-930	930-1,000	1,000
Nigeria	800	800-850	850-900	900-950	950

The consumers were in a dilemma. If the price range in the agreement was to have any relevance to reality, then that range must be increased to show that the Council was not blind to the actual prices which had risen too rapidly in the last six months. But to accept such a change in the price range would mean the acceptance by the Council of a "follow-the-market" policy which in itself might mean following always a de-stabilised market. They argued that the proposed new range would be inoperable; even on the new lower sectors the buffer stock (although currently full of money) would not find itself able to buy and thus to obtain the metal to check the current price inflation. Price increases now might lead to large scale over-production within three or four years; and in that event the range then might have to be reduced. Japan was firm on no increase in the price range at a time when the ratification of the agreement was before its parliament. But Belgium, Denmark, the Netherlands and Mexico were not against an increase, and a vote on the question: Are the present floor and ceiling prices appropriate to achieve the objectives of the agreement? received an almost unanimous negative from both consumers and producers. It proved difficult, however, to settle the degree of inappropriateness. A vote on the range of floor £800 and ceiling £1,000 gave an overwhelming consumer rejection.

The producers had, however, proved their case. This acceptance of a change in principle was translated into figures in January, 1962, when the consumers initially proposed a range of £780 floor to £930 ceiling and then more generously accepted a range of £790 floor to £965 ceiling (with a mid-point of £877).[3]

This new price range was to last for nearly two years. In that period the actual market price showed very wide fluctuations, mainly as a result of fears about disposals from the U.S.A. surplus strategic stockpile. The market moved between a minimum of £845 a ton and a maximum of £1,035, that is, in the Council's middle and upper sector and above the ceiling, but only for a short time in the lower sector and at no time anywhere near the Council's floor price. The period was, in general, one of massive

[3] Floor £790, lower sector £790 to £850, middle sector £850 to £910, upper sector £910 to £965, ceiling £965, mid-point £877 a ton.

under-supply of tin, but the Council found itself first buying and then selling metal, although in neither case in any substantial quantities.

The buffer stock in action, 1962-63

Through the first quarter of 1962 the new price range was in fact ineffective in producing any action by the Council. The market, smelling a shortage which even the Council could not deny to itself, kept the price near to or above the Council's ceiling, but the Council, having no metal in its buffer stock, could do nothing to ease the price. The position, not of shortage but of anticipated supply, changed sharply in the second quarter of the year. The world market had been uneasy for almost a year at the possibility of surplus tin from the U.S. strategic stockpile becoming available to the consumer. By February, 1962 the U.S.A. was asking the Tin Council for its views on disposals; by April the Council was seeking a common viewpoint to present to the U.S.A.; in May it was expressing publicly its deep concern on the problem; in June the U.S. Congress gave its approval to the disposal of 50,000 tons. By July the London tin price was £100 a ton lower than it had been ten months earlier when the U.S. administration had first asked Congress for that tonnage. It had now, in about a quarter, slid down through the upper and middle sectors of the agreement's new price range. In late July the discussions between the Council and the U.S. on the rate and pricing of disposals proved inconclusive. The market price threatened to slide again through the lower sector. The buffer stock had very ample cash resources (it held about £15.5 millions) and it promptly supported the price by buying at the highest possible point at which it could take metal, that is, at the top of the lower sector. This was to be its real floor price for the next nine months.

The market was now thoroughly frightened, even although, when the U.S. programme of disposals was translated into figures in the autumn, it should have been clear to it that the tonnage was in no sense excessive and that the U.S. selling practices would not involve price cutting. The Cuban missile crisis in October, 1962, which would normally have been expected to galvanise so sensitive a commodity, had no long-term effect on the tin price. By the end of 1962 the buffer stock Manager had bought 3,270 tons of metal. By that time the U.S.A. had sold 1,400 tons.[4]

Selling out the buffer stock, 1963

The statistical shortage was to continue at an even higher rate through the year 1963 and the U.S. was to continue to release too small quantities and to release them too spasmodically. In the first quarter of 1963 the price remained little above the point at which the buffer stock had rather obviously made it clear that it would buy, but in the second quarter, in face of an inadequate flow from the stockpile, it steadily shifted upwards in the middle sector. There was not enough room for it to move, as buffer stock policy had in effect pinned it almost at the beginning of that middle sector. By early June the price had reached the point (£910) at which the buffer stock Manager was permitted (but not compelled) to sell. Immediately the Manager began to sell and sold over half his stock at this low price before realising that he was offering jam on a platter to speculators from New York. His withdrawal from more sales did not mean a rocketing price; the price was being determined by U.S.A. sales, and the highest actual price reached in June at £927 was still very much below the ceiling price of £965 at which he would have been obliged to sell. The Manager was now left with only 1,335 tons, a quantity

[4]The Tin Council's annual report for 1962-63 points out that 1,765 tons was bought by the buffer stock before the U.S.A. sold any of its tin and 1,050 tons was bought in the period during which the U.S.A. sold its first 1,400 tons. It would seem that there was no strict correlation between the tonnage arising from U.S. sales and the tonnage bought by the buffer stock. The buffer stock was buying at its highest possible buying point. This naturally increasing the tonnage that it took in. It is fairly certain that one reason for this prompt buying was to emphasise the argument then being used by the Council that U.S. sales would be at the expense of the producers' money in the buffer stock.

[5]They stressed *inter alia* that the provision of new dredges costing $M125 millions would give an output of 6,000 tons of tin a year.

[6]Bolivia asked for a price range of £890 to £1,065, Indonesia for £875 to £1,050, Thailand for £850 to £1,030 and the Congo for a range that would give an average price of £1,000 (figures quoted in December, 1963).

insufficient to dampen any market rise but large enough to be a constant temptation to speculators.

The inevitable happened. The market was now fully conscious of the shortage and exaggerated its immediate importance. An increase in the proposed tonnage of U.S. disposals confirmed its worst fears. There was panic buying in the U.S.A. from the stockpile in July and August. Curiously enough, this buying was not at panic prices; the New York price in August was not much more than one-twentieth above what it had been a year before. But the impetus could not be stopped. Sales from the I.T.C. buffer stock began to be heavy again towards the end of October and the handful of tons then remaining (245 tons) was distributed through the market in November. At the end of 1963 the buffer stock held no metal, but £16.5 millions in cash. It could have no effect on prices for the remainder of the second agreement; it could not under the terms of the agreement return any of this money to the contributing countries who were lagging so behind-hand in production; and it concentrated on being merely an investment trust operating in a narrow investing field.

The new price ranges of 1963-64

The price range of £790 to £965 established in January, 1962 had not been unreasonable and, in practice, the monthly average price in London had through nearly two years fallen within it. The average market price for 1962-63 at £903 a ton had been close to the mid-point of £877 in the Council's range. The average level of price did not, therefore, justify another increase in the price range. Nor did the trend. The market price had swung down through 1962 and had then been fairly flat until March, 1963; it had then swung up, but even by October, 1963 was not yet as high as it had been two years earlier. But all these cold-blooded facts denied what could not be denied – the smell of shortage in the air. The producers brought forward another price claim in late 1963. The Malaysians used much the same arguments on declining grade of ground, the cost of installing new dredges and the need for those dredges to increase production.[5] More impartial figures backed the producers' request.[6] World mine production in 1963 was to prove to be no higher than it had been in 1962 and less than 4 per cent higher than in 1961 or in 1960. The market price in November, 1963 broke through the ceiling and reached for the latter month over £1,000 a ton, for the first time since the Korean boom of 1951. In December, 1963 the Council raised its floor price to £850 – an increase of a modest 8 per cent – and its ceiling price to £1,000 – an increase of under a timorous 3 per cent.[7]

This gesture by the Council proved to be meaningless in the atmosphere of 1964 where no one could have anticipated the astonishing rise in market prices. The programme in March, 1964 for the disposal of 20,000 tons a year from the U.S. stockpile should statistically have met the world shortage, but statistics seemed to have here, as often enough at other times, little relation to the price. Very heavy buying from the stockpile in July-August pushed the price up; relatively little buying in September-October pushed it up further. In October the price was at a level (£1,584 a ton) which was beginning to make the Korean boom seem pale.

In the middle of 1964, with the market price well above the ceiling price so recently established by the Council, Nigeria brought up the price issue again. This time the producers shifted their argument to sounder and wider grounds of principle. They departed from their traditional concentration on rising operational costs and lower grade of ground and they turned to the need for a higher price range – and especially a higher floor price – as a stimulus to the miners in order to meet a shortage which everyone now admitted. The producers' memorandum in November, 1964 argued that there was an obligation on the Council so to foster production that, when the U.S. stockpile tonnages were disposed of, enough tin would then be available from normal mining sources. Investment in mining had always carried a high risk. There was a necessary time-lag in the linked

---

[7] A new price scale floor £850, lower sector £850 to £900, middle sector £900 to £960, upper sector £960 to £1,000.

[8] Floor £1,000, lower sector £1,000 to £1,050, middle sector £1,050 to £1,150, upper sector £1,150 to £1,200, ceiling £1,200, mid-point £1,100 (November, 1964-July, 1966).

processes of study, exploration and extraction which meant that an immediate market price was of little importance. The basis of investment was a floor price since this was the only guarantee of a minimum price on which long-term planning could be based. In the nearer future the U.S. disposals might be continued even if the market turned to weakness from its present strength and a higher floor price would impose a necessary brake on the consequences of those disposals on investment.

Malaysia and Thailand, the cheaper producers, wanted a new floor of £1,000; Bolivia and Nigeria, the dearer producers, one as high as £1,150 with a ceiling perhaps some 30 per cent higher.

Consumers could scarcely object to the producers' insistence on the long-term importance of the floor price; they could not deny that the actual market price had risen by no less than £500 a ton in the last twelve months. In a gallant attempt to look realistic in its price policy, the Council accepted in November, 1964 its greatest step yet in price changing. It raised its floor and ceiling by about one-fifth, the floor to £1,000 and the ceiling to £1,200.[8] The Council had no means of enforcing its new ceiling, since this was far below the current market prices, since its buffer stock had no metal to sell and since it had no intention of doing anything under the provisions of the agreement relating to action in the event of a shortage.

The recovery of the mining industry which had been so long delayed became clear, first in Indonesia and then in Bolivia, in 1964. In neither case was the recovery continued into 1965, but by then it had spread into Malaysia, Nigeria and Thailand. By 1966 overall mine production was at last back to where it had been in the five years prior to the export control crisis of 1958-60. The high consumption of 1964 was to flatten out in 1965-66 although it was still about 40,000 tons above the pre-control level. Sales from the U.S. stockpile reached a peak in 1964 and then began to drift down. These forces took time to influence the price. The effect of the U.S. supplies on the market price in the spring of 1965 was sharp but fairly short-lived and the price rose again to peaks of nearly £1,530 for the months of May and September, 1965. But the back of the price boom had been broken by the end of the year. Early in 1966 the market price began to fall and continued to fall almost without interruption for a year. It was not, however, until July, 1966 that yet another upward revision of the Council's price range was to bring the declining market price again within the scope of possible action by the Council.

XV
The Tin Shortage and U.S. Disposals from Strategic Stockpile, 1961-68

THE ending of export control in the autumn of 1960 was not followed by any sudden resilient upsurge in tin production. Export restrictions had been too prolonged and too severe. There were also some special factors. In the Congo Republic the official ending of restrictions coincided with the confusion of an entirely inexperienced new régime. In Indonesia the break with Dutch management in the middle 1950s was followed by a fall in production almost unchecked until 1966-67. In Bolivia the benefits of the massive supply of capital aid to the nationalised industry, mainly from the U.S.A., were not to be reaped until after 1963. The 1956 level of production was not restored in Thailand until 1961, in Malaysia and in Nigeria until 1965, in Bolivia until 1967; it had by no means been reached as late as 1972 in Indonesia or in Congo-Zaïre. The world as a whole had to wait until 1966-67 to produce what it had produced in 1956. Unless tin consumption contracted (and, in fact, it did the opposite and expanded) the world was open to a period of intense shortage and mounting prices. The only balancing force lay in the U.S. strategic stockpile of enormous, if unknown, tonnage, but this was frozen.

The essence of the second agreement of 1961-66 was, first, the acceptance of the fact of shortage, both by the Council and the United States; secondly, the acceptance by both parties that this shortage, or part of this shortage, could be met by releases from the strategic stockpile; thirdly, the fight as to the principles on which such releases should be made and the machinery for sales; and, fourthly, an intense controversy as to the effect of the releases on the world tin price and therefore on the economic health and export earnings from tin of the developing countries, to whom any disposal from the strategic stockpile was clearly of the greatest importance.

FIGURE 14

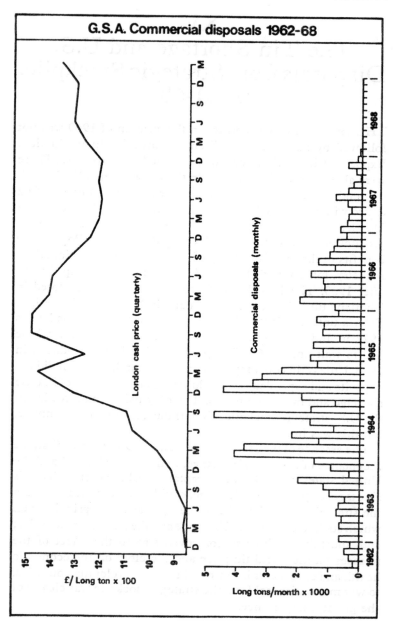

G.S.A. Commercial disposals 1962-68

There was no doubt as to the fact of a shortage of tin, although, of course, vigorous dispute as to its arithmetical extent. In February, 1961 the Statistical Working Party of the Council estimated that the simple shortage of tin in 1961 (that is, the difference between mine production of tin and the commercial consumption of tin metal) would be no less than 26,000 tons. This was about one-sixth of total world consumption. Even after allowing for a continued intake of tin from the U.S.S.R., the net statistical shortage would still be 11,600 tons. This estimate, as did so many of the Working Party's estimates, erred sharply in optimism. In May, the Working Party revised its figures, cut the estimated output from the mines and the anticipated intake from the U.S.S.R. and made allowance for additional supplies of 4,000 tons due to be sold from the U.S. special stockpile held at the Texas City smelter; but even so the net shortage had now become 15,700 tons.

For 1962, the Working Party anticipated, for no very good reason, a sharp rise in mine production, but still a net deficit of 17,000 tons, again almost immediately revised to 20,000 tons. In July the deficit had steadied down to 19,100 tons. In October, as a result of new and reduced figures for consumption, the net shortage was radically, and incorrectly, reduced to 11,000 tons. The circumstances in which the estimates of consumption were made had a peculiar difficulty. A complicated trade in high-grade tin alloys involving the Netherlands, Federal Germany and the U.S.A. was active, and it was not easy to disentangle the figures so as to arrive at a true assessment of the amount of virgin tin metal being used. But by April, 1962, in looking backward, the Statistical Committee was down to a net shortage of only 4,000 tons for 1961 and, in looking forward, to a net shortage for 1962 of only 5,000 tons. It cannot be said that the confidence within the Council (the estimates were not made public to the world but were known to the Americans) as to the reliability of its own forecasts was improved by these statistical wanderings.

The question of disposals by the U.S.A. became acute in April, 1961. In that month the chairman of the Tin Council (G. Péter) was told by officials of the U.S. governmental General Services Administration (G.S.A.) in Washington that they had become seriously concerned at the shortage of tin production for 1961 which the Council itself was estimating (before allowing for any U.S.A. releases from the stockpile) at nearly 20,000 tons. The U.S. held a special stock of 3,933 tons of tin metal at the Texas City smelter. This was not part of the strategic stockpile and, therefore, its disposal required no sanction from Congress. The U.S. proposed to sell it to ease the shortage. M. Péter stressed the need to co-ordinate release of this Texas City stock with the price structure in the Tin Agreement; but he stressed also that he was not empowered to discuss the very much more important question of releases from the strategic stockpile. In informal talks in London the Americans referred again to the likelihood of a tin shortage in 1961; more important, they referred in general terms to the strategic stockpile; and they appeared to be prepared to be approached by the Council on the question of the disposal of part of this stockpile.

Within the Council there could be no serious logical objections to the sale of the small and specific Texas City stock. The consumers were aware that the then ceiling price in the Tin Agreement was being threatened. The producers were prepared to agree that the Texas tonnage could usefully bridge the gap in supply. On 1 June, M. Péter, speaking of the Texas City stock alone, told the U.S. representative in London (Mr. George Jacobs) that the Council hoped the U.S.A. would take steps to

---

[1] 500 tons were offered for sale on 18 July; only 350 tons were sold.

[2] In practice, sales from these were slow. By 31 March, 1962, actual sales totalled 1,100 tons by Canada. Italy was selling about 50 tons a month from the autumn of 1960. Both countries limited sales to the internal market. The Italian stock seems to have been held nominally by the Italian railways.

[3] The State Department also knew of the political importance of Bolivia. On 6 October, 1961, in a letter to President Paz Estenssoro of Bolivia, President Kennedy had said:

"The course of action which we have suggested is the sale of small lots of tin over a period of several years. This tin would come from the 50,000 tons of tin which we have now in excess of our strategic requirements. We do not intend to depress the price of tin through these sales. They would be initiated at a time of world-wide shortage and would have the effect of discouraging tin consumers from substituting other materials for their normal tin consumptions. In this way we can protect the long-run stability and continued prosperity of the tin market."

sell 3,933 tons; that the Council as a whole would ask the U.S.A. to refrain from selling should the L.M.E. tin price fall below the mid-point in the upper sector of the price range in the Agreement; and that Bolivia and Indonesia had made the reservation that sales should not be made below the ceiling price in the Agreement. The U.S.A., in late June, finalised its plans for the Texas City stock. It proposed to offer 900 tons for sale as soon as possible without unduly disturbing the market and the balance of 3,000 tons as circumstances or the market determined from time to time.[1]

Meanwhile, the buffer stock Manager of the Council had run into serious trouble on his home ground. His holdings of tin metal, adequate enough at nearly 10,000 tons in early June, had, to his great surprise, melted away almost overnight under an attack by speculators. On 21 June he had to announce that his stock was exhausted. There seemed very little reason to expect the new Agreement to be able to find immediately the necessary metal resources in its buffer stock to defend the ceiling price on behalf of consumers. There were only three sources of supply, apart from an extension of normal production. Two of these sources were too small to affect the position (the 3,000 tons Canadian governmental stockpile and the 5,000 tons Italian governmental stockpile).[2] The third source was the U.S. strategic stockpile, of enormous size and potential impact. For the next six years the work of the Tin Council was dominated by this U.S. disposals problem.

The beginning of discussions on disposals

The U.S. administration also had its problems. The tonnage regarded as necessary for the strategic stockpile had been arrived at after the most detailed discussion of the U.S.A.'s potential needs in time of war, and any proposed reduction of the stockpile holdings would require a parallel study and a parallel approval by all interested parties. To use part of the strategic stockpile to meet a world tin shortage (perhaps a temporary shortage) was not really within the general objectives of stockpiling policy, which was a matter of defensive security. To make a profit on the sale of surplus tin (once it had been established as a matter of policy that sales should be made) was the necessary act of a good administration but should not be regarded as a factor determining policy. Congress remembered acutely the actions of the pre-war International Tin Committee in squeezing the U.S. tin consumer; it remembered with greater acuteness the "gouging prices" which the U.S. taxpayer had paid to build up the stockpile in the 1950s; and it was not now likely to show much sympathy for the fears of the producing countries as to disposals. The State Department was well aware that disposals affected tin-producing countries in areas where important U.S. foreign policies were concerned—in Bolivia, in the Congo and in south-east Asia; it was important that the onus for initiating disposals should not lie or should be made not to lie with the U.S.A.[3]

The magnitude of the problem was clarified in March, 1962. Senator Stuart Symington (chairman of the Armed Services Committee of the U.S. Senate) announced that the stockpile held 348,000 tons. The amount surplus to strategic requirements was 164,000 tons and was worth $442 millions. This surplus represented one year of total world production (outside China and the U.S.S.R.). The U.S. Congress was asked for authority to dispose of an initial 50,000 tons out of the 164,000 tons.

The Council was well aware that its discussions on the special Texas City stock had been merely a preliminary to the much wider fight on the strategic stockpile. As early as October, 1961 it had limited its own range of action. A meeting of Heads of Delegations had at that time unanimously agreed on the principle that a "cut-off" price (that is, a price below which the U.S.A. should not sell) should be fixed at the appropriate time for disposals, but could not agree what the "cut-off" price should be; and the meeting of the Council in April, 1962, which had before it a request for its views from the U.S.A. on disposals, was faced with the need to define a policy—it was hoped, a common policy—on the matter.

[4] The G.S.A. (General Services Administration) was the U.S. disposing agent, concerned with the mechanics of disposal and the best price. Overall policy on disposals was provided by the Office of the President on Emergency Planning. Discussion on priciples and details with interested governments and the Tin Council was through the U.S. State Department.

The producer members of the Council stood very firmly on the principle of a minimum "cut off" price for U.S. disposals. That price, below which no disposals should be made, should be equated to the ceiling price in the Tin Agreement, that is, to the price which showed when market opinion was accepting the fact of a shortage and which was the level that would stimulate production to meet consumption in the long run. Sales by the U.S.A. at prices below this ceiling price would prevent a body of new production coming into life and might destroy part of the marginal output which was struggling even on the Council's current ceiling price of £965 a ton. In the second place, the tonnages to be disposed of by the G.S.A.[4] over any particular period should not exceed the shortages estimated by the Tin Council. This second part of the producers' case was bound to put the consuming members of the Council in an awkward dilemma. They had to accept, whatever their reservations on particular countries, the general conclusions as to the statistical shortage which the Council approved.

The producers maintained, in the third place, that G.S.A. disposals should be made in an orderly way so as not to disrupt the market. This sentiment was unexceptionable. No intergovernmental commodity agreement whose aim was to prevent undue market fluctuation could accept otherwise. But the principle, if inevitably accepted, still left to argument actual or potential results of any of the steps taken by a disposing government.

The consumers were not prepared to accept the first of the producers' assumptions that a disposals price below the ceiling price was inconsistent with the purposes of the Tin Agreement. The agreement envisaged for the achievement of its objectives the establishment of a price range to prevent price fluctuations. The ceiling price was a decisive figure in the sense that, on the shortage reflected by that price, the U.S.A. should have freedom to sell any necessary tonnages at or above the ceiling price. This would help consumers to meet their immediate shortage of supplies; they could not rely upon the slow effect of higher market prices on stimulating production in the future. A large majority of the consumers felt that the ceiling price could not be regarded as the cut-off price for U.S. sales.

Nevertheless, the Council in April, 1962 accepted the inevitable and expressed to the U.S.A. its willingness to enter into stockpile discussions. A month later a meeting of Heads of Delegations expressed to the U.S. Secretary of State their deep alarm at the statements made to the U.S. congressional sub-committee and received a reassurance again that the U.S.A. would take no action to disrupt markets and would fully consult with the Council and substantially interested governments.

The Council met in late May in an atmosphere not noticeably lightened by a report that the U.S. Congress had ceased for the time being to consider legislative action approving the release of the 50,000 tons. Some consuming members of the Council repeated that the U.S.A. was free to dispose of tin at prices not below the ceiling price in the agreement; that the U.S.A. should be allowed the possibility of making prudent sales in the upper sector of the agreement range, yet so controlled that the market was not unduly disturbed; but that the U.S.A. should not sell in the agreement's middle sector, since this would not be in harmony with the practice of the agreement. A working party of the Council failed to reach a compromise on the stand to be taken.

The breathing space for the Council was short. In June, 1962 the U.S. Congress completed the legislative action authorising disposal of the 50,000 tons, news which brought the tin price down sharply. This tin was to be made available for sale, on the

[5]Press release, U.S. State Department, 21 June, 1962.

[6]Press release by the Office of Emergency Planning, 25 June, 1962, and report of the meeting in *Comtel-Reuter*, same date.

[7]The report of the Statistical Committee may be summarised as:

		in thousand long tons	
	1961	1962	1963
Available:			
from mine production	136·4	143·5	149·5
from other sources (U.S.S.R., Italian, Canadian and Texas City stocks and from the I.T.C. buffer stock)	19·9	7·1	6·0
Total	156·3	150·6	155·5
Requirements	159·5	155·1	157·8

basis of prevailing market prices, in small quantities over a period of several years, in order to prevent market disruption. The rate and condition of sale would be determined by the General Services Administration in concurrence with other interested agencies and departments of the U.S. government. Before a final decision on the rate and conditions was reached, consultations with substantially interested governments were to be completed. The rate and conditions of sale would be so devised as to protect the interests of producers, processors and consumers.[5]

Further justification for the U.S. releases was made by Edward McDermott, Director of Emergency Planning,[6] before the American Tin Trade Association, a body generally opposed in principle to the Tin Agreement and immediately concerned with seeing more tin on sale. He was well aware of the concern in the metal trades that the U.S. government would dispose of its surpluses of strategic and critical materials in such a way as to depress prices and disrupt markets. This tin disposal would take perhaps up to 10 years, depending on market conditions. The U.S. government knew the importance of tin to many tin-producing countries and would, in accordance with established policy, take into account the effect of any disposals on the world tin market. No further quantities of government-owned tin would be sold until the present 50,000 tons was disposed of. They had entered a period of increased demand for tin which current production could not satisfy.

The Council at its meeting in July, 1962 faced its most severe test since the recent export control crisis. It found some comfort in the statistical position. The Statistical Committee had discovered an almost balanced position for 1961 between the tonnages available for consumption (including the 10,080 tons which had been seized from the buffer stock in June) and the requirements for consumption; it assessed the 1962 position now at a shortage of only 4,500 tons; and it believed that the 1963 position would show a shortage of only 2,300 tons. Neither the 1962 nor 1963 forecasts took any sales from the U.S. stockpile into account. Obviously, these figures would not justify any substantial releases from the stockpile.[7]

The U.S. plan of disposal, put forward for an initial period of six months, was before the Coucil at this meeting.[8] Whatever the Statistical Committee was prepared to believe, the market certainly thought otherwise and was ripe for a price explosion. The consumers were now thoroughly alarmed. They reiterated their position, namely, unrestricted sales above the ceiling price, moderate sales in the upper sector of £910-965, nominal sales in the area £880,910 (that is, above the mid-point), no sales below £880. But to meet the wishes of the producers they agreed to put forward to the U.S.A., as a basis of discussion, that the total amount of releases from the stockpile in the initial period should be related to the estimated deficit for that period, as agreed between the Council and the U.S.A. When the price was in the upper sector releases by the U.S.A. should be of a moderate amount within the agreed total; when the price was above the ceiling there need be no limitation on sales. With this programme, the Council agreed to send a massive delegation to Washington, with the acting-chairman (Sr. Manuel Barrau of Bolivia) in charge.

The 1962 discussions

This consultation with the State Department was the first of a series over the next four years. So long as the tin shortage existed—as it did until 1966—the Tin Council was fighting a very difficult battle in admitting the shortage but in denying its magnitude. The arguments on both sides in July, 1962 were to be

[8] The plan was for the disposal of 300 tons a week for six months (as to 260 tons in the open market, weekly and non-cumulative, and as to 40 tons weekly cumulative to government agencies or on foreign aid programmes). Sales would be by competitive bidding at prices reasonably consistent with prevailing market prices. If sales exerted substantial downward pressure on prices the offerings would be reduced or suspended. The total amount of tin would be, at the maximum, 7,800 tons.

[9] The U.S.A. started with its initial proposal of 260 tons commercial sales a week (plus 40 tons for governmental and foreign aid). The Council counter-proposed 100 tons a week when the price was above £910 and 60 tons a week below £910; this had no response and the delegation then proposed a sliding scale of nothing when the price was below £850 (the floor), rising in steps to 260 tons when the price was £960. The Americans ended by being prepared to consider the possible scale of 160 tons a week below a price of £910.

[10] 160 tons commercial disposals, 40 tons other.

repeated many times. The Council forecast a shortage for 1962 of about 100 tons a week, well below the tonnages proposed by the Americans. Prices for disposal should in the Council's view be related to the range in its own agreement, and in particular to the upper sector of £910-£965 where the Council could itself sell tin. In the middle sector of £850 to £910 the market should be generally free of abnormal factors, whether buffer stock action or stockpile sales. If strategic sales pushed the price into the lower sector below £850 where the Council might buy, their Manager might find himself absorbing tin, which in effect arose from the American disposals, in order to check a downward spiral.

The Americans held that the decision of Congress on disposals, the pressure in the States of consumers' public opinion under the shortage, and the U.S. balance of payments position made a substantial disposal essential. The U.S.A. could not in principle accept a "cut off" price of £850, although admittedly they would not be happy to see the price much below that point. The average cost of the tin in the stockpile (about 108·7 cents a pound) was below the current price, but the question of making a profit was not relevant. If prices trended significantly downwards through the lower sector the U.S.A. would reduce or even suspend sales. If the consultations failed there would be an increased opposition within the U.S.A. to a possible accession to the tin agreement.

Haggling took place on the tonnages to be offered by the Americans, but not on the principles.[9] The parties could not agree and publicly said so. In August the U.S.A. announced a programme of disposals at 200 tons a week.[10]

The market had been affected by two contradictory factors—its fear of the consequences of any disposals (whatever the tonnage) from the stockpile and its inherent belief in a forthcoming shortage; in consequence price fluctuations, although serious, were not catastrophic. At the time of the U.S. administration's request to Congress in September, 1961 for the release of the 50,000 tons from the stockpile the London price had been well above the ceiling price in the agreement; that request pushed the price down and until early March, 1962 it remained in general not far below the ceiling. The price movements between September,

1961 and March, 1962 still showed fears of a shortage as the major influence on the price; in that period the range between the maximum and minimum daily prices reported was only between £994 and £940 or of the order of five per cent.

The market changed very abruptly when in March, 1962 the magnitude of the total American surplus – 164,000 tons – was revealed. The price fell during April and early May almost without interruption through the upper sector of the agreement. On the Congress approval of the disposal of the 50,000 tons it dropped through the whole of the middle sector and entered, at £845, the first stages of the lower sector in June – a drop of £115 a ton in three months. There was every sign that the slide would continue and the buffer stock Manager made it clear that he would buy – although he did not need to buy – to support the price at this top of the lower scale. The failure of the discussions between the Council and the Americans weakened the price again,[11] the Manager began buying and continued to buy through August,

[11]The price was reported to have fallen on a single session of the London Metal Exchange to £825 before the Manager brought it back at once to £849. This very short-term fall (which did not become an official L.M.E. price) was perhaps not unwelcome to the Manager; it justified his later buying always at the top of the lower sector; and it dramatically highlighted the fears of the producers that American selling policy would mean the Manager's using their money to buy tin emerging ultimately from the stockpile.

[12]The test was the New York price. The prompt price in New York as quoted by *The American Metal Market* was the only reliable barometer of prevailing market prices in the U.S.A. It was not a full market price in the sense of the L.M.E. price. The buyer, bidding for G.S.A. tin in relation to the New York price, had to take account of such factors as the geographical location of the G.S.A. tin offered (none was in New York City, but all in a series of depots over the U.S.A.), the time of delivery, the terms of payment, the closing hours of the tender to the G.S.A., etc.

[13]To the outsider one of the striking features of the London market is the surprisingly small amount of physical tin which it carries in stock as the basis of its transactions. The stocks reported in the L.M.E. warehouses in normal times (that is, outside the times of buffer stock operations) may be only between 800 and 2,000 tons or perhaps only two to four percent of total Western Europe annual consumption. This seems to be an open invitation to short-term (not necessarily long-term) squeezes on the market; but buffer stock buying may convert the squeeze from short-term into long-term since the buffer stock may have to hold for a long time its purchases bought for a short-term purpose.

[14]Consisting of Manuel Barrau (Bolivia), J. M. Rochon (Canada), United Kingdom and Malaysia (Yeo Beng Poh).

September and early October whenever the price fell immediately out of the middle sector. In the quarter July-September (mainly in the two latter months) he bought 1,805 tons.

The G.S.A. offered its first 200 tons for weekly bidding in mid-September. All bids were below the stockpile criterion[12] of reasonable relationship to the market price and were rejected; so were the bids for the second 200 tons. The demand was disappointing and in the first six weeks of the programme a total of only 340 tons or one day of total use in the U.S.A. was sold.

The buying and retention of tin by the Manager, although not large in itself, had taken off the London market much of the loose tin which that market requires as a basis for its normal operations[13] and the resultant squeeze on the market came in October. It came at a time of international political crisis – the Cuban missile confrontation between the U.S.S.R. and the U.S.A. – which would normally be guaranteed to have a galvanising effect on the tin price but, in this case, most surprisingly did not.

At Bangkok in November, with its Manager still buying, the Tin Council re-considered its policy towards the next round of. the American programme. It accepted that the tin shortage for 1962 was now 10,400 tons and for 1963 would be over 6,000 tons. A working party[14] drew attention to the changing habits of the Americans in holding private stocks (these had fallen by one-third in a year) in view of the immediate availability of tin from the stockpile. An increase in the American demand would not be reflected in the world price since it would be met not from the open market but from the stockpile; in that event, the price position in 1963 might repeat that of 1962 – a statistical shortage but with the price continually depressed by the prospect of stockpile supplies. At a time of statistical shortage the buffer stock had found itself buying tin in complete contradiction to the essential point of the agreement. This abnormal position would be accentuated if the buffer stock, whose object was to deal with temporary and unforeseen market circumstances, had to turn to deal with the very long-term factor of U.S. disposals.

As for the forthcoming programme, the producers wanted a limit of 100 tons a week (roughly the estimated deficit for 1963), the consumers wanted 200 tons a week, but the delegation to the United States was told to stress the principles of a minimum "cut off" price, the statistical shortage and the need to prevent buffer stock buying.[15]

The mission[16] to Washington in December, 1962 found little change in the attitude of the Americans, who were possibly disappointed at the slow rate of their actual sales of tin, and the existing programme of 200 tons a week was continued until March, 1963. American sales were no higher, however, in the

[15]Entry into negotiations with the U.S.A. was approved by all consumers and by four producers. Bolivia was against, unless there was a serious statistical shortage and the price was above £965. The Congo abstained.

[16]The delegation consisted of Malaysia, United Kingdom, Canada, the new Chairman (H. Allen), the Chairman of the Statistical Committee (A. M. Brooke) and the Secretary.

[17]In the "trial programme" at 200 tons a week from September, 1962 to March, 1963 the G.S.A. offered 5,600 tons for sale, received bids (some not very serious) for 11,440 tons and sold 2,720 tons.

[18]At 2,500 tons in each quarter in approximately weekly instalments of 200 tons, with sealed bids for tender, on the basis of the New York prompt price of the day of sale.

[19]In the quarter, mainly at the beginning, he sold 1,940 tons out of a total holding of 3,275 tons.

[20]The Council invited the Americans to meet a working party, open to all members of the Council, in London. The U.S. representatives included Miss Marion Worthing of the State Department and John Croston of the G.S.A. Miss Worthing had a remarkable range of experience on tin and other metals in the State Department, an outstanding intelligence and a deep sympathy for the problems of the tin producing countries. John Croston was amongst the world experts on the geological and economic problems of tin.

[21]A working party of the Council in 1963 considered a proposal which had been floated (perhaps before enough thinking) on the U.S. stockpile. The buffer stock Manager should be used in disposing of some of the G.S.A. stockpile. He might be an agent, disposing of an agreed tonnage, or a principal buying from the stockpile and selling on his own account. In any case, his operations would be in or above the upper sector of the price range. The short-term possibilities of the Council's thus acquiring a corner on the market were tempting; the long-term implications of the Council's taking over any substantial part of the stockpile (without the rest of the stockpile being frozen for ever) and thus inheriting all the stockpile problems were not examined carefully enough. The Council delegates who were later canvassed by the Chairman and the Manager in favour of this idea gave it no support.

first quarter of 1963 and the price was even lower, but the buffer stock had to buy no more tin.[17]

The disposals programme for 1963-64

In March, 1963 the U.S.A. considered its longer-term programme for the year from April, 1963 – a total of 10,000 tons.[18] The proposals contained no reference to a "cut off" price. They were slightly modified in view of the objections from the Council and the published programme referred only to 2,500 tons (at 200 tons a week) over the quarter April-June. Sales would be reduced or temporarily suspended during periods of significant relative weakness – a repetition of a principle which the U.S.A. had earlier accepted.

Sales by the G.S.A. in the second quarter of 1963 were not alarmingly high, but it was clear that the price had now definitely turned upwards. In April-May the London price moved right through the middle sector; in June it moved into the beginning of the upper sector. The buffer stock Manager sold a substantial part of his stocks very quickly at this point,[19] but failed to hold the price rise. The Americans were in some dilemma. A rising price was not increasing the demand on the stockpile; the G.S.A. selling methods might not be flexible enough.

The Council, realising what little effect the U.S. sales were having on a price which was in the upper sector, was now reflecting the alarm of the consumer members. Its working party thought first in terms of an immediate lift in the G.S.A. disposals to 260 tons a week until March, 1964 (making an annual rate of 13,000 tons, in line with the now accepted statistical shortage) or even 300 tons a week, subject to price qualifications. The producers still insisted on no sales below the ceiling price. The discussions with the Americans in London in June[20] were not productive. The suggestion of the Manager selling tonnages from the strategic stockpile was dismissed.[21] The Americans proposed a supplemental 400 tons a week in replacement of the current 200 tons immediately or an aggregate of 2,500 tons (additional to the 200 tons a week) to be spread over three months.

There was more pressure from the Council[22] but without success; and the U.S. programme for June-September, 1963 was lifted to a maximum of 400 tons a week.

The first of the 400 tons in this *ad interim* programme was promptly fully sold and for the third quarter of the year the G.S.A. was to sell a total of 2,710 tons. This gave an opportunity for the G.S.A. to examine its selling policy, to discover that its tender method (based on the New York prompt price on each Wednesday) might make large profits for the market[23] and to change to the more flexible system of receiving daily bids. Encouraged by the higher sales and newer method the U.S.A. continued for the next six months October-March, 1964 at the 400 tons a week figure. With the London price well up in its own upper sector the Council could scarcely claim that even the larger disposals were killing the price. In fact, an immediate criticism of the disposals came from within the U.S.A. The American Tin Trade Association naturally objected to the new method of daily selling, pointed out that the G.S.A. was eating inequitably into the private sales market and asked for a reduction in the weekly programme.[24]

On the Council the producers were restive. On principle, they felt that the consultations with the Americans had not the character of consultations between equals.[25] The Council could not openly criticise the American actions without breaking market confidence in the tin price; this led to the actual producers in member countries believing that the Council was endorsing the American line. The Council had not yet secured concrete U.S.

[22]The Council asked for the lump total of 2,500 tons to be reduced to 2,000, the period of application to be subject to review by both parties after three months and a limitation on weekly sales. At this meeting the consumers, to show solidarity, said nothing.

[23]Bidders on a Wednesday tender would, on a rising market, make profits since the next sealed bids for G.S.A. tin could not come in for a week. The market was certainly a rising one. There were the usual and unprovable reports that the New York price could be rigged to provide an unnaturally low level as the basis for bids on the Wednesdays.

[24]*Comtel-Reuter*, 12 August, 1963.

[25]The word has a different and stronger meaning in its Spanish translation.

[26]£889 for 1961, £896 for 1962 and £888 for 1963 (to September). But the margin between the highest and lowest daily prices had been £214 in 1961, £130 in 1962 and £82 in 1963 (to September).

assurances as regards the minimum disposals price or maximum tonnage disposals over the longer term. The Council should not, they argued, allow the G.S.A. to upsurp the functions of the Council as the highest body responsible for tin affairs. If no definite understanding were reached with the U.S. on the often-stated principles (including the "cut off" price) the producers would ask for an increase of £60 a ton in the current price range. Such an increase would at once allow the buffer stock Manager to defend at a higher price against the long-range depressing effect of the disposals.

The Americans now extended their supplemental programme (that is, at 400 tons a week) for the six months October, 1963 to March, 1964. This level was well above any figure representing the statistical shortage which the Council had contemplated, but the market was now backing the American assessment.

The intensity of the discussions between the Council and the U.S.A., the forebodings of the producers, the anticipations of shortage and the occasional wild price movements over 1962-63 had, in fact, hidden a very surprising longer-term stability in the price. Over 1961, 1962 and 1963 the variations in the annual average prices proved to be negligible.[26] But now the price began to gallop. In October the London price moved right through the upper sector to the ceiling and then through the ceiling; the buffer stock Manager sold the balance (1,145 tons) of his stock and by the beginning of December the price was approaching £1,000 a ton.

A price boom had started which was to push the Council out of the field of influencing prices for the next two years. The G.S.A. had to abandon its careful long-term and moderate programme and to accept the need for *ad hoc* decisions to meet the clamour of the American consumers for tin at almost any price.

U.S. sales through 1964

In December, 1963 the G.S.A. raised its offerings to a weekly limit of 600 tons for two months to February, 1964. In January it sold 1,590 tons; in the first fortnight of February another 830 tons; in the clear fright of the third week of February 850 tons;

and in the panic of the last week of February no less than 2,515 tons.[27] The whole of the tonnage on offer had been absorbed by the end of February and the G.S.A. offered a further 5,000 tons up to mid-March. But the panic continued with sales of 2,415 tons in the first week of March.

The Tin Council remained a passive witness to events and prices which it could not now influence. Its own assessment of the shortage (now 3,500 tons for the first half and 5,400 tons for the second half of 1964) was of little relevance to the daily experience of the G.S.A., which had already sold in the first quarter of 1964 more tonnage than the Council's anticipated shortage for the whole year. In an attempt to bring back its own price range into contact with reality the Council in December, 1963 had lifted its floor and ceiling slightly and in vain. It was due to meet the U.S.A. in February in Washington to discuss the American programme for the year from April, 1964. The consumers on the Council were restless, were anxious to see more tin and had even suggested asking the G.S.A. to adopt a system of forward sales – a step which would have eaten into the already reduced Malaysian exports to the U.S.A. They were successful in getting the Council to include a reference to the need of American releases to meet the reasonable requirements of consumers in the terms of reference of the mission. That mission in Washington (February, 1964) was beginning to run out of arguments. It was asked to stress the natural play of economic forces to permit the gap to be filled, if possible, by stimulating produc-

[27]By late January, 1964 the London price was £1,040 a ton and by end-January £1,100. On 24 and 25 February the New York price rose by 8 cents a pound (£64 a ton) and the G.S.A. sold a daily record of 920 tons at 154 cents a pound. The next morning (26 February) the London price was up another £60 a ton to £1,248, but in the afternoon heavy selling brough it back by £60. The G.S.A. sold 755 tons and had only 715 tons left available for sale. One 27 February rumour (correctly) said that the G.S.A. would offer more tin. The London price fell £100 a ton to £1,148. On 28 February the London price was down again to £1,120 and the New York price to 135.6 cents (£1,085 a ton).

[28]Nothing in or below the lower sector, 200 tons a week in the middle sector, 400 or 500 tons a week in the upper sector, a sufficient quantity at or above the ceiling. Bolivia did not agree to this scale.

[29]That is, 21,145 tons sold and 28,855 still unsold out of the 50,000 tons approved by Congress in 1962 and 98,000 tons for which Congress approval would be sought.

[30]Press communiqué, 20 March, 1964.

tion; the general level of mine production should not be disrupted so as to create a crisis when disposals finished; the U.S.A. releases should be related to a range of prices;[28] the American unloadings should not drive the Council to buy tin for the buffer stock or to have to apply export control; the Americans should accept the obligations of a reduction in offers and a "cut off" price when the market price went down; the U.S. methods of sale should not disrupt normal trading practices; the short-term release programmes might encourage speculation in prices, would handicap the raising of capital for the tin industry and might therefore aggravate the future production position.

The Americans listened politely, could not accept the principle of a sliding scale of tonnages or a "cut off" price and pointed to the rapid growth of G.S.A. sales as a protection for consumers everywhere.

The American long-range programme, 1964

In March, 1964 the administration announced its long-range programme which, together with previous actual disposals, would clear out the whole of the 148,000 surplus from the tin stockpile.[29] This total excess would be disposed of over a period of approximately six to eight years. It could be reviewed from time to time. The plan had been prepared after a thorough review of the current world tin situation; it took into account authoritative forecasts of anticipated production and the consumption trends expected in the next few years. The disposal amounts could be absorbed by regular market channels and without causing hardships in domestic and friendly foreign economies. Disposals would be made in a manner aimed at minimising market difficulties and would take into account the investment of capital in exploration and development of new tin supplies. The plan would allow disposals to be phased so as to protect and foster the health and growth of the world's tin mining industry. Offers would be accepted only if they were reasonably consistent with the prevailing market prices; sales would be reduced or suspended during any periods of significant relative price weakness.[30]

The one year plan, within this over-all programme, anticipated a disposal of 20,000 tons in the year from April, 1964 – an
amount to meet demands and to re-build stocks.

The American assurances were generous and the Tin Council
welcomed the plan.

Sales in 1964–65

The proposed 20,000 tons yearly figure seemed justified in the
April-June quarter of 1964 when actual sales were 4,400 tons.
But this level of sales did not depress prices. The New York price
rose 16 cents a pound in June and another 9 cents in July; by
mid-July the London price was £1,295 and the Japanese on the
Tin Council referred (vainly) to the possibilities for action under
the shortage clause in the agreement. But the demand within the
United States was rising far beyond the anticipations of the State
Department. In the six months to September the G.S.A. sold
14,000 tons of tin out of its yearly anticipation of 20,000 tons.
The State Department, at its request, talked with the Tin Council in London. In October the G.S.A. estimated that the demand
upon it for the next six months would be 18,000 tons, and the
total made available in the yearly plan to March, 1965 was raised
to 32,000 tons. This figure was substantially in excess of current
Tin Council estimates of shortages.[31]

The increase of the tonnage available in April had no effect in
checking the rising price. That price jumped again in New York
– 24 cents in September and another 19 cents for October. The
maximum daily prices reported in London (£1,715) and in New
York (217 cents) were well in excess of the peaks of the Korean
scramble of 1951.

The magnitude of the G.S.A. October addition to sales available had an immediate, if temporary, effect on the price and on
the clamour for tin. The American consumer, in taking 29,000
tons from the G.S.A. during the calendar year 1964, had over-

---

[31] The Council in October, 1964 estimated a shortage between mine production
and metal consumption of 19,000 tons for 1965.

[32] Total disposals during 1964 at about 31,000 tons were equal to the combined
mine output during that year from Nigeria, the Congo and Indonesia.

bought, and the price shot down after October as rapidly as it had shot up (14 cents in November, 27 cents in December, 5 cents in January, 1965).

During 1964 – the heaviest year for sales from the strategic stockpile[32] – the American consumer was buying in the most spasmodic manner, a fact which in itself accentuated the heavy price swing; but the rate of buying, when it took place, was so intense that it guaranteed higher and higher prices.

The purchases seemed to have no systematic relation to the normal pattern of American consumption. Much of the G.S.A. tonnage was sold in large blocks during a relatively short number of days. Between January and November, 1964, for example, no less than 19,165 tons from the stockpile were sold on only 55 market days; on the other hand there were another 95 days on which the amount sold each day was 25 tons or less. A spasm of heavy buying might take place only on one day (for example, 1,105 tons on 15 September to be followed by nine days of digestion, with only 160 tons sold in that period). It might last one day (for example, 445 tons on 8 October and be followed by 15 days of quietness with a total of only 250 tons sold). In the September example, the heavy buying was after 12 days of relative quietness but in which prices were rising sharply; in the October example the purchase was after 13 days of relative quietness, also with rising prices. But the heavy buying might last a number of consecutive or almost consecutive market days (for example, purchases of 3,345 tons in eight consecutive market days in February were followed promptly by a total of 2,560 tons in the next immediate five days in March; 1,125 tons were sold in four consecutive market days in May, 1,025 tons in three consecutive days in July, 3,035 tons in 11 consecutive days in August and 1,370 tons in four consecutive days in November).

The heaviest buying was not necessarily in periods of the highest prices and did not necessarily force the price higher. The purchases totalling 5,905 tons in consecutive days during February-March began after a period of fair stability in the price, and were made in the first five days at rapidly rising bids, in the next six days at rapidly falling bids and in the next two days at rising prices. The minimum bid accepted in this period was

131.50 cents and the maximum 154 cents a pound. In August the buying of the 3,035 tons was at a price which was almost flat throughout the buying period (minimum price 159 cents, maximum 160.50 cents). The four consecutive days of heavy buying in November (1,370 tons) – which followed fifteen days of quiet buying during which the price had risen as high as 215.50 cents – was at prices substantially lower (180.50 to 190.875 cents) than on these previous days.

It was the duty of the G.S.A. towards the American public to secure a reasonable market price for its sales. In general, the G.S.A. accepted bids which were usually about 0.50 cents a pound below the New York price, a reasonable enough variation in view of the less favourable conditions on which the stockpile tin was offered. But on occasions with fairly substantial tonnages[33] this G.S.A. principle was not effective in practice. It is clear that on those latter occasions the G.S.A. did not by any means obtain a fair relationship with the New York price and it is probable that over 1964 as a whole the differential in favour of the

[33] For example, 475 tons on 8 July at 5.50 cents below the New York price, 920 tons on 25 February at 5 cents, 1,105 tons on 18 September at 3.875 cents below.

[34] Exports of tin metal over 1964-66 from the U.S.A. to countries other than those in receipt of aid tonnages were not abnormal. Re-exports, which may have come in part via the G.S.A. sales, were 1,800 tons in 1964, but unimportant in other years.

[35] 4,269 tons in the first quarter, 4,055 tons in the second, 3,460 in the third and 4,710 tons in the fourth quarter.

[36] Also with Mr. E. L. McGinnis of the U.S. Embassy in London.

[37] 28,000 tons for commercial sale exclusive of government and foreign aid against 32,000 tons including other uses for 1964-65. The other use in 1965 has been 2,150 tons.

[38] The New York prompt price had been collected in New York by *The American Metal Market* only as an indicative price. The G.S.A. would now determine its basis price daily by averaging the price ex-smelter Penang for Straits tin (for 45-day delivery) and the London quotation for "futures asked" (second session) for standard tin (for ex-English warehouses for 20-day delivery) as adjusted to an ex-docks basis New York and adding a premium for prompt delivery. In the event of unusual circumstances or conditions in London or Penang, the G.S.A. would utilise the more representative of the two markets or would establish (as it did later) an appropriate market. G.S.A. communique, dated 26 February, 1965.

buyers was more than the 0.5 cents a pound which the G.S.A. had envisaged. There was, therefore, perhaps some justification for the fears of the producers on the Tin Council that the result, although not necessarily the intention, of the G.S.A. actions was to shade the tin price within the U.S.A. Fortunately, practically all the G.S.A. tin was consumed within the U.S.A. and was not shipped into outside competitive markets, otherwise the outcry would have been justifiably much greater.[34]

In the early days of disposals most of the buying from the G.S.A. was done by the New York brokers (sometimes in very substantial amounts) but the entry at a later date of the tinplate manufacturers directly into the field showed both their acceptance of relatively long-term shortage (otherwise they would not have weakened their buying links with the Penang market) and their confidence in the long-term maintenance of the G.S.A. plans. At the same time, since their buying programmes were related to their own large and long-term consumption needs, they gave the G.S.A. a much more stable market. This was reflected in the much steadier rate of disposals in the year to March, 1966.[35]

The programme for 1965–66

The Council talked again with Miss Worthing of the State Department[36] in London during February, 1965 on the American plan for Year Two (the year from April, 1965). The Americans were thinking in terms of a new method of pricing sales (which the consumers approved) and of a tonnage of 32,000 tons (which the producers believed to be too high and where the consumers preferred an unstated figure). The outcome was a published figure of 28,000 tons[37] and a basic selling price calculated by the G.S.A. and not reliant on possible local manipulation in New York.[38]

The tonnage made available much more than covered all the tin that the U.S. consumer proved willing to take. World mine production was rising and world metal consumption was falling – both gently. But the American consumer remembered his own high consumption of 1964 (the highest for five years), the high world consumption (the highest for over 20 years) and the world shortage, present for the last six years, culminating in a deficit of 24,000 tons for 1964 and certain to continue through at least 1965.

The price began to rise again early in 1965. The figure of 28,000 tons from the stockpile for the year from March, 1965 was regarded as not enough. The price rise in May to 192 cents a pound and stayed around that very high figure for the next five months.

It was not until October, when world mine production was showing for the first time in many years that it could overtake consumption, even if only temporarily, that the price began to turn downwards. Even so, the average price in New York for the full year 1965 proved to be over 20 cents a pound above, and in London £173 a ton above, the 1964 prices.

TABLE 14

Disposals of tin metal from the U.S.A. surplus stockpile.

Year	Sold commercially	Other disposals, including A.I.D.	Total	As percentage of world consumption
	tons	tons	tons	%
1962	1,400	—	1,400	1
1963	9,325	1,301	10,626	17
1964	28,994	2,153	31,147	18
1965	20,169	1,564	21,733	13
1966	14,350	1,926	16,276	10
1967	4,865	1,281	6,146	3
1968	35	3,460	3,495	2
1969	—	2,048	2,048	1
1970	—	3,069	3,069	2
1971	—	1,736	1,736	1
1972	—	361	361	0
1973	19,262	344	19,606	9

[39]The delegation of the U.S. was headed by Anthony M. Solomon of the State Department. It included E. R. Getzin, Miss Marion Worthing, John Harlan (G.S.A) and Stanley Nehmer (Department of Commerce). The Council delegation included H. W. Allen (Chairman), Khaisri Chatikavanij (Thailand), Toru Udo (Japan), Mohammed Noor Hassam (Malaysia), S. D. Wilks (United Kingdom), M. A. Brooke (Statistical Committee), W. Fox (Secretary) and R. T. Adnan (Buffer Stock Manager).

The disposals plan for 1966–67

Commercial sales from the stock in the year to March, 1966 were running at the rate of over 16,000 tons a year. Miss Worthing and Mr. McGinnis met the Council in London in February, 1966 to consider Year Three (the year to March, 1967). The U.S.A. estimated the statistical shortfall for that year at 20,000 tons; the Council at 12,500 tons. The U.S.A. was thinking in terms of maintaining for Year Three the same total as for the current year (that is, 28,000 tons including an unspecified cushion against eventualities). The U.S.A. stood firm on this figure but called it an availability, providing programme flexibility.

The shortage in the world was passing its worst. On the production side the bulk of the recovery was being carried by Malaysia and Thailand who added 7,000 tons to their joint output in 1965 and another 9,000 tons in 1966; world consumption was beginning to slacken; the Council and the U.S. estimates of the shortage in 1966 had been both grossly pessimistic; and the provision of and availability by G.S.A. of 28,000 tons proved to be unrealistic. In Year Three the G.S.A. was to sell in 1967 only one-third of what it had sold in 1966. The tin price was moving always downwards through 1966 – from £1,425 a ton in January to £1,210 in December. Perhaps more important for the Council were the changes which it made in its own price range, first in November, 1964 so as to give a ceiling price of £1,200 and an upper sector of £1,150-£1,200, and then in the middle of 1966 to a floor of £1,100 and a ceiling of £1,400. This latter change, made when the falling market price was £1,275 brought the Council back as a force to be considered; and the price trend – a fall of no less than £150 in the immediately previous six months – indicated that it might be in the market soon as a buyer of tin. In the circumstances, with a new agreement just begun and under the psychological pressure of a meeting in a producing country (in Congo-Zaïre in September), the Council asked the United States for a statement of policy on G.S.A. sales, particularly in the light of possible buffer stock buying to support a falling price.

The meeting with the United States in Washington in October 1966 was impressive.[39] The position of the United States seemed to be changing. Actual disposals had begun to tail off from the middle of 1966 and the clear indications were that world production and consumption would be moving into balance for 1967. By August the G.S.A. had sold slightly more than half the total surplus[40] and was therefore well within the six to eight years which the U.S.A. had stated, at least publicly, would be the long-term disposal period. The price, for the first time for years, showed signs of collapsing. The State Department had entered into assurances which now might have to be honoured. The political factors had also swung in favour of the producers. The U.S.A. was involved in the internal struggle for the control of the raw materials of Congo-Zaïre; there had started a parallel struggle in Indonesia after the overthrow of the Left there under Sukarno; Bolivia seemed wide open to subversion. Above all, there was the war in Vietnam for which the Thai airfields were valuable. Much public opinion within the U.S. had shifted to oppose any further running down of strategic stocks. In such an atmosphere the few million dollars immediately involved in pressing on with a tin disposals programme unpopular in very sensitive international areas was of no importance.

The Council concentrated on its simplest, strongest and most relevant point – that the U.S. should not sell tin while the buffer stock of the Council was buying. The Americans stressed that they had shown very great flexibility in their pricing system to avoid market disruption, but they were now prepared to enter into a commitment that would meet the Council at least half-way. The U.S. agreed in principle to moderate its tin sales programme if this should be inconsistent with the contingent operations authorised under the International Tin Agreement.[41] As those contingent operations meant, amongst other things, the

[40]Disposals (including aid and governmental) 76,000 tons, balance left 72,000 tons.

[41]Department of State communiqué, 28 October, 1966 reprinted in full in the *Annual Report,* 1966-67, of the International Tin Council.

[42]For example, in the earlier period the tinplate firms, in their own names, bought at least 2,500 tons of the total 4,685 tons sold in August, 1964; 500 tons out of the 1,850 tons sold in November, 1964; and 1,175 tons out of the 4,535 tons sold in December, 1964. The biggest dealer from the New York market was C. Tennant, who took 730 tons in August, 370 tons in November and 1,740 tons in December.

purchases of tin at prices within the lower sector of the price range and the possible authorisation of sales and purchases at prices within the middle sector, as well at the application of export control, this American concession was important and the commitment was wide. Co-operation between the two parties (by direct and confidential talks between the director of the G.S.A. and the Manager of the buffer stock) were arranged to secure co-ordination. The market read into the last point perhaps more than the co-ordination could mean in practice.

The G.S.A. had in effect fixed its own price in 1967 at 154 cents a pound, but its trade was beginning to die. Through its earlier stages it had been heavily, if not willingly, supported by the solid purchases from the tinplate makers.[42] By the end of 1964 these were drawing over one-third of their total requirements of tin from the stockpile. The most ardent upholders of private enterprise in the U.S.A. (including the United States Steel Corporation) had become the largest clients of the state organisation. But these large buyers were drifting out of buying through 1965 (although still remaining as occasional customers into 1966) and this in itself helped to reduce the G.S.A. turnover. The peak of sales of 9,245 tons for the first quarter of 1965 had dropped to 3,325 tons for the last quarter of that year. Through 1966, after some initial spurt, the downwards movement continued to only 2,210 tons for the last quarter. The G.S.A. found that even Tennants stopped using it and it was left to feed the small clients. The G.S.A. had changed its character as a seller. From September, 1966 its price had stopped moving in relation to London or Penang; it had clearly decided (as it was entitled to do) on an arbitrary fixed price which remained almost unchanged at around 154 cents a pound until its activities as a seller petered out at the end of 1967. By the last quarter of 1967 its sales were down to a total of only 735 tons.

The end of U.S.A. selling

The G.S.A. programme was nominally continued for a further year from April, 1967 but without a tonnage being stated. After discussions with the Tin Council in London in June, 1967 the programme (an administrative availability of 28,000 tons for 1966-67) was extended until 30 June, 1967 and, later, until further notice. But sales had in effect stopped at the end of 1967 and were formally withdrawn in July, 1968.

The State Department (perhaps stimulated by the G.S.A. which had a more immediate concern with the volume of tin being sold) was still thinking of ways in which disposals of tin might be maintained without breach of the assurances which had been given in October, 1966. In December, 1968 the State Department[43] discussed with the Council the possible revival of U.S. disposals, pending a review of selling policy. The Americans had looked at two methods of disposal which would not interfere with the tin agreement. The first was for the sale of tin to firms on contract to the American government; this method would need the G.S.A. to quote a daily price and would involve very many transactions; it would, however, cover 10,000 tons of tin a year for two years. The second was to barter tin from the stockpile against other commodities of which the stockpile was short. This method might be equally complicated and would also involve a daily price quotation; it would cover about 7,500 tons a year for two years; but it had the advantage that the necessary transactions would minimise the effect of these disposals on the market and therefore on the price. The tin acquired in the first proposal would be used by firms in their own plants; the barter arrangements under the second proposal would be through private firms and would not be inter-governmental. The Tin Council made the obvious point that either method would mean the supply of tin still coming from the stockpile and not from the traditional producing suppliers.

The Americans carried their barter suggestion so far as to sound out opinion within the U.S. tin traders. The attitude of these traders was discouraging and the idea died, at least for the time being.

In June, 1970 the State Department came back again,[44] this time with a more concrete and ominous proposal. The Americans were planning to resume disposals in July, 1970 on a com-

[43] Mr. Katz of the State Department and Mr. J. McGrath in London.
[44] Mr. J. McGrath (State Department) to the Council in London.

mercial basis at the prevailing market price (subject to the assurances of October, 1966) of 6,000 tons of tin in the coming year.
They added an interesting proviso. If the Council wished, it
could buy any of this tin, provided it paid the current market
price. This idea (even if we assume it to be serious) had lost the
attraction, if any, which it had ever had in the eyes of the Tin
Council. The Council was now engaged in rapidly running
down its own buffer stock in the hope of ending the third agreement in 1971 with no metal at all. If it wanted to buy tin it could
buy on its own price range from £1,380 downwards. The current
market price was well above this figure, had been rising generally over the past year and looked like rising more. The idea died.
The other and more important U.S. proposal to start disposing
again at 6,000 tons a year ran into serious trouble with Bolivia in
the spring of 1971 and was also dropped.

But, in fact, the Americans had by now drawn the teeth from
their surplus stock by their own action. In March, 1969 they
announced the result of their long re-assessment of stockpile
objectives. The objective for tin (that is, the amount not available
for disposal) was raised from 200,000 tons to 232,000 tons. This
changed the threat from the surplus entirely. The surplus had
had 57,120 tons left available; now it had only 25,131 tons. This
was equivalent to under two months' world supply, whereas the
initial surplus of 148,000 tons had been equivalent to 11 months.
The position was even better than that. The shipments on
foreign aid and for governmental usage from the surplus stockpile had averaged nearly 2,000 tons a year over the last four years;
if the shipments were to continue at this rate the surplus available for commercial sales would fall to the order of perhaps less
than 15,000 tons. In fact, the total available for all purposes had
sunk to 18,500 tons by early 1973. This figure presented no real
menace to the market, at least until the next revision of the
American objective.

In its discussions with the State Department, the Council
always stressed that the tonnage of disposals should be in relation to the statistical shortage. It happened in result that the total
commercially sold during the four most active years, 1963-66
(80,000 tons) was not far removed from the total real statistical

shortage of tin for that same period, but the actual sales in any one year did not coincide with the statistical shortage for that year. The Council's own estimates for the statistical shortage were so variable and unreliable in the period as not to provide a standard for the U.S. to follow. In any case, the U.S. was meeting a demand which was sometimes far removed from the mere statistics of a shortage. In general, the U.S. in the early stages under-provided and in the later stages over-provided tonnages for disposal, and the initial steps in disposal were taken too leisurely.

On the question of the actual "cut off" price for disposals, to which the U.S. always remained firmly opposed, the consumers and producers on the Council could not agree. A "cut off" price related to a price scale in the agreement would have moved at the choice of the members of the agreement (in fact, that scale was moved four times, on each occasion upwards, in the period 1962-67) and would have taken the disposal prices of its own tin out of the hands of the U.S.A. – an intolerable political position.

The Council, more realistically, concentrated on the argument that U.S. sales should not force the buffer stock into buying tin to support the world price, a powerful and emotive argument. Over January, 1962 to July, 1966 the point at which the buffer stock could begin buying was raised from £850 to £1,200. There was in the period only one time – in the second half of 1962 – when disposals might have had such an effect on price as to force buffer stock buying. At all other times, the disposals price was above – and often far above – the buffer stock buying range.

As regards the industry, the tonnages of disposals did not have the deplorable effect of checking production. World mine production rose by 26,000 tons in 1962-66 (at about 5 per cent per annum) as compared with about 2 per cent per annum in the next five years. The price of tin showed intense variations during the period of disposals but the movement was very sharply upwards – over £500 a ton between 1962 and 1965. Even in 1966, after the total disposals of 81,000 tons, the annual price was nearly £400 a ton higher than in 1962.

For the consumers the disposals were an almost unmixed benefit in the sense that they prevented the price of tin resting long on the famine prices shown by the actual highest daily prices (£1,715 and £1,625) temporarily reached in 1964 and 1965. The other side of the picture was the wild gyration of prices within the year– a range between highest and lowest daily prices of £186 in 1963, of £695 in 1964, of £435 in 1965 and of £261 in 1966. This, however, did not check the upward movement of consumption.

For the Council as a whole the dispute had one undoubted benefit. By its agreement of October, 1966 the United States had accepted one basic principle of the tin agreement – the operations in defence of the price. It had also accepted, generously and without any pressure, the direct interest not only of the producing countries but also of the Council (and therefore of other consuming countries) in the policies and practices of the U.S.A. on disposals of stocks owned by the U.S.A., and it had accepted the application of a body of principles which it would find difficult not to continue to accept in the future when the even bigger question of unfreezing the frozen strategic stock was raised. There still remained in that frozen stock 232,000 tons of tin or well over one year of total world production.

XVI

The making of the Third Agreement, 1965

THE third tin agreement was due to be considered at a United Nations Tin Conference in New York in March-April, 1965. In preparation, the Council, although intensely pre-occupied with the question of American disposals, appointed as early as the end of 1963 a committee to prepare a new draft agreement without commitment. That committee, operating through three working parties, produced a fairly well-clothed "skeleton" agreement which went back to the Council and then on to the United Nations meeting. This draft document raised few contentious points and avoided new principles.

The producer members of the second agreement met in Madrid in early 1965, knowing the essence of the skeleton third agreement. They agreed unanimously to enter it, but with some qualifications. The new agreement should reflect the new spirit of UNCTAD; the right of the buffer stock Manager to operate in the middle sector of the price range should be recognised; there should be a new price range (in place of the present £1,000-£1,200), with a spread of some £300 between floor and ceiling; and the size of the buffer stock should remain unchanged at 20,000 tons. The UNCTAD conference of 1964, in one of the greatest outpourings of talk of the twentieth century, had laid down the signposts for development. These signposts were to mark clearly the road which the developed countries should follow in aiding the evolution of the developing countries; they also marked, but very much less clearly, the roads

[1] The second agreement had referred to "maintaining and expanding their import purchasing power".

[2] Article XXIV, para I. The same Article provided for signature "on behalf of governments of independent States represented at . . . the Tin Conference held in 1965"; thus it excluded from signature the German Democratic Republic and the People's Republic of China (which were not present at the conference) but included the Federal Republic of Germany (which was not a member of the United Nations but was invited to and attended the conference).

which the developing countries might themselves follow if they were willing. In the field of commodity problems, UNCTAD emphasised the need for inter-governmental commodity agreements. In those agreements there would be one common factor – the need to use the agreement to promote the export earnings from that commodity of the producing developing countries. To this admirable sentiment all developing countries naturally agreed; the developed countries, given the atmosphere of 1964, could not disagree in principle.

Fortunately, the third tin agreement required little adaptation in its objectives, and no change in its machinery, to accommodate the new wind from UNCTAD. The new agreement accepted without much discussion what UNCTAD had accepted; its modified objective became "(the need to) make arrangements which will help to maintain and increase the export earnings for tin, especially those of the developing, producing countries"[1] and it included what UNCTAD had had to include, the qualification "while at the same time taking into account the interests of consumers in importing countries". It made relatively little change in other directions. It left the size of the buffer stock at 20,000 tons; it made no change in the price range but, as in the second agreement, left that responsibility directly to the Council.

The risk to the third agreement came, surprisingly enough, from the side of producers, who had hitherto been unanimous in their support for its continuation. The first and less important risk was from Indonesia. In a spasm of outraged dignity, the government had temporarily quitted the United Nations and some of its associated agencies. Indonesia was not officially represented at the tin conference in New York and her influence could be exercised only through a back-stage non-delegation. Her signature was, however, ensured by the provision in the agreement of rights to sign by those who had been members of the second agreement even where they were not represented at the United Nations tin conference.[2]

Very much more startling was the reaction of Malaysia, the biggest producer of tin in the world, without whom there could be no agreement. Some strong forces had been working within the Malaysian administration against the principle of an agreement. The opinions had been embittered by the action of the United States in releasing tin from the stockpile, thus preventing a killing on the market by producers with tin available for sale, and by the failure of the Tin Council to prevent these stockpile sales. These views used emphatic language which managed to combine the Council and the Americans in one conspiracy. They regarded the Council and its buffer stock not only as a complete failure but as a tool of the U.S. government and the General Services Administration. An instrument – the Council – designed to stabilise the tin price had in recent years resulted in wider price fluctuations than had been seen for many years. Malaysia had had to sell at prices slashed by U.S. action on stockpile sales. Malaysia was using up resources at low prices; the other producers, like Nigeria and Indonesia, were now selling less but would reap a golden harvest in future when Malaysia's reserves were used up. Many mines in Malaya were tied up with European merchanting and speculative interests and advisers on the Tin Council so influenced could have no objectivity. There was no substitute for tin and the world would always want it. They should press the U.S.A. for a higher "cut-off" price on its tin disposals; they could withdraw from the Tin Council and stop being so reasonable; they might revert to the machinery of an international tin study group.[3]

The most representative bodies in the trade (the All-Malaya Chinese Association, the F.M.S. Chamber of Mines and the Malayan Chamber of Mines in London) were, and were known to be, in favour of a third agreement and the Malaysian government had at the recent UNCTAD conference reiterated its deep belief in the system of international commodity agreements with

[3] N. Cleaveland (Pacific Tin Consolidated Corporation, the only U.S. tin mining company operating in Malaya) was more moderate in his opposition. The Tin Agreement had not stabilised price or employment; adequate supplies of tin were not available; the overall efficiency in mining had not been improved. A basic change in the industry had made serious chronic under-production instead of periods of over-production the major problem. The international financial price-controlling facilities had been a serious burden on the producers. The Council's price range was entirely unrealistic. The possibility of over-production was not foreseeable in the remote future. They might consider an agreement limited to producers.

particular praise for the tin agreement. Nevertheless in December, 1965 the Malaysian Cabinet managed to swallow the anti-American and anti-international arguments circulating within parts of the administration, stood on its head and declared:

"Having weighed the advantages and disadvantages of the Third (Tin) Agreement . . . to Malaysia, and taking into account that, since the drafting of the Agreement in April, the world supply and demand of tin has shown evidence that a burdensome surplus of tin is unlikely to arise, the Cabinet decided that Malaysia should not participate in the Third Agreement.

"The most important basis for the establishment of a commodity agreement is that there should be a prospect of burdensome supply of the commodity. In the case of tin, the Cabinet considered that the agreement is unlikely to be effective in arresting any immediate trend in world tin price in view of the strong demand for tin. Further, the establishment of a buffer stock as envisaged in the agreement will not serve any effective purpose as there is no surplus tin for the buffer stock Manager to buy in order to stabilise the price of tin within the price range proposed in the agreement.

"Malaysia also does not consider the price range which consumer countries have agreed after prolonged negotiations realistic in view of the present and future prospects of the world tin supply and demand position.

"Malaysia has always supported and still supports the principle of commodity arrangements where the circumstances justify such arrangement and provided that such commodity arrangements are supported by all producers and consumers.

"In respect of the Third . . . Agreement there are a number of important tin consuming countries who have declared their sympathy and support of the principles of international commodity arrangements and have even encouraged developing countries to get together and establish such arrangement but have not themselves come forward to subscribe to the Third . . . Agreement. In withholding her participation in the Third . . . Agreement Malaysia is acting on the basis that she is not entirely convinced of the circumstances justifying the renewal of the current tin agreement."[4]

This damaging denial of everything in the tin agreement for which Malaysia had hitherto stood alarmed all the other potential members of the third agreement. As producers they could see disappearing an international body which, whatever its weaknesses, represented at least some barrier to the flood of surplus tin. These producers and a delegation from the Tin Council put pressure on Kuala Lumpur. The pro-agreement forces within the Malaysian administration re-asserted themselves. Within a few days the Malaysian Cabinet, nominally for the sake of the other developing producing countries, re-asserted its old principles and declared its intention of signing.

The temporary aberration of Malaysia may have had another and longer-term effect. It is possible – but perhaps not very probable – that the United States, in spite of her inactive part in the New York conference in 1965, was seriously considering membership of the Agreement. If so, the initial failure of Malaysia to sign and the temper of the initial Malaysian declaration may have been decisive in deciding that the U.S. administration took no further action towards membership.

[4] Press statement of the government of Malaysia, 16 December, 1965.

XVII

Prices and Export Control Under the Third Agreement, 1966-71

THE agreement of 1966-71 was the first in which the Tin Council was able to operate without the massive and perhaps overwhelming influence of abnormal outside factors. The U.S.S.R. had been a dominant external force under the agreement of 1956-61, uncontrollable until the settlement with the Council in 1959. So far as most of the period of the agreement of 1966-71 was concerned Russia was drawing tin from the free market at a rate which was not determining but was important enough to prevent a surplus developing too soon. The U.S.A. disposals, in tonnage, timing and price, had been decisive in directing the work and thinking of the agreement of 1961-66; but those disposals had, in effect, ceased to influence the market price or market psychology since the arrangement with the State Department in 1966.

The five years of the third agreement of 1966-71 were pleasanter and easier in other fields. Mine production was high and rising, but not too fast. Consumption was also high and rising. For the first part of the agreement, consumption was slightly above production; for the second part, production was above consumption; towards the end the statistical position was in balance. In neither case was the gap excessive; the highest surplus of production in any year was 7,000 tons, the biggest shortage 5,000 tons. In U.S. dollar terms, the range between the highest and lowest annual prices in 1966-71 was under 20 per cent; the range between minimum and maximum daily prices, although large in monetary value, was small in percentages (ranging only between 5 and under 20 per cent in the year). The Council's floor price was not at any time seriously threatened, although, following the usual tradition, the ceiling price, when challenged, could not be held. The buffer stock was operating – buying for two years, selling for two years and then buying again for a year.

Export control was put into operation, but only for one year (and, rather curiously, while the buffer stock was selling). Sterling (the currency in which the agreement prices were expressed) was often under pressure and often weak; it was the weakness of sterling, not any change in the supply and demand position, which provoked the largest shift yet seen in the Council's price range. It was also a period in which all the actions of the Council – buffer stock buying, buffer stock selling and export control – seemed destined to push up prices generally.

The change in the price range

Under the first agreement there had been only one change in the price range, a change confined to the floor price with the consequential adjustment in the sectors. Under the second, the three price changes affected not only all the elements in the price range but also made the degree of change much more substantial; the floor and ceiling prices were raised by nearly 40 per cent. In the third agreement another factor came into the calculation – the devaluation of the pound sterling. In the initial price change in that agreement made in 1966, before sterling devaluation, the floor price was raised by one-tenth; in the second change in 1967, made consequent on devaluation, both floor and ceiling prices moved up one-sixth; in the third change made in 1970, but not for currency reasons, the floor was raised again by another 7 per cent.

Under the first agreement, the mid-point in the price range rose only by £45 a ton, under the second agreement by £295, and under the third agreement by £425 a long ton.

At the first meeting of the third Council in July, 1966 the producers requested a revision of the price range, to give a floor price of £1,200 and a ceiling of £1,500, with a mid-point of £1,350. This claim had the merit that the mid-point was close to the average price on the market over the previous six months. That price had, however been moving downwards and currently was well in what would have been the lower and therefore buying sector of the proposed range, even though there was still a small shortage and the U.S.A. was still selling tin from the stockpile to

meet that shortage. It was no longer opportune for the producers
to concentrate on the argument of the need for a higher floor
price to ensure future production. But Thailand pointed out that
it was running out of good mining ground in the traditional
mining areas and had to look forward to opening up new areas in
the sea (where dredges had higher costs and a shorter life) or in
the north of the country (where rich tin land was confined to
difficult narrow valleys). Bolivia referred to the steady weakening of her grade of ground and the steady rise in mining costs.
Malaysia felt that there was no fear of forcing over-production at
the proposed new prices; it was still necessary to bring out all the
marginal output. If the market price fell below £1,250 a ton (as,
in fact, it was about to do for the next twelve months) production
would go down (in fact, it was to rise). Indonesia stated that its
new mining policy (attracting foreign investment by
production-sharing schemes with loans repayable from the
revenue from increased production, and allocating to the domestic tin industry a higher quota of its earnings of foreign
exchange) was dependent on a higher price level. In Nigeria
mining costs had risen 11 per cent in the past four years.

The consumers were cautious. Canada, although well aware
of mining problems and mining thinking, was concerned that
the high level of tin prices would mean substitution by other
materials with a degree of price stability (presumably,
aluminium and nickel). They should not resort to buffer stock
operations or export control at price levels so high as to be
dangerous to the future of the industry. If the producers needed
capital for exploration and development, buffer stock funds
which were not being used could be returned to them. The
United Kingdom pointed to the implications of raising the price
range; an increase in the floor price would reduce the tonnage of
metal which the Manager could buy and at the same time would
bring nearer the point at which he would begin to buy. Japan,
which had steadily become more important as the mouthpiece of
the pure consumers, also emphasised the dangers of substitution; but Australia (naturally) and Czechoslovakia (more unex-

¹ The new range was floor £1,100; lower sector £1,100 to £1,200; middle sector
  £1,200 to £1,300; ceiling £1,400; mid-point £1,250.

pectedly) favoured the producers' claim. The usual working party produced a compromise fairly favourable to the producers – a floor of £1,100 and a ceiling of £1,400. This widened the price range to 27 per cent as compared with the 20 per cent or so hitherto accepted as normal.

The new range[1] came into effect (July, 1966) at a time when the market price had been moving down rapidly (by about £150 a ton in the first six months of 1966). By August the monthly market price had dropped to the mid-point in the new scale (£1,250) and by December it was near the beginning (£1,200) of the lower sector. The tin market seemed to be at the start of a cyclic change. The high prices of 1964-65 were giving rapid results in output in 1966-67. Malaysia, in spite of its gloomy forecasts on the quality of mining ground, increased its output by about one-seventh, high-cost Bolivia similarly and Thailand even more. For 1966 world mine production was perhaps 20,000 tons more than it had been two years earlier. The industry was saved from an immediate surplus only by the continued inefficiency of the Indonesian mines and by the hold-up of concentrates in Thailand as a result of a smelter breakdown. Even so, in the second half of 1966, world mine production had at last overtaken metal consumption for the first time in over eight years.

In spite of the signs of change American consumers had continued to buy steadily and on a substantial scale from the U.S. surplus stockpile through 1966. In October, 1966, after discussions with the Tin Council, the U.S. agreed to modify its tin sales programme if this should be inconsistent with the contingent operations (that is, buffer stock and other operations) authorised under the tin agreement, but this public acceptance of concern did not mean the ending of stockpile sales. For the first half of 1967 they were still 4,600 tons.

The market price fell in January, 1967 to the top of the lower sector. The buffer stock, which had been out of operating action for over three years but which had called up an initial £10 millions under the new agreement, at once began buying on this high point, although the rate of downward price movement in the previous three months – under 2 per cent – had been very slight. If the object of the buffer stock was to embarrass the U.S. into the cessation of stockpile sales it had no effect. The U.S. programme for disposals was continued at least nominally after March and again after June.

The period of price stability between February and July, 1967 – when the London price moved by under 2 per cent – was the product in part of small buffer stock buying (1,500 tons), in part of a fairly stable consumption/supply situation (with mine supply slightly below metal consumption) and in part of the maintenance over almost the whole period of a fixed price for U.S. disposals. A sudden market flurry in June, consequent on the Egypt-Israel war, was met in the U.S.A. by not very substantial buying from the stockpile.

The fundamental change in the market position – the slow shift towards a surplus – was now being hidden by price movements which reflected not a surplus or shortage of metal but a deep-rooted suspicion of the quality of sterling. By 1967 almost every important tin producer in the world was reporting a mine production higher than for the previous year. In the same year every important metal consumer – with the exception of Japan – was reporting a consumption that was significantly or slightly lower.

More directly relevant from the point of timing and price in 1967 was the great increase in the flow of metal on the market from the smelters in the United Kingdom and Thailand. The major fact which preserved an uneasy degree of balance in the first nine months of the year 1967 was the outflow of tin to the U.S.S.R. and also to the other European socialist centrally-planned countries.

The production surplus was reflected in the market price in the late summer. In the quarter July-September, 1967 the buffer stock Manager was buying, not on a substantial scale (only some 2,000 tons), but still at almost the top of his permitted buying sector.

The tin problem was now merged in the currency problem. In times of currency instability it is a truism that loose money tends to run for safety into commodities. In the long war between 1965 and 1967 against the pound sterling there were constant reports of a shift from sterling into tin as into gold and the other non-

ferrous metals.[2] It was not, however, until November, 1967 that
the pound was broken.

Prices and the devaluation of sterling, 1967

The Council's price range had always been expressed in sterling. Occasional informal comments had been made during the
renewals of the agreement that a more stable currency might be
used in an agreement devoted to stability. During the 1950s and
early 1960s the only alternative stable currency that could be
considered as a realistic candidate was the American dollar.
One-third of the world's tin production was taken by the U.S.A.
and the tin mining companies everywhere had substantial outgoings in buying dollar mining equipment. The U.S. authorities
would have been only too willing to see New York helped to
transform itself into a real world tin market. But the prestige of
sterling and perhaps the over-riding importance of the world tin
market in London were at stake. The matter was less important
to other consumers (since none of them could claim any essential currency stability or hope to rival London as a metal market)
or to producers (who had qualms about supporting the claims of
the currency of a country not a member of the tin agreement).
There had been no serious support for a change in such a long-
established habit.

But the possibility of a devaluation of sterling had been taken
into account. In the first agreement the Council was obliged to
review the price range if it were considered that movement in the
relative values of currencies made such a review necessary, and
was empowered (initially through its Chairman) to suspend the
operations of the buffer stock "if such a suspension is necessary

---

[2] I speak with some hesitation here. Money moved into tin (as into other metals)
on the hint of devaluation. But there is no evidence of the volume of that money
moving or of the real effect on prices. The currency speculator has always been
an easy target; all the easier because he is always anonymous; and it is too
convenient to drag him in when no other reason can be imagined for price
movements. He was a favourite target for the Council's buffer stock Manager,
who has my every sympathy.

[3] To a floor of £1,283, a lower sector of £1,283 to £1,400, a middle sector of
£1,400 to £1,516, an upper sector of £1,516 to £1,633 and a ceiling of £1,633
(operative 22 November, 1967 to 16 January, 1968).

to prevent buying or selling of tin by the Manager to an extent likely to prejudice the purposes of this Agreement". This proviso, modified in detail but not in principle, remained in all future agreements.

In late November, 1967 the pound sterling was devalued by one-seventh. In face of the anticipated consequential sharp rise in prices (which would mean a new market price above the existing ceiling price in the Council's range and therefore an obligatory selling of the stock) the Chairman of the Council provisionally restricted the operations of the buffer stock so that the stock did not sell tin. The Council, on 22 November, fixed a new provisional range which simply raised the existing range in all its aspects by one-sixth[3]. The buffer stock was brought back into operation. The anxiety of the Council to resume operations was natural enough, but it was based on what proved to be a natural miscalculation of the effects of devaluation. The new sterling market price moved up not by the full 16.6 per cent of devaluation, but only by about 10 per cent. The lowest price before devaluation at which the buffer stock had been buying was about £1,180 a ton, well above the floor and towards the top end of the lower sector. After devaluation and until the end of the year 1967 the new market price was around £1,350 (that is, roughly half-way down the lower sector in the new range). In January, 1968 the average market price was down to £1,323 or two-thirds of the way down the lower sector and unpleasantly near to the new floor. As currency speculators were disappointed in the failure of the tin price to realise the full benefits of devaluation they began to unload on the immediate and still substantial profits of the current new market prices. In the first three weeks of January, 1968 the buffer stock Manager bought no less than 2,400 tons when the London price was around £1,320 or moving towards the new floor price.

The volume taken by the buffer stock in January was on a scale approaching some of the buying in the more hectic weeks of the 1957-58 crisis. The total of 7,165 tons held in stock by 18 January, 1968 was the highest for eight years. The total buying power of the £20 millions subscribed to the buffer stock in the third agreement had been reduced to some 16,000 tons by the increases in the floor price in July, 1966 and in November, 1967 so that by January, 1968, with 7,160 tons already bought, the Manager had enough money to buy only about another 10,000 tons[4].

In late January, 1968, after two months of the provisional price range, the Council considered the fixing of the range for a longer period[5]. It was perhaps unfortunate that it did not take the opportunity, in the light of its knowledge at that time of the intake to the buffer stock, to consider the price not only for currency reasons but also in the light of Article VI – whether the floor and ceiling prices were appropriate for the attainment of the objectives of the agreement. The straight devaluation change had produced a mid-point in the new scale of £1,458 a ton, a price well above the market price both currently and for the year to come, and had therefore almost guaranteed the buying of tin by the Manager within the lower sector. But the Council had at no time as yet seriously thought of the need for a reduction in its price range and could scarcely be expected to do so now, and the provisional price range was accepted with slight modifications[6].

Export control in 1968-69

The agreements had always taken steps to ensure that obstacles would be placed in the way of any hasty entry into the field of export control. The first agreement provided that the entry into export control (that is, the fixing of a total permissible export amount) should not become effective until at least 10,000 tons of metal were held in the buffer stock or the Council found that this quantity was likely to be held before the end of a current control

---

[4] This excluded the possibility of drawing on the bank overdraft of £10 millions from Hambros Bank Ltd. In fact, the Council never showed any willingness in using the overdraft which, in any case, would now buy perhaps 7,000 tons or after October, 1970 only about 6,500 tons.

[5] The Secretary strongly supported a shift of price strictly in accordance with the rate of devaluation, on a thin interpretation that the agreement permitted nothing else but really to ensure that devaluation was not used to result in an effective real reduction in the price range. It would have been better, however, to have examined more carefully the real movement of market prices.

[6] Floor price of £1,280, lower sector £1,280 to £1,400, middle sector £1,400 to £1,515, upper sector £1,515 to £1,630, ceiling £1,630 (January, 1968 to January, 1970).

period (as, for example, the Council found in 1959). This fixed probably too high a tonnage target, and the second agreement modified the restriction to provide for no export control unless the buffer stock was likely to hold 10,000 tons at the beginning of the control period. This figure was reduced to 5,000 tons if export control were being newly entered into. But, in a very sweeping extension of authority, that agreement permitted the Council, by a two-thirds distributed majority, to ignore these figures of 10,000 and 5,000 tons and therefore, in effect, to introduce export control at any time, irrespective of the position of the buffer stock. This amended version was continued into the third and fourth agreements.

Through 1967 there had been, the Council believed correctly, a rough balance between mine production and commercial consumption. Admittedly, smelter production of metal had been unexpectedly high. But a major long-term depressing effect on the market – sales from the U.S. stockpile – had become far less acute. Commercial sales from that source had fallen to little over 1,500 tons in the second half of the year and could reasonably be expected to wither further. Another major stimulating factor on the market – the intake of tin by the U.S.S.R. from the free markets – continued to hold good. The Russian tin, which had flooded so disastrously on to the Western European market, stopped coming abruptly by 1961. The movement was then reversed, although not in the same prodigious amounts. In 1963 the Russians were importing over 3,000 and in 1964 over 4,000 tons of tin from the free market countries. As Chinese supplies to the U.S.S.R. faded, the Russian intakes from other and less socialist sources increased. They were over 5,000 tons by 1965 and remained of this order for the next seven years. At the same time, partly as Russian supplies to them fell off, the other socialist centrally-planned countries also turned more to the traditional sources of supply. By 1967 supplies to all the socialist countries were taking over 5 per cent of all available metal production and had become almost a balancing factor warding off any surplus.

In January, 1968 the buffer stock Manager, influenced perhaps by his buying of tin at a faster rate than ever before, waxed

FIGURE 15

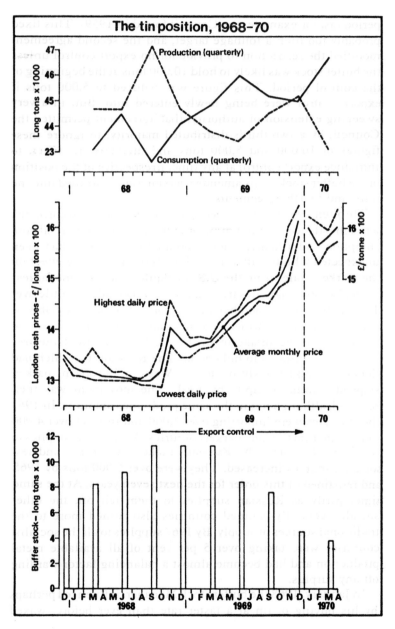

The tin position, 1968-70

eloquent, in spite of the statistical assessment, on the need for export control. There was, in his opinion, an ever-increasing total level of stocks, an insufficient consumer demand, high interest rates which made it impossible for consumers and traders to continue to finance stocks and nothing short of war would turn the tables and induce the market to improve. Current market prices were, in his opinion, dangerously near the floor and the way to improve them was not to increase the financial resources of the buffer stock but to apply a limited degree of export control, as a preventive measure and as a deterrent against unwanted speculation. But the consumers and many producers were by no means convinced. The Council had noted that on its own assessment there was a slight shortage due for the first half of 1968 and it contented itself with calling up from the producers the second half (£10 millions in cash) of their contribution to the buffer stock. In an attempt to steady a market which had been upset by the talk of export control the Council expressed, publicly and unnecessarily, its full determination to use as required all the resources available under the agreement (including the £10 millions bank overdraft) to achieve its objectives, but not at present export control.

The increase in mine production continued through 1968 to give the highest world total since the production boom of 1940-41. The capital poured into the Bolivian industry under the Triangular Plan at last brought Bolivia back to the level it had enjoyed before the export control period of 1958. As compared with that pre-control era Malaysia was now producing a quarter more and Thailand twice as much. There were still some laggards.

The Congo-Zaïre continued to mine at little more than half its earlier rate; Indonesia, although recovering, had by no means drawn herself out of the slough. Of more immediate interest in the context of higher production was the clear emergence of Australian tin as a revitalised industry. It was, however, easy to over-stress the magnitude of the upward swing in the production chart. That level in 1968 was still not yet one-tenth above what it had been a dozen years earlier. World consumption of tin metal in the late 1960s had reached the highest annual figure for nearly thirty years, but recently the line on the chart has shown signs of flattening out.

The possibility of a surplus of any important character was limited by two factors. Purchases by the U.S.S.R. and other socialist countries were being maintained on a fairly high scale. More important, disposals from the U.S. stockpile for commercial sale had stopped effectively at the end of 1967. Official suspension was declared in July, 1968 and thereafter U.S. disposals could be ignored (except for polemic reasons) for the rest of the third agreement.

The market had its spasms of erratic behaviour, but not on supply grounds. The success of the attack against the pound sterling encouraged speculation in the gold price, and in March, 1968 such speculation was reflected in a short-lived rise in the tin price. The price, however, failed to move out of the lower sector of the Council's price range, and through the summer was moving slowly down towards, but not to, the new floor price. To keep it from falling further the buffer stock Manager needed to buy only just over 2,000 tons between mid-January and the end of June, 1968, that is, less than he had bought in early January alone. But this was enough, when added to his slow cumulative buying through the year 1967; and his sudden rush of buying in January, 1968, made his total holdings 11,290 tons by September. The buying power left to him at the end of September, 1968 was only 5,000 tons or (allowing for the unlikely probability of drawing on the Hambro overdraft) about 11,500 tons.

The position now developing was quite different from the approach to the first and only previous application of export control post-war in 1957. Then, there had been no continuous expansion of production, but in fact a slight shrinkage over a number of years; there had been many years in which consumption had fallen below – and sometimes very well below – mine production; the price weakness had been caused first by the withdrawal of a long-standing buyer – the U.S. strategic stockpile – and secondly by the entry of a new seller – the U.S.S.R. – into the market. Then, the price had moved down the full gamut

---

[7] The actual tonnage quota for the 104 days from 19 September to 31 December, 1968 was 42,950 tons. Thereafter, quotas were for quarterly periods.

of the price range from ceiling to floor within a short twelve months and the buffer stock had used up most of its full buying power within a period of only six months. These circumstances might reasonably enough be accepted as abnormal. The picture presented in 1967-68 was one of a creeping surplus where there existed no abnormal outside factors, except the currency one, where due warning was being given from the falling price in the market and where the buffer stock had full time for reflection. The buffer stock of the third agreement in September, 1968, still held resources which might cover an anticipated surplus for some time; the buffer stock of the first agreement, after twelve months of buying, in June, 1958 had held no resources at all against a continuing heavy and prolonged surplus of supply. Perhaps the only common feature in the two periods was that in both cases an upwards change in the Council's floor price and lower sector had eaten into the ultimate buying power of the buffer stock, had therefore weakened its authority to hold the price and had brought nearer the point at which the Manager would think it necessary to appeal for export control.

In September, 1968 largely, but not entirely, on the price argument – the average price for the previous month at £1,297 had been dangerously near the floor of £1,280 – the Council agreed on export control. On the Council's statistical assessment, that limitation, even if justified, could be imposed only in a very mild degree if the object were solely to bring the level of mine production back to the level of metal consumption; but could be imposed in much sharper degree if the object of the exercise was to create an artificial shortage for some time during which the buffer stock could sell some of its metal holdings. The rate actually accepted for the first period of control – a quarterly rate of 37,500 tons[7] – represented a limitation of about 4 per cent or only one-tenth of the rate of cut thought necessary in the early periods of export control in late 1957.

Export control continued for another year and for too long through 1969. An overall revival in consumption came almost at the beginning of control. By the first quarter of 1969 consumption was in excess of mine production; and for the full year 1969 was to be about 5,000 tons higher. The market price reacted very sharply even to the relatively modest degree of export control. It rose from £1,300 in September, 1968 right through the lower sector and temporarily into the middle sector at £1,405 during November, the most rapid fluctuation in the market price in such a short period (except on devaluation) within the last four years.

On the entry of export control, the buffer stock Manager had been given (as he had been given in 1957-60) the authority to sell in the middle range in order to prevent export control pushing up the price too rapidly. That authority was continued while export control was maintained. By May-June, 1969 the price was moving under export control towards the middle of the upper sector and the buffer stock Manager was selling about 2,800 tons of tin a quarter in order to restrain it. If the objective of export control was to achieve a market price around the mid-point of the Council's price range or to adjust the position of the buffer stock so that it was balanced in cash and metal, control proved successful in its first nine months; if the objective was to enable the Manager to sell metal at the higher prices resulting from export control so that his only problem at the end of the agreement in June, 1971 would be to distribute a bigger kitty completely full of cash this objective was being successful in the later stages of (and indeed after the end of) export control. But the two objectives could scarcely be squared. By September, 1970 the buffer stock had sold 3,600 tons during control; it was down to 7,600 tons and the market price was still rising. The Council felt obliged to raise the quota for the fourth quarter of 1969 to 41,500 tons (that is, above the level operating before export control). A number of producing countries was feeling the strain of living up even to their quota. For the second and third quarters of 1969 shortfalls from Nigeria and in the fourth quarter from Nigeria and Malaysia had to be redistributed to other producing members to ensure that the total permissible export amount should be fulfilled.

The price position even with both export control and buffer stock in operation was becoming almost a shambles. In November the price rose into the upper sector and almost to the ceiling – a jump of between highest and lowest price in the month of £116. In December it was to break the ceiling. In early

December the Council accepted the absurdity of maintaining export control, even at a nominal level, when the market price was at a level exceeded in previous years only in the great shortage of 1964. It ended export control and adopted a diametrically opposite policy of calling upon the producing countries to increase their exports as much as possible.

The episode reflected little credit on the stabilising policy of the Council. During the fifteen months of control the London tin price (monthly average basis) rose by £316 a ton or by about one-quarter, a rise higher almost than any movement shown since the Council came into existence, except in 1964-65 when the Council was outside the field of influencing prices.

The manner of ending control and the failure of the buffer stock, in spite of heavy sales, to prevent the ceiling being broken were guaranteed to keep the market price high during the next three or four months. Further sales by the buffer stock below the ceiling did not weaken the price. By March, 1970 the buffer stock was down to 3,870 tons, a figure ominously near to the tonnage which had ended in 1963 in the pockets of speculators. The market felt, wrongly, that the basic deficit in supplies was not yet being met; it swallowed rumours that the forthcoming tin conference on the fourth agreement would end up with a higher price range; it looked at the unprecedently high price of copper. In April the London price again went through the ceiling (now £1,605 a metric ton); the turnover on the London Metal Exchange on that day was 3,190 tons; and the buffer stock sold nearly 3,000 tons. The Manager was left with only 1,000 tons; under a fairly transparent device the Chairman, in suspending selling operations, was able to say that some tin was still left at his disposal by reason of the nature and location of his holdings.

The downward reaction after this market victory was made more rapid and more heavy by the failure of the Geneva conference to arrive at a new and higher price range. By June the minimum daily price was £152 a ton below the maximum price of May. But the price was still high and the buffer stock, with its handful of stock saved from the speculators, was perhaps glad to be almost quiescent for six months.

The new prices of 1970

Since the last price change in January, 1968 the producers had seen an almost continuous rise in actual market prices, which had resulted in an average price through 1970 on the mid-point of the new range. The consumers were in a weak position. Their usage was running at an almost record level although market prices were double what they had been ten years ago, and they could scarcely maintain with any strength that the present price was damaging consumption. Timing was important for the producers. The current agreements permitted any member to bring up the price question at any Council meeting; they also compelled the Council to consider (but not necessarily to settle) the problem at its first meeting. In the first two agreements a decision on a price increase had not been made until nine and six months after entry into force; the third agreement had acted at its first meeting. The producers had received no change at the conference in April-May, 1970 on the draft fourth agreement, and they were aware that the backbone of consumers might be stronger with the future entry into membership of Federal Germany.

In May, 1970, before the new agreement, the producers asked for a new range of £1,360 to £1,750 a metric ton. Their supporting economic argument was fairly thin. Their costs were being affected by inflation, but their simple and fundamental case – that market prices had been rising constantly for the past two

[8] Floor £1,350, lower sector £1,350 to £1,460 middle sector £1,460 to £1,540, upper sector £1,540 to £1,650, ceiling £1,650, mid-point £1,500 (metric tons).

[9] In examining short-term price movements, account should be taken of one apparent operational weakness in a buffer stock. There is almost inevitably a tendency for a buffer stock Manager to shift from imposing an immediate stop on a price fall into imposing a longer-term straight line price on the market. This tendency will be helped if he has authority to both buy and sell, and can corner on the market. When the buffer stock stops buying there may be sudden destabilising jump in the market price. For example, during the six months, September, 1971-February, 1972, in which the buffer stock bought a total of over 4,000 tons of tin at differing prices within the lower sector, the market price was kept within a fairly narrow range – from between a highest of £1,433 and a lowest of £1,398 or a move of 2·5 per cent. In March, 1972 the price jumped and showed for that one month alone a range between £1,520 and £1,444 or a move of 5 per cent.

years – could not be gainsaid; nor could it be overlooked that their signatures were still required for the fourth agreement. In October the Council accepted a modified increase – a rise of about 7 per cent in the floor to £1,350 a metric ton, but only 3 per cent in the ceiling to £1,650 a metric ton. The mid-point in the new agreement was still in relation to current market prices.

The Council had shown a tendency to narrow the middle sector in its price range – from 10 per cent (of the floor price) in 1964 to under 6 per cent in 1970; this narrowness of the middle sector and the tendency apparent in the buffer stock to act quickly as soon as the price dropped into the lower sector meant that relatively slight market movements brought the buffer stock into action. By the end of the year 1970 a very small amount of buffer stock buying in the higher part of the lower sector and in the first quarter of 1971 a little more buying on a similar level brought the price back into the middle sector. On the production side, the long upward movement, checked in very small degree by the export limitations of 1969, had been resumed. In particular, Indonesia had at last shown a real revival. Consumption, stationary through 1970 – even dropping in Japan – had moved up again in 1971. There was through 1971 a most unusual and almost exact parallelism between world mine production and world consumption, both on the highest levels for thirty years or more. Stocks of free metal (that is, outside the ownership of the Tin Council) did not change much over the year. The future prospects, so far ahead as the market looked, seemed buoyant.

Nevertheless, the slow slide in the prices in sterling but not in dollars, which had begun after the increased price range of October, 1970 continued without interruption through much of 1971. By the middle of that year the market price was in the lower sector. In the last quarter of 1971 heavy buying, at a quarterly rate higher than at any other time under the third agreement, still left the price above the middle of the lower sector. In 1971, the range between highest and lowest daily prices had been limited to £99 a ton or 7 per cent in sterling (as compared to a much higher 16 cents or 10 per cent in the U.S. dollar quotation).[9] In 1970, when export control operated for the whole and the buffer stock for part of the year, the range had been £206 or 27 cents.

XVIII

The Fourth Agreement

THE fourth tin agreement, operating for five years from 1971, continued the basic principles of export control and buffer stock without change. It did, however, make changes in its practices to increase the flexibility which, since the discussions of 1950 and 1953, had been built into successive agreements. But more important was the relationship, outside the agreement, which was now developed on the buffer stock side with the International Monetary Fund.

By 1969-70 the International Monetary Fund had fully accepted its obligation to help in financing buffer stocks in an international commodity agreement and was prepared to apply it to the tin agreement, the only agreement with the necessary qualifications. Those qualifications were simple enough. Governments desiring help were required to be members of the Fund; they should be able to prove balance of payments difficulties; as members of an international agreement they should be under obligation to make payment to a stabilising buffer stock; and they should make an initial small contribution to the buffer stock from their own resources. They would then be allowed to use their drawing rights on the Fund to obtain the greater part of their further payments into the buffer stock. This Fund commitment guaranteed that there would be no problem in raising the total £27 millions required to be called up by the buffer stock in the tin agreement. Five of the initial producing members of

---

[1] Bolivia, Indonesia, Malaysia, Nigeria and Thailand, but not Congo-Zaïre or Australia.

[2] No producing country asked for a distribution of surplus funds. If requests had been made it is possible that a solution could have been found, for example, loans from the buffer stock to the government concerned.

[3] If the buffer stock holds excess cash (that is, above the initial compulsory contribution of 7,500 metric tons=£10,125,000 *plus* any voluntary contributions) the Council can authorise refunds out of such excess (but not necessarily all the excess) to the producing countries (but not to others). But any producing country can refuse to take its share of the excess and may leave it with the Manager. It is to be hoped that a fifth tin agreement will be simpler.

the fourth agreement[1] drew on the Fund in the first year. The Fund may do more, and may apply the same standards of generous drawings at low interest if the compulsory contributors to the buffer stock were to strengthen that stock by making additional voluntary contributions.

One minor change in the buffer stock mechanism of the fourth agreement was introduced to make it more palatable to the International Monetary Fund. Under previous agreements, contributions, once made to a buffer stock, were not refunded until the end of the agreement, even where those funds were in substantial excess of the likely requirements for some time ahead. This restriction had been notably irksome during the second agreement when the buffer stock was investing in relatively long-term securities within the United Kingdom whilst mine production of tin was stagnant in face of rising consumption.[2] The International Monetary Fund could scarcely ignore this contradiction. The new agreement provided, therefore, for the principle of the distribution of excess cash in the buffer stock during the life of the agreement, but this principle was so closely hedged with qualifications as to lose most of its point.[3]

These facilities for buffer stock financing from the Fund emphasised the growing importance attached to the rôle of the buffer stock in dealing with short-term and even medium-term price fluctuations. It was, perhaps, ironic that within eighteen months of the beginning of the agreement the Tin Council turned to export control.

The buffer stock and voluntary contributions

The authorised tonnage in the buffer stock of the fourth agreement was slightly lower (20,000 metric tons against 20,000 long tons in the previous agreement, with an equivalent value of £27 millions against £20 millions). Its initial buying power was substantially less. The stock of 2,692 metric tons at the end of the third agreement was bought for the fourth agreement for £3·9 millions at the current price for June, 1971 (about £100 a ton above the floor price). This purchase reduced the buying power of the initial stock to 17,000 long tons or less than five weeks of world consumption, even on the assumption that all future buying would be on the basis of the floor price. Fortunately, the provisions in the agreement for further finance were now being translated into reality. In 1972 the Council approved a standby credit of £8 millions (as it had approved, but not used, a credit for £10 millions under the third agreement). More important was the reality of voluntary contributions to the buffer stock from consuming countries. For twenty years the producing countries had pressed for consumers to make such payments. Their argument was simple. The agreement was providing equal benefits (protection for producers at the floor price, protection for consumers at the ceiling price); it provided a stability of price and production as valuable to consumers as to producers. It was unfair that the whole weight of financing should fall on producers. This argument failed to stress that the first three agreements had, for one good reason or another, failed to protect the consumer at the ceiling price, and the appeal drew no response.

The opinions of consumers began to be modified in the late 1960s in the Netherlands and in France, for reasons not necessarily connected with the tin agreement. These two countries, as others, were concerned with the need to maintain their general development aid to the developing countries, including the tin producing countries; they were also concerned with the efficiency of that aid. Development aid, as normally given, had too often seemed like pouring money down the sink for extravagant or useless purposes. An international commodity agreement conferred a general economic benefit – which could be measured – on developing countries as a whole. A contribution to the buffer stock of any such agreement was not open to abuse and could be justified on general grounds. Further, the buffer stock was in itself a good financial proposition. All buffer stocks in tin, both before and after the Second World War, had made a profit for their participants and it was reasonable to expect that, under general inflation, long-term rising prices would make them con-

[4] Unless otherwise instructed by the Council (Article 25, c(i)).

[5] Brazil was the backbone of the International Coffee Agreement, covering both producers and consumers and operating a quota system for exports; she was also the backbone of the coffee producers who in 1971-72 took over for themselves in effect, the quota powers of the agreement.

tinue to be profitable. Development aid given in this form with a fairly sound guarantee of profit and a share in the ownership of a metal, more desirable perhaps than paper money, had now more attractions.

In 1972 the Netherlands made a voluntary contribution of 5 million guilders (about £0·6 millions) to the Council's buffer stock. France followed with £1·2 millions. In both cases, these sums contributed were related to the votes held by them as consumers on the Tin Council. The precedent created for the future of the buffer stock finances was as important as the precedent created by the International Monetary Fund on financing.

Curiously enough, the fourth agreement destroyed formally the theory of equality of benefit for producers and consumers. It left unchanged the obligation on the buffer stock Manager to use all his resources in defence of the floor price. But it abolished his theoretical obligation to sell at the ceiling price. When the market price was at or above the ceiling price the Manager[4] should sell at the market price (not, as previously, at the ceiling price). It is not wise – although it is often done – to include in an international agreement a provision which cannot be applied practicably or which members will not apply because of the consequences; but it is regrettable, at least in the eyes of the outside world, that the principle of equality of rights of producers on the floor price and of consumers on the ceiling price was not diplomatically maintained.

The major "outsider": Brazil

In 1971 the seven producing members of the agreement covered 91 per cent of world mine production (outside the U.S.S.R. and China). The major non-member was Brazil. There, 'the opening up of a new tinfield in Rondônia had brought the Brazilian production to some 4,000 tons of tin by 1970, with hopes for a very much higher increase in the fairly near future. The Brazilians attended the Geneva conference of 1970 on the fourth tin agreement. They made it clear that Brazil had no real objection to the principle of international commodity agreements.[5] But Rondônia was due to develop most rapidly. A tin production of 10,000 tons a year (that is, well above the current standards of Nigeria and the Congo-Zaïre) might reasonably be expected in the quinquennium 1975-80. This anticipated mine production would more than meet the domestic requirements even of an expanding local economy. Brazil could not be expected to cripple the expansion of a new and infant tin mining industry by accepting in its growing stages any quota restrictions.

The other producers were, of course, anxious to bring Brazil within the agreement. A whole series of formulae which would reduce the restrictions on Brazilian exports to the minimum was worked out, but the Brazilians insisted that their tin export trade should be well established before the question of restriction arose. The other producers had faced the problems of the outside producers even before the Second World War; they had then made the mistake of bribing these outsiders, especially Thailand and the Belgian Congo, into the agreements on far too generous terms. They were not prepared now to repeat the experience for the benefit of Brazil. The negotiations broke down on this question of timing and tonnage conditions.

The shift from the sterling price basis

Mine production, metal production and metal consumption balanced at a high level through 1971. This happy position should have reflected itself in at least a stability of price; but, in fact, the price stability of the first half of the year (expressed in sterling in London, in dollars in New York and in Malaysian dollars in Penang) was broken on currency grounds in the second half of the year when the New York price rose but the London and Penang prices fell. For October the lowest London price was £1,398 a metric ton or at the very beginning of the lower sector in the Council's price range. The buffer stock Manager found himself buying around this price, in the fourth quarter of 1971, no less than 3,200 tons of tin – the largest quarterly tonnage taken in by the buffer stock since the flood of early 1968

---

[6] The fourth agreement read: "...the market price of tin shall be the price of cash tin on the London Metal Exchange or such other price or prices as the Council may from time to time determine". Article 25 (c).

which had prompted then his urgent request for export control.

The sag in the sterling price through 1971 was reversed in 1972, but sterling was still regarded with suspicion. On the change-over of sterling to a floating basis in the middle of the year, the London price rose temporarily into the upper range of the agreement but the Malaysian price actually fell.

The fourth agreement had envisaged two possibilities – that the Council for its price range might have to shift from sterling (although the only likely earlier alternative in mind was the dollar in New York) or that, to frighten speculators, the Council might do what the U.S. authorities had done in the later stages in selling their strategic stock, that is, fix its own calculated price.[6]

The disadvantages of the shaky condition of sterling made the other disadvantages of the London market more obvious. London was believed to be a more speculative market than was Penang. The turnover on the London Metal Exchange had been increasing rapidly in recent years (from 122,000 long tons in 1968 to 170,000 metric tons in 1972), an increase out of all proportion to the expansion of world production or even of consumption in Western Europe. The physical basis of the market had weakened with the relative decline in the output of the tin smelters in the United Kingdom. The more admirably London fulfilled its function as a hedging market, the more likely it was to draw speculation.

The New York market had made gallant but unsuccessful attempts in the past to become the major world market; now its currency was as suspect as sterling. Penang was strictly a market for physical tin. It had no facilities for hedging. In 1972, it priced 81,000 tons of tin metal or approaching half the world's consumption. Over half of the U.S. intake and three-quarters of the Japanese intake was bought there. The simplicity of the Penang market mechanism, as compared with the complexity of the London market, possibly made it more open to control by the buffer stock. Whether the Penang market price was the most important in the world (as the second buffer stock Manager believed) or whether the three markets in Penang, London and New York merely complemented one another was not agreed among experts. But, immediately, the Malaysian dollar in

Penang had a stability which neither London nor New York held.

In July, 1972, following the shift of the pound sterling from a fixed to a floating parity, the Tin Council changed its price range, for the first time in the history of any international agreement on tin, from a sterling to a Malaysian dollar basis.[7]

The buffer stock and export control, 1972

The tin price (in Malaysian dollars) was held through 1972 at roughly the same level as it had stood through 1971, but the cost to the buffer stock was even heavier. The statistical position, especially in production, was changing sharply. World mine production, after a rest in 1971, moved up again in the second half of 1972. The recovery in Indonesia added one-sixth to its tin production in the two years, but the outstanding expansion came from Australia. By 1971 Australian output had risen to 9,300 tons and by 1972 to 11,900 tons. World mine production for 1972 was 197,000 long tons, the highest level for thirty years.

There was no corresponding advance in world tin metal consumption. The flattening-out of consumption in Australia meant that all the new mine production there was in excess of the home market and in 1972 the Australians were putting over 6,000 tons of tin a year on the world market. The creeping surplus which had been seen in 1967-68 was being repeated.

The buffer stock had found it necessary to buy no more than 1,500 tons of tin in the first half of 1972, but the change to the

[7] Floor price $M583/picul, lower sector $M583 to $M633, middle sector $M633 to $M668, upper sector $M668 to $M718, ceiling price $M718, mid-point $M650·50 ex-works Penang (1 picul=133·3 lb of tin).

[8] Some of this tin was bought in Penang. Buffer stock holdings were 1,213 tons in December, 1970; 6,059 tons in December, 1971; 12,504 tons in mid-January, 1973. Stocks in London Metal Exchange warehouses were reported at the same dates as 2,130 tons, 6,059 tons and 6,725 tons (all metric tons). It had never been a habit of the buffer stock to hold tin in New York.

[9] That is, the £27 millions compulsory contributions, the £1·9 millions in voluntary contributions from France and the Netherlands, and the overdraft arrangements for £8 millions.

[10] At a tonnage equivalent to 44,700 metric tons a quarter.

[11] The buffer stock sold over 5,000 tons in the summer quarter (stock on June 30, 1973 10,069 tons; on September 16 4,820 tons).

Malaysian dollar price in Penang pushed it into much greater activity. In the third quarter of 1972 the Manager bought 2,000 tons, and in the fourth quarter (to mid-January, 1973) another 2,300 tons[8] were acquired at prices near the top of the lower sector. He now held 12,504 metric tons of metal, the highest total in buffer stock ownership since July, 1959. He had used half his maximum potential cash resources[9] and what was left, even if used to the full capacity, was unlikely to bring in another 12,500 tons.

In January, 1973 the Council declared an export control period.[10] But this was on the basis of holding the level of exports as they then stood, not on the basis of reducing that level. The Council thus applied the principle of ensuring adequate supplies for consumers and accepted the possible continuation of buffer stock buying to take up any further surplus. It permitted the Manager to buy and sell in the middle range of its price scale but, in ordering him to be a net seller in the upper part of that sector, implied that his buying activity in the lower part might so push up the actual market price as to increase the producers' total earnings from tin, even on an unchanged tonnage.

But the Council was running into its worst monetary crisis. The simultaneous sickness in 1973 of sterling and the U.S. dollar pushed tin prices into a whirlpool paralleled only by the price panics of the Korean war of 1950-51 and the U.S. strategic disposals of 1964-65. In the summer the flight from currencies into the metal coincided with some increases in real consumption, especially in the United States, and the market price rose to $U.S. 2.50 a pound in New York and to a price of £2,100 in London.

The Malaysian price, to which the Council's scale was now linked, was held under temporary control by buffer stock buying. The fourth agreement was, however, now about to show the old and, perhaps, incorrigible weakness of being unable to protect the ceiling price when the market was being dominated strongly by forces other than supply forces. Very heavy sales of buffer stock tin [11] showed that the buffer stock by itself could not hold the line against a price explosion of this magnitude. The Council promptly shifted course, abandoned export control in September and repeated its old practice of shifting its own price range up to follow the new market prices [12].

The U.S. stockpile is opened

But the price trouble was as nothing in comparison with the bombshell now due to explode. The U.S.A., restless for the past five years against the suspension of sales from the strategic stockpile of tin, was about to fulfil the worst expectations of the Council. In April, the American President announced most drastic reductions in the frozen stockpile. The stockpile objective was reduced from 232,000 tons to 40,000 tons. A total of 210,000 tons of tin was available for sale [13].

The nominal reasons for the decision to dispose were, first, the need to counter inflation by pushing down raw material prices and, secondly, the need to re-assess the objective of the stockpile. It was true that, since the U.S. Joint Chiefs of Staff had laid down objectives in 1944 and later reduced them, the U.S. administration had persistently bought too much and stocked too much tin. The Tin Council now faced a flow of tin from the stockpile which, at the most optimistic view, might have a long-term debilitating effect on the prosperity of the industry. That industry – with its long and unique record of international co-operation aimed at securing at least a degree of stability in the world tin price and enforcing some degree of order in a market notoriously disorderly – deserved something better than to have thrown upon it the consequences of the massive disposal of a strategic stock which had been built up unnecessarily and perhaps even irresponsibly 20 and 30 years earlier.

The tin price booms again, 1973-74

The Tin Council was soon to lose its influence on the tin price. The curtailment under export control of mine production in the first half of 1973 was about 5,000 tons, but at the same time

[12] New range: floor 635; lower sector 635-675; middle sector 675-720; upper sector 720-760; ceiling 760. All in Malaysian dollars per picul.

[13] Including tonnage unsold from the pre-1968 releases.

a sudden and even unexpected revival in tin consumption, particularly in the United States and Japan, took up over 8,000 tons.

The basic price disturbance came from the resurrection in the United States of disposals from the strategic stockpile. With the war ended against the communists in Vietnam, the U.S. administration no longer felt the pressure from the neighbouring tin producing countries who had always wanted the stockpile frozen in American hands. There were also other forces within the United States. The administration was now re-considering the size to which the stockpile could safely be reduced and was re-examining the strong presidential support for the current holdings of tin in stock being converted into cash assets to strengthen the national revenue. The maintenance of a large tin stockpile could no longer be strategically justified and the sale of that stockpile, whose value at current prices was well over a thousand million dollars, could not be rejected.

The proposals for disposals had, outside the United States, reactions similar to the reactions earlier in and after 1962. The tin producing countries repeated in 1973 that both the threat and practice of disposals would have an unacceptable, downwards effect on world tin prices. The tin consuming countries remembered that, while sales from the strategic stockpile over the years 1962-64 had helped to meet the world industrial requirement for consumption, they had also given the tin price an unexpected upward movement. The discussions between the Tin Council and the U.S.A. were decided by Washington with little more than a re-commitment of the principles already acceptable in 1968.

The world market was moving rapidly out of hand. The British pound and the American dollar, but not yet the Malaysian dollar, were drifting into a currency whirlpool. The volume of industrial consumption of tin was moving slowly into the year 1974 but the price of tin in the United States began to rise earlier and rapidly. The U.S. strategic stockpile found itself selling at unexpected tonnages and unexpected prices nearly 18,000 tons of tin in the second half of 1973. This disposal was more than twice the annual rate in 1964 and in 1965.

The Tin Council had similar trouble. Over the third quarter of 1973 its buffer stock sold over 5,000 tons of tin, the fastest rate of selling for many years. It was necessary to suspend the buffer stock sales. As a safeguard against a very rapid rise in the London price (to £2,155 in September) the Tin Council pushed its price scale to a higher level (on 21 September, 1973 to a floor of M$635 a picul and a ceiling of M$760 a picul) but the increase of only 9 per cent on the floor price and 6 per cent on the ceiling showed itself inadequate in November when the Penang price moved through the new ceiling.

The tin prices no longer reflected the balance of tin production and consumption or of supply and demand; they were reflecting the weakness of London and New York currencies. By the end of 1973 the peak market prices of the two markets were double the lowest prices at which they had stood at the beginning of the year. The jump in prices was accepted as phenomenal. This was a ready, but perhaps not always a justified, acceptance. In the first post-war price crisis—the Korean war of 1950-51—the rate of upward movement from the lowest price to the highest price was 180 per cent in London, 149 per cent in New York and 153 per cent in Penang/Singapore. These currencies in the three main markets were linked. In the second price boom of 1964-65, with the three markets still in general currency relation, the margin between highest and lowest prices was over 60 per cent. In the third period, covering 1972 and 1973, London, with the weakest currency, showed a rise from lowest to highest price at 127 per cent, New York with a less shaky currency, a rise of 106 per cent and Penang, almost a strong currency, a rise of only 69 per cent.

The Tin Council had seen its price arrangements largely inoperative in the price boom of 1964-65 during the second international tin agreement. At the end of 1973, half-way through the life of its fourth international agreement, the international body accepted again a temporary paralysis of action in the face of the lates swings and even gyrations of world tin prices.

XIX

An Assessment of Tin Control

WHAT have been the achievements of tin control over the years?
Has the tin industry enjoyed greater stability of production and
revenue than might otherwise have been expected? These questions need to be answered against the background of the stated
objectives of those concerned.

The most important of the objectives in the formidable list set
out in the latest of the agreements are: to ensure a degree of
stabilisation – long-term and short-term – in the price; to obtain
some long-term adjustment between production and consumption; to increase the export earnings of the producing countries
without detriment to the interests of the consuming countries;
and to mitigate the difficulties for producers during surpluses of
production and for consumers during shortages of production. It
is a general objective that the desirable price should be
remunerative to producers and fair to consumers. By implication, the price of tin should not move so much out of harmony
with the prices of other competitive materials as to distort the
pattern of consumption or to weaken the market for producers.

In looking at what has happened within the industry to see
how far the agreements have been successful, I have taken three
groups of periods for comparison. The first covers the 15 years
from 1924 to 1939. This first period was marked by two peaks of
production and consumption, by one intense trough of price and
production, and by a fairly long cyclical swing (four years up,
four years down and then four years up). It was also subject to
control (through export limitation or the use of a buffer stock) for
its latter nine years. The second period is the six years 1950-55.
Here production was fairly high and fairly constant, although
below earlier and later peaks, and consumption was fairly stagnant. The market was free and without any international control;
but, in fact, it was dominated by the U.S.A. demand, both at the
beginning in the Korean panic and later by stockpile buying.
The third period is the 17 years 1956-72, covered by the post-war

388

FIGURE 15

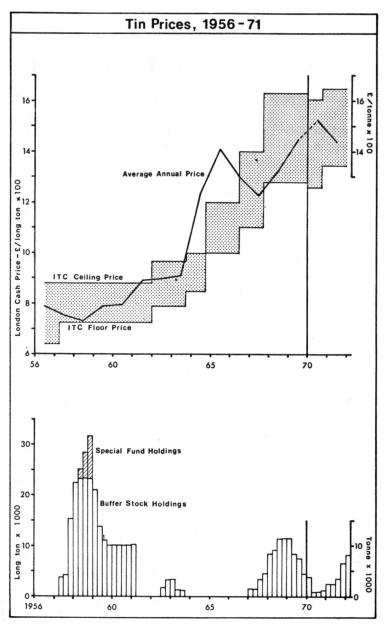

tin agreements. In the first of those agreements there was a gross over-supply of tin; in the second agreement a gross under-supply; in the third a production and a consumption both moving upwards to high levels and in rough harmony. Abnormal supply forces from the U.S.S.R. and then from the U.S.A. affected the first and second agreements but not the third. Control of the market by the Tin Council was strong in the agreement of 1956-61, almost non-existent in the agreement of 1961-66 and effective enough (but more in buffer stock than in export limitation) in the agreement of 1966-71.

How far have short-term price fluctuations been weakened? Possibly one simple test statistically is to examine the movements between the highest and lowest daily prices quoted during a year. In the years 1924-30, that is, before the acceptance of the pre-war control system, the range was never less than 22.8 per cent or higher than 72.7 per cent.[1] In the period 1931-39, the range was between 10 per cent and 72 per cent, so that international control, applicable throughout those years, seems to have reduced the short-term movements relatively little. In the free and uncontrolled period of 1950-55, the influence of Korea was shown in remarkable swings in both 1950 and 1951 (well over 100 per cent), but even other successive years had swings as low as 9·5 per cent and as high as 73 per cent. In the first agreement (1956-60) the greatest swing was only 23 per cent and the lowest 5 per cent; in the second agreement (1961-65) 68 per cent greatest and 15 per cent lowest; and in the third agreement (1966-70) back to figures of 22 per cent greatest and 12 per cent lowest.

If we look only at the more effective control periods in these agreements, we find that the greatest margin between maximum and minimum daily prices in a year was only 20 per cent in 1958-60 and 22 per cent in 1967-71. In general, except where, as in 1964-65 outside forces (the fears and facts of U.S. sales) had overwhelmed the market, it can be properly claimed that the post-war agreements have substantially smoothed down short-term fluctuations.

[1] That is, in relation to the lowest daily price.

The degree to which longer-term fluctuations have been checked may be measured by comparing the price movement from year to year on annual averages.[2] In the period 1924-39 the smallest movement (under 1 per cent in 1927) was in a year outside control; the next smallest (2 per cent) in the middle of control (1935); the largest movement (44 per cent) was during control, the next largest (28 and 30 per cent) outside control. It is possible that annual variations under control were not much better in the pre-control years. In the period 1950-55 the range was from a lowest annual change of under 2 per cent to a highest change of 45 per cent – a position very little different from the controlled years of the 1930s. Under the first agreement, not surprisingly in view of the very tight degree of control, the changes were at their least (from a minimum of 1·4 per cent to a maximum of over 7 per cent); under the second agreement, when prices rocketed, the changes were much as they had been in the 1930s (from a minimum of 1 per cent to a maximum of 36 per cent); and under the third agreement they moved back to a fairly narrow range (from a minimum of 6 per cent to a maximum of 9

TABLE 15
Short-term and longer-term price movements in tin, 1924-72

Period	Character of period	Highest daily price in period £	Lowest daily price in period £	Range %	Highest annual price in period £	Lowest annual price in period £	Range %
1924-30	Free of control	321·1	104·6	207	291·2	142·0	105
1931-39	Controlled	311·3	100·3	210	242·3	118·5	104
1950-55	Free of control	1,620·0	566·0	187	1,077·0	719·0	50
1956-60	Controlled	890·0	640·0	39	797·0	735·0	11
1961-65	Controlled	1,715·0	779·0	120	1,413·0	797·0	77
1966-72	Controlled	1,664·0	1,180·0	41	1,554·0	1,228·0	27

Note: Prices in £ a long ton.

[2] The change between annual average prices of two years expressed as percentage of price in the lower year.

[3] A mathematical formula as the basis of changes in the agreement's price range might have saved much discussions on the Council and given better results in practice, but the Council, perhaps with sound political sense, never considered this alternative.

per cent). In general, and naturally when the Tin Council was strongest, the year-by-year range was smaller.

It is more difficult to assess the effect of the agreements on the stability in the tin price over longer periods. The producing and consuming members of the agreements were perhaps interpreting the word stability in differing senses. To the producers, stability, speaking plainly, meant a guarantee (iron-clad with their money in the buffer stock and their sacrifices in export limitation) of a floor price, the continuous upward shift of that floor with the consequential continuous upward move of the price range in the agreement, and no guarantee of a ceiling price in which they expressed no long-term interest. To the consumers, stability meant a recognition of the greater economic interest of the producers in the agreement, an acceptance of the value of the floor price as a means of ensuring the maintenance of mine production needed for future consumers' needs, and a check on the upward movement of the price to prevent it becoming onerous on their raw material costs. The consumers were also bound to realise that, as their own manufacturing costs were being de-stabilised by monetary inflation, they could scarcely expect their own influence on raw materials to be other than to stress the economics of moderation in inflation.

It has constantly to be stressed that the members of the Tin Council were voluntary members of a voluntary organisation with a limited life. Their delegates were not present as an impartial body of economists but as civil servants, each aiming at the national interest of his government; they were supported by advisers who were inevitably very much alive to the economic interests of the group from which they were drawn. The Council's decisions on changes in its price range were naturally often determined by political reasons[3] or, more simply, by following with a time-lag the actual movement of prices in the market. This latter practice has satisfied both parties reasonably enough, possibly because so far it has not had to tackle the very thorny problem of negotiating a price movement downwards. It is, of course, not a long-term price policy.

Under the agreements, the price changes were irregular in timing. In the 17 years 1956-72 prices were changed only eight times[4]– once in the first agreement, three times in the second, three times in the third and once early in the fourth agreement. Most of the changes in the second agreement had no relation to the reality of market prices outside the Council. In the first agreement, when real price problems were most serious, the only change was one of 14 per cent in the floor price; in the second agreement the total and irrelevant change in the floor price was 37 per cent; in the third agreement the total and relevant change in the floor (including the change necessary on the devaluation of sterling) was 35 per cent – all upwards. It is probable that the Council, although an advocate of stabilisation, changed its own price range in the second and third agreements too often and too little under pressure from unstable market prices.

What was happening to actual tin prices over a much longer period? During 1924-39 the tin price averaged £215 a ton. The first great jump came on the reopening of the metal markets after the war when the sterling price and the dollar price doubled. During 1950-55 the sterling price at £830 was nearly four times the pre-war quotation. During the 17 agreement years the sterling price almost doubled again, from £788 in 1956 to about £1,531 in 1972. During the first agreement the sterling price and dollar prices showed almost no change between 1956 and 1960; during the second agreement both the sterling and dollar price rose by nearly 60 per cent; and during the third agreement the sterling price rose by about one-fifth and the dollar price by one-sixteenth. If we can ignore the influence of the changing exchange value of sterling and concentrate on the dollar price, this showed in 17 years of the agreements 1956-72 a rise of 75 per cent or about a simple 4·5 per cent a year against a rise over the 16 years 1940-55 of 90 per cent or 5·5 per cent per annum and almost no change between 1924 and 1939. An average price of 46·6 cents over 1924-39 had become an average price of 104·4 cents during 1950-55, of 99·1 cents for the first agreement of 1956-60,

---

[4] Ignoring nominal changes.

[5] A rough calculation based on tonnage of tin in concentrates produced and the New York price less an arbitrary 5 per cent for smelting and freight charges. The International Monetary Fund estimated the export earnings of the six developing members of the tin agreement at $U.S.543 millions in 1968-69, $U.S.567 millions in 1970-71 and at $U.S.626 millions in 1971-72.

of 136·1 cents for 1961-65 and of 164·1 cents over 1966-72. The movements in these agreement years do not suggest long-term stability in price, but we may be asking for the moon in expecting such stability.

Export earnings from tin

Clearly, the general upward move in prices under the agreements, coupled with the general increase on production under the third agreement, has benefited the import earnings of practically all the producing countries. During 1924-39 the annual export earnings from tin may have averaged around $U.S.150 millions. In 1950-55 they had risen to about $365 millions, but dropped to about $295 millions during the depression of the first agreement during 1956-60. They jumped back to about $415 millions for the second agreement and again by another half to $610 millions for the third agreement[5]. The Tin Council cannot claim the credit for the increase during the second agreement, but it was certainly responsible for maintaining the 1956-60 earnings on such a high level in view of the then over-supply position and it can claim much of the credit of the rise for the years 1966-71.

Has the international control produced a greater tendency for production and consumption to keep in a better natural balance? During 1924-39 when there was a surplus, that surplus averaged about 13,000 tons in the year; when there was a shortage of supply, that shortage averaged about 11,500 tons. Under the post-1956 agreements the average annual surpluses were slightly smaller at 10,500 tons but the shortages were substantially larger at 19,000 tons. The relative balance in the third agreement merely indicates the coincidence of high production and high consumption. The agreements in themselves have not seriously pretended to be able to do anything to stimulate the consumption of tin.

Have the tin agreements been responsible for any shift in the degree of importance of the areas of tin production? The Havana Charter was followed by the first agreement of 1954 in taking as its objective "to promote the progressively more economic production of tin"; fortunately, this phrase, so full of implications for destroying some producing areas, has not been carried into the fourth agreement. It would have been impossible to achieve without the minute examination of costs not only for each country but for each type of mining in each country; it could not have taken into account collapses in production due to political causes, as in Burma, Congo-Zaïre and Indonesia.

In fact, over 16 years of the agreements, the two countries generally regarded territorially as the cheaper producers – Thailand and Malaysia – have increased their shares in world production, although within each the gravel pump mines, generally accepted as the form of production with higher operating costs, have been responsible for a greater proportion; the dredges, generally accepted as the form of production with lower operating costs, have been responsible for a smaller proportion of the national output. Indonesia, whose costs in real currency are unknown, has declined in relative importance, but Bolivia, whose costs in lode mining are known to be high, has increased in relative importance. The tin world has perhaps been fortunate that the death in Burma and the semi-death in Congo-Zaïre of substantial tonnages of tin supply and the very slow revival in Indonesia have given time not only for the other producing areas of the world to adjust their own output to fill the gaps but also to permit the absorption within the market of new tonnages from

TABLE 16

Changes in world mine production of tin, 1937-71

| | in percentages of world | | |
Decreases:	1937	1956	1971
Congo-Zaïre	4·4(a)	7·6	3·5
Nigeria	5·2	5·5	3·9
Burma	2·5	0·5	0·3
Indonesia	18·2	17·9	10·6
Increases:			
Bolivia	12·2	16·0	16·1
Malaysia	37·5	37·0	40·3
Thailand	7·7	7·4	11·6
Australia	1·6	1·3	5·0

(a) Including Rwanda-Burundi.

TABLE 17

The U.S.A. prices of non-ferrous metals, 1956-71

Index 1965-67 = 100

Period	Aluminium	Copper	Lead	Nickel	Tin	Zinc
1950-55	82·8	76·3	100·8	64·4	63·2	95·6
1956	105·5	115·1	106·5	79·8	61·3	94·5
1956-60	109·1	88·7	88·1	88·4	60·0	83·5
1961-65	97·5	87·0	81·2	96·9	82·5	88·6
1966-71	108·1	94·6	96·2	128·7	98·0	102·5
1971	117·6	142·8	92·4	162·8	101·3	72·5

Source: International Tin Council *Statistical Year Book*.

Brazil and Australia. To expect the Tin Council to be able (or even willing) to change the world pattern of production is expecting too much.

Is it possible to see whether the tin agreements have given the tin industry something in longer-term price stability, which the other non-ferrous metal industries have not obtained? Clearly, one thing peculiar to tin has been the provision of and finance for a floor price beneath which the actual market has not been allowed to fall, except for a few brief days, during a period of 17 years. This has undoubtedly maintained the capital supply for the current requirements of the industry. On the other hand, it does not seem to have attracted the general flow of new capital that might have been expected, and there are no signs that the other non-ferrous metals, without the advantages of a floor price, have found any difficulty in raising capital to maintain or expand production.

Aluminium, without a price agreement or an international organisation, has maintained since 1937-39 an astonishing price stability and an astonishing expansion in the rate of production. Nickel, with a tight-fisted private control of production by one producer (which has been dented in recent years), was able to hold prices with relatively little movement until the late 1960s. Copper, lead and zinc, where the price explosions took place mainly in the years just after the Second World War and where there was (for copper and zinc) a further rise in the late 1960s, have ended in 1966-72 in much the same relationship as they were in 1937-39 (in dollar prices). During the three tin agreements, tin moved up in price by more than one-half as compared in the same period[6] with a slight fall in the price of lead, a rise of one-tenth in zinc, and even in aluminium a rise of one-third. Only nickel rose more than tin (due almost entirely to the shortage which broke the price domination of International Nickel around 1970).

It is possible that the fairly rapid relative price rise under the tin agreements merely reflects a position where tin in the 1950s was underpriced as compared with the other non-ferrous metals. In 1937 tin (dollar price) was nearly three times the price of aluminium, less than twice the nickel price, four times the copper price, and between nine and eleven times the lead and zinc price; in 1972 the tin price ratio had become six times aluminium, over three times the copper price, ten times the lead and zinc price but only a quarter higher than the nickel price.

The distortion in the price relationship over 35 years has been relatively slight, except in the case of aluminium, and there the gap in favour of aluminium is so wide as to forecast the greatest future threat to the expansion of the consumption of tin. The general problems of all non-ferrous metals, even when they are inter-competitive, are very much the same; their particular problems differ very widely. Tin has its particular problems of the very large number of unimportant and even tiny producers, the relatively small size even of those units which can be regarded as important, the limitations on tin deposits by geology and in volume, the concentration of mine production in areas far removed from the centres of consumption and the narrow basis and relatively slow development of that consumption. Nevertheless, the factors which called the tin buffer stock into existence as long ago as the 1930s and revived it in the agreements from 1956 onwards exist fundamentally in all non-ferrous metals – wide short-term and long-term fluctuations in prices, great depressions in production and consumption, excessive dependence on the upswings and downswings of industrial activity in the developed countries – and are in no sense peculiar to tin. So far as the use of a buffer stock in tin has eased some of those problems

[6] Comparing averages of the uncontrolled period 1950-55 with the average of 1966-71.

or the use of export control (as in 1958-60) has preserved part of the world tin mining industry from extinction the weapons of the tin agreement have been justified. The other non-ferrous metals – in particular tungsten, zinc, lead and copper – would do well to examine whether their own special problems are open to the same solution.

Postscript

A more personal view

I HAVE spent nearly a quarter of a century in the administration of the two international bodies – the International Tin Study Group and the International Tin Council – which have been concerned with creating and operating a control machinery for tin. I have seen, sometimes many times over, so many of the world's tin problems – the accumulation of information on production, markets and prices, the continuous argument on the principles of co-operation, the detailed operation of the agreements in practice, the results of export control and buffer stock, the long and repetitive discussions with the U.S.A. on sales from the strategic stockpile, the organisation of production conferences inside and outside the United Kingdom – that I have found it difficult to see the wood for the trees. By trade and in principle I have been committed to world international co-operation in tin and I am likely to have become a biased party in judgement on the question: How valuable has this co-operation proved to be?

I may as well admit another bias. In the conflict of interests between those who buy and use tin and those who mine tin, most international civil servants will find themselves swaying towards the producers. This is in part because general international philosophy and liberal sentiments have moved in favour of the poorer developing countries, who provide the greatest part of the world's tin, and away from the richer developed countries, who have exploited the natural resources of the poorer countries and who have eaten up those resources in making mountainous supplies of consumers' products. I am not sure that I look at it this way. I have my doubts as to whether the conflict is quite so simple.

The developing countries as a whole do not seem to me to be better or purer merely because they are poorer and less efficient; and, indeed, I have seen enough evidence to know that within developing countries the gap between rich and poor (including those working in tin) may be wider than the average gap between developed and developing countries. I appreciate the political arguments in favour of the shift of ownership and development of resources, including resources in tin, from outside owners to indigenous owners, even if the shift is not an economic proposition in the eyes of economists. But I have noted that the harshest modern exploitation of labour in tin was in Bolivia under local private ownership and that even nationalised tin industries in Bolivia and Indonesia have often been severe with labour; and I still have to be persuaded that the tin miners in the developing countries are likely to get a very much better deal from their indigenous mine owners or from the new middle class bureaucracies of the new independent states than they have received in the past from foreign owners or even foreign governments. But there is some rough justice, which may not be very economic, in evening out the balance between countries which nationally have little and countries which nationally have more, if only as a preliminary to some further levelling out within the developing countries between those who have little and those who have much.

I have accepted, with some natural reluctance, that an international political and economic organisation concerned with tin, as it might be with any other commodity, is likely to find itself limiting, fairly strictly, what it can achieve. The limit is determined not by the bland, meaningless, comprehensive or contradictory principles on which the member countries have agreed in the international tin agreements, but by the degree to which the machinery set out in the text of the agreements is allowed to operate or, at times, indeed is allowed not to operate as the result of a general consensus among members whose interests may normally be expected to conflict.

Over the past 15 years or so, members of most of the international commodity agreements—in wheat or sugar or coffee —have reached fundamental disagreement, and have on many occasions tacitly ignored the application of the most important price or export control provisions of their agreements or have seen the authority underlying these provisions transferred out of the hands of the nominal governing body. I have never expected, and do not expect now, that officials of the Tin Council can advance matters step by step much beyond the minimum degree of common agreement necessary to ensure the continuance of the international tin agreements from one quinquennium into the next. But it is at least encouraging to remember that a sense of unity in tin has developed from time and experience in tin agreements, as opposed to the chequered and perhaps less successful histories of other commodity agreements.

I still regret that neither the International Tin Study Group nor the International Tin Council has at any time made a study of one fundamental question which I have always found the most interesting in tin, as in the other non-ferrous metals dealt in on the London Metal Exchange. This is the problem created by the inability of this market to arrive at prices which reflect accurately the current relationship of supply and demand for physical metal. This Exchange has also developed as a hedging market and, perhaps to an even greater extent, as a centre for purely speculative dealing. It has not yet proved possible to examine the degree to which the distortion of prices resulting from hedging and speculative activities may ultimately be subject to some control or how far, and how desirably, the market may be limited exclusively to physical dealings.

Nevertheless, the London Metal Exchange seems at all times to show undue complacence about the more speculative uses which have been made of the tin market. It may well be asked whether more authoritative or representative bodies, whether speaking for tin consuming or tin producing governments or for the International Tin Council, might not put enough pressure on the Exchange to ensure that its internal affairs are so arranged as to minimise the impact on prices of massive speculative dealings. It is certain that, while no action is taken, things will remain as they are and the ability of the International Tin Council to stabilise prices through the Metal Exchange will always be limited.

Forty years ago, when I was working with the Labour Research Department, I came to accept socialist thinking about profits, wages and ownership in British coal mining, engineering, printing and building. When, early in World War II, I came to be concerned with non-ferrous metals and their contracts, I found, however, that I was working within a framework which did not concern itself with wages or principles of nationalisation in the mining industries of undeveloped territorities. Nor is this a matter which, at any time or by any government, has been raised for discussion in the Tin Study Group or the Tin Council, and each country has always acted quite independently on the question of nationalising its tin mining industry. In retrospect, my regret is that where nationalisation has occurred, as it has done to a high degree in Bolivia and Indonesia, the elimination of private ownership has shown has shown suprisingly little result in the context of a greater community of interest between labour and a nationalising government.

One last thought, which stays with me, concerns the principles and operation of the buffer stock system as it has been revealed within the international tin agreements. I have observed over the years the slow and, indeed, almost the reluctant approach by so many members of the tin agreements to considering the need for providing in adequate quantities the money and metal without which the buffer stock cannot function effectively. It has, however, been some consolation that, in the more recent life of the tin industry, the International Monetary Fund, operating as an international agency of the United Nations, has made a positive contribution to the financing of the buffer stock on tin.

APPENDIX A
The essential tin figures, 1920-72

Period	World mine production 000 long tons	World metal consumption 000 long tons	Average annual price £/long ton	Range between highest and lowest daily prices		Range in annual prices	
				£/long ton	%(a)	£/long ton	%(b)
1920-24	126	121	214	280	131	138	65
1925-29	165	160	254	147	58	87	34
1930-34	126	133	164	144	88	112	68
1935-39	171	199	218	158	72	53	25
1940-44	167	121	274	70	25	44	16
1945-49	120	112	442	455	103	303	69
1950-55	167	135	830	1,042	123	357	43
1956-60	140	148	772	250	32	62	8
1961-65	144	164	1,069	936	88	524	49
1966-70	176	176	1,369	487	35	331	24
1956	166	150	787	167	21	48	6
1957	163	143	755	74	10	33	4
1958	116	136	735	125	17	20	3
1959	119	148	785	54	7	50	6
1960	136	163	797	43	5	11	1
1961	136	158	889	214	24	92	10
1962	142	161	896	130	15	8	1
1963	141	162	910	185	20	13	1
1964	147	171	1,239	695	56	330	27
1965	152	167	1,413	435	31	173	12
1966	167	171	1,296	261	20	117	9
1967	172	169	1,228	190	16	73	6
1968	181	174	1,323	168	13	100	8
1969	177	182	1,451	299	21	128	9
1970	183	179	1,554	209	13	103	6
1971	183	184	1,461	101	7	93	6
1972	194	189	1,531	205	13	70	5
1973	184	206	1,992	1,615	81	461	30

Production, consumption and price figures from International Tin Council: *Statistical Year Book*.

(a) As percentage of average annual price of period.

(b) Similarly for quinquennial periods. For individual years 1956 and onwards percentage change compared with the immediately preceding year.

APPENDIX B
Votes in the four agreements, 1957-71

Country	1956-57 Tonnage	Votes	1961-62 Tonnage	Votes	1966-67 Tonnage	Votes	1971 Tonnage	Votes
Consumers								
Australia	1,442	32	3,500	50	4,556	57	—	—
Belgium	1,739	38	2,370	35	3,067	40	2,945	29
Canada	3,770	77	3,797	54	4,885	60	4,313	40
Denmark	3,913	79	4,201	59	573	12	740	11
Ecuador	4	5	—		—		—	—
France	8,405	165	11,023	146	10,737	127	10,409	90
India	3,698	75	4,157	58	4,427	55	4,512	42
Israel	103	7	—		—		—	—
Italy	2,730	56	4,050	57	5,705	70	6,500	58
Netherlands	2,477	52	2,960	43	3,553	45	4,857	45
Spain	463	14	833	16	1,614	23	2,301	24
Turkey	784	20	773	15	780	14	—	—
United Kingdom	19,763	380	21,179	277	19,715	228	17,476	147
Austria	—	—	610	13	642	12	596	10
Japan	—	—	11,841	157	16,997	197	24,396	204
Korea	—	—	—	—	289	8	257	7
Mexico	—	—	1,133	20	1,200	19	—	—
Czechoslovakia	—	—	—	—	2,502	33	3,548	34
China (Taiwan)	—	—	—	—	—	—	339	8
Bulgaria	—	—	—	—	—	—	670	10
Federal Germany	—	—	—	—	—	—	13,029	111
Hungary	—	—	—	—	—	—	1,195	15
Poland	—	—	—	—	—	—	3,597	34
U.S.S.R.	—	—	—	—	—	—	7,400	65
Yugoslavia	—	—	—	—	—	—	1,335	16

Producers	Percentage	Votes	Percentage	Votes	Percentage	Votes	Percentage	Votes
Belgian Congo and Rwanda-Burundi	8·72	90	—	—	—	—	—	—
Bolivia	21·50	213	18·21	182	16·92	172	18·36	182
Indonesia	21·50	213	19·73	197	10·79	111	11·65	117
Malaysia	36·61	360	38·45	378	46·74	450	45·02	440
Nigeria	5·38	58	6·32	66	7·01	74	4·86	52
Thailand	6·29	66	9·11	93	13·98	143	13·29	133
Congo-Zaïre	—	—	8·18	84	4·56	50	3·93	43
Australia	—	—	—	—	—	—	2·89	33

SELECTED BIBLIOGRAPHY

International Tin Committee: *International tin control and buffer stocks*. 1941. (Tin Producers' Association).

M. J. Schut: *Tinrestrictie en tinprijs*. 1940. Nederlandsch Economisch Instituut.

K. E. Knorr: *Tin under control*. 1945. Stanford University.

United Nations: *Conference on trade and employment: Final act and related documents*. 1948. United Nations.

U.S. Department of Commerce: *Materials survey: Tin*. Prepared for the National Security Resources Board by the Division of Mineral Economics, Pennsylvania State College. 1953.

E. S. Hedges: *Tin in social and economic history*. 1964. London.

Yip Yat Hoong: *The development of the tin mining industry of Malaya*. 1969 (an excellent book, with very detailed bibliography).

Liaqat Ali: *The principle of buffer stock and its mechanism and operation in the International Tin Agreement*. In *Zeitschrift des Instituts für Weltwirtschaft an der Univ. Kiel, Band, 96, heft I*. 1966.

W. Fox: *The work and problems of the International Tin Council*. In W. Fox (editor): *A second technical conference on tin, Bangkok, 1969, vol. 3*.

P. Legoux: *De quelques codes d'investissements*. 1968. Paris.

W. T. Dunne (editor): *Tin in your industry*. Tin Industry Research and Development Board. 1968.

W. Fox (editor): *A technical conference on tin, London, 1967*. 2 vols. 1968. (530 references). International Tin Council.

C. L. Sainsbury: *Tin resources of the world*. U.S. Geological Survey Bulletin 1301. Washington. 1969.

W. Fox (editor): *A second technical conference on tin, Bangkok, 1969*. 3 vols. (58 papers and 550 references). 1970. International Tin Council.

T. Geer: *The post-war tin agreements: A case of success in price stabilisation of primary commodities*. In *Schweizerische Zeitschrift für Volkwirtschaft und Statistik*, no. 2. 1970.

United Nations: *Mineral resources development with particular reference to the developing countries.* U.N. Department of Social and Economic Affairs. 1970.

W. Fox: *Tin and the international agreements.* In *The pricing and marketing of metals.* The Institution of Mining and Metallurgy. 1972.

A. Legoux: *Backwardation et contango.* Annales des mines. February, 1970.

J. E. Denyer: *The production of tin.* Paper to the I.T.C. conference on consumption, March, 1972. International Tin Council.

W. Fox: *The reserves and availability of tin and the world consumer.* Paper at the I.T.C. conference on consumption, 1972. International Tin Council.

The International Tin Study Group publications:

J. F. Houwert: *Review of the world tin position over the year 1946.* The Hague. 1947.

Review of the world tin position, 1947-48.

Tin 1949-50: A review of the world tin position.

Tin 1950-51.

Tin (1951-53).

Tin 1954.

Statistical Bulletin, monthly.

Statistical Year Book, issues for 1949, 1952 and 1954.

Supplement to the Statistical Year Book, 1953 and 1955.

The United Nations conference on tin, Geneva, 1950.

The United Nations conference on tin, Geneva, 1953.

A statement on the position and prospects of the tin industry, 1950.

Tin and the Tin Study Group, 1948-52.

Tin and the Paley report. 1952.

Notes on tin. Monthly, duplicated.

The International Tin Council publications:

A draft international agreement on tin: Paris and Geneva, 1950.

A revised international agreement on tin. Produced by a drafting committee in London, August, 1953.

The first, second, third, fourth international tin agreements.

A new international tin agreement: A working draft (for the second agreement). 1960.

A draft for the Third International Tin Agreement.

G. Péter: *The work of the International Tin Council, 1956-59.*

Monthly Statistical Bulletin. Continued from the I.T.S.G.

Statistical Year Book: Tin, tinplate and canning. Issues for 1959, 1960, 1962, 1964, 1966 and 1968. (Invaluable for statistics).

Statistical Supplement (to the Statistical Year Book). Issues for 1963, 1965, 1967 and 1970.

Annual reports. Yearly from 1956-57 to 1970-71.

W. Robertson: *Report on the world tin position with projections for 1965 and 1970.*

A. La Spada: *Patterns of world tin consumption.*

A. La Spada: *Prospects for world tin consumption up to 1975.*

Australia:

Bureau of Mineral Resources: *Australian mineral industry reviews.*

Brazil:

J. B. Kloosterman: *A two-fold analogy between the Nigerian and the Amazonian tin provinces.* In *A second technical conference on tin, Bangkok. 1969.*

Bolivia:

Negociaciones para un contrata de estaño. Asociacion de industriales mineros de Bolivia. La Paz. 1952.

Oscar Davila Michel (editor): *Primer simposio internacional de concentracion del estaño.* 1966. Oruro (Universidad Tecnica).

Compendium on the loan application for the third phase of the rehabilitation plan. Corporacion Minera de Bolivia. 1964. La Paz.

Reservas mineralogicas de Bolivia. Corporacion Minera de Bolivia. 1967. La Paz.

D. J. Fox: *Tin and the Bolivian economy.* 1970. Latin American Publications Fund.

Amado Canelas O.: *Mita y realidad de la Corporacion Minera de Bolivia.* 1966. La Paz.

Congo-Zaïre:

Maniema: *Le pays de l'étain.* 1953 (Symétain Company).

P. Anthoine, P. Evrard, C. Kharkevitch and G. Schaar: *The*

Symétain tin deposits: Geology and mining. In *A technical conference on tin, London, 1967.*

The work of the Géomines at Manono. In *A technical conference on tin, London, 1967.*

Indonesia:

G. A. Gilmore: *Mining investment and Indonesia. Mining Magazine,* October, 1971.

Indonesia builds anew. Tin International, January and February, 1972.

Prospects for tin mining in Indonesia: Prospects for bids. Duplicated. 1967. Ministry of Mines, Djakarta.

Malaysia:

Massive bibliography in Yip Yat Hoong: *The development of the tin mining industry of Malaya.*

Tin mining in Malaya (photographs). Tin Industry Research and Development Board, Malaya.

Nigeria:

R. Lukman: *Some problems associated with the valuation, mining and processing of alluvial cassiterite on the Plateau tinfield of Nigeria.* In *A second technical conference on tin, Bangkok, 1969.*

Spain and Portugal:

D. J. Fox: *Tin mining in Spain and Portugal.* In *A second technical conference on tin, Bangkok, 1969.*

Thailand:

Rachan Kanjan-Vanit, Pow Kham-Ourai and W. Champion: *Off-shore dredging of tin deposits in south Thailand.*

Poolsawasdi Suwarnarangse: *Tin mining problems in Thailand.*

Pajon Sinlapajan: *Tin dredging in Thailand.* These three in *A second technical conference on tin, Bangkok, 1969.*

Work and problems on tin in Thailand. In *A technical conference on Tin, London, 1967.*

United Kingdom:

A. K. Hamilton Jenkin: *Mines and miners in Cornwall.* In a series.

U.S.A.:

Disposing of excess strategic stockpile materials. Report to the

President by the Executive Stockpile Committee. 1963.

C. L. Sainsbury and others: *The circum-Pacific tin belt in north America*. In *A second technical conference on tin, Bangkok, 1969*.

Trends in the use of tin. A report of the National Materials Advisory Board. Washington. 1970.

Periodicals:

Notes on tin. Issued, duplicated, at about monthly intervals by the International Tin Study Group and then by the International Tin Council.

Mining Journal (weekly), *Mining Magazine* (monthly) and *Mining Annual Review*.

Tin International (monthly).

INDEX

The following abbreviations are used

ITC = International Tin Committee (prior to 1946).
ITA = International Tin Agreement (from 1956).
ITCouncil = International Tin Council (from 1956).